Seismology and plate tectonics

David Gubbins

Professor of Geophysics,
Department of Earth Sciences, University of Leeds

Published by the Press Syndicate of the University of Cambridge
The Pitt Building, Trumpington Street, Cambridge CB2 1RP
40 West 20th Street, New York, NY 10011–4211, USA
10 Stamford Road, Oakleigh, Victoria 3166, Australia

First published 1990
Reprinted (with corrections) 1992

Printed in Great Britain at the University Press, Cambridge

British Library cataloguing in publication data

Gubbins, David
Seismology & plate tectonics.
1. Plate tectonics 2. Seismology
I. Title
551.1′36

Library of Congress cataloguing in publication data available

ISBN 0 521 37141 4 hardback
ISBN 0 521 37995 4 paperback

Contents

to my family,
Stella, Matthew,
Katherine and Clare

Preface

This book grew from a course taught with Dan McKenzie to final year physicists in Cambridge. Our task was to provide undergraduates who already possessed a good grounding in physics and mathematical methods with a broad exposure to geophysics in just 24 lectures. Fortunately it is possible to bring a student in geophysics to the forefront of the subject quite quickly: within 24 lectures it is possible to explain research techniques that are in current use and were used to make discoveries that are now less than 20 years old, which is impossible in most other branches of physics. In order to provide the course with structure, and to introduce an essential quantitative element that would allow us to examine it with numerical problems as well as purely descriptive accounts, we decided to concentrate on the development of seismology since the introduction of the World Wide Standardised Seismic Network in about 1960 and its role in the development of plate tectonic theory, the major advance in earth science this century.

The seismogram, a recording of earth movement after an earthquake, usually measured at large distance from the earthquake source, shows a rich variety of different types of elastic waves. This variety is created by changes in properties of the earth at depth, and seismology remains the main tool for discovering the structure of the earth's interior. Reading seismograms is an excellent way to learn about the earth, and therefore several examples of seismograms have been included in the book, together with practical exercises for locating earthquakes, identifying different seismic phases and the free oscillations, and determining the nature of the earthquake source. These practicals form a central part of the book and have been an important element of the course. Modern digital records are provided for Practical 3 but all others are analogue recordings; it is true that digital seismology will be the main research technique in the future, but reading analogue records will retain its superior pedagogic value. The *WWSSN* records have been reduced and can be reproduced at full size for reading by enlarging in a standard copying machine.

The resulting book is one mainly on basic seismology, with rather more emphasis on the seismogram than other texts (BULLEN & BOLT, for example), and the extraction of information from seismic recordings for specific geophysical problems by describing the practical techniques by which the original research was done. Plate tectonics is described in Chapter 7 and forms a rather separate conclusion to the book, drawing on many of the techniques described in the earlier chapters. This chapter has little geological content; the aim of the theory is to provide a simple framework for understanding large scale tectonics and is best kept as simple as possible at first reading. I have had to repeat some material to be found in other texts in order to make this book self contained; I hope the result is not too repetitive. Seismology is treated only briefly in STACEY'S book, which is an excellent introduction to global geophysics for physicists, and perhaps this book will serve to complement it. The mathematical level of the book is comparable with that of BULLEN & BOLT, and may be used as a gentle introduction to theoretical seismology such as is found in AKI & RICHARDS.

This book has been several years in the writing and I have received invaluable help from many colleagues. Dan McKenzie first suggested writing the book and contributed to the initial planning; he also designed Practicals 1, 2 and 4 (although of course any mistakes found in this version of the practicals are entirely my own responsibility) and provided the proof of Euler's theorem. It was very unfortunate that he was not able to continue with the work; his contribution would not only have improved the book but also increased my enjoyment in writing it. He read and commented on a late draft.

The chapter on plate tectonics owes much to a course designed by Geoff King, which I took on those frequent occasions when a large earthquake took him away from Cambridge during term. Much of my knowledge of low frequency seismology, given in Chapter 5, was learnt from Guy Masters and Freeman Gilbert during a visit to the Institute of Geophysicis and Planetary Physics in La Jolla as a Green Scholar.

I am grateful to several other colleagues who read and commented on early drafts: Robin Adams, Jim Ansell, Bob Engdahl, Phillip England, Brian Kennett, Dick Walcott, the staff of the ISC, and a number of undergraduates who took the course. Where I have used figures from previously published work, this is acknowledged in the caption. I am grateful to Jim Taggart and the staff at *NEIC* for providing microfiche copies of the seismograms for reproduction.

1

Introduction

Geophysics is the application of physics to problems in earth science. Part of the subject involves the development of geophysical methods to solve geological problems. Once a technique has been developed the exciting research element of the work transfers to the geologist. He may be using magnetic properties of rocks to infer past deformations, or earthquake sequences to infer current deformation; in either case the novel aspect of his work is the deformation and not the well-established geophysical tools he is using. There has been a flow of ideas, theories and techniques from geophysics into traditional geology; this flow is continuing today. The more interesting part of geophysics is the development of new theories for earth processes. The physicist brings new skills and a new outlook to the subject. Provided he takes the trouble to familiarise himself with the observations, which is time-consuming but not particularly difficult, he is often able to make significant contributions to the subject. He cannot compete with the traditional geologist when detailed geological knowledge is required (geological mapping of an area for example). On the other hand the study of the deep earth is likely to remain the province of the geophysicist (and geochemist) because the study of surface rocks provides little information on the behaviour of the earth's interior.

Probably the most exciting current geophysical research is on geodynamics: the study of movement within the earth. We can now describe a great many of these movements, but we know very little about the underlying driving forces. Slow creep in the solid parts of the earth gives rise to plate motions, or continental drift as it is sometimes called. The surface movements have been well mapped and they form a framework for understanding geological processes, but the underlying convection and the forces driving it are still poorly understood. Deeper down flow in the liquid interior generates the earth's magnetic field by a process which is even less well understood, but the evolution of the magnetic field over geological time and the techniques for extracting past field directions from rocks are well established. Fluid dynamics is also being applied to classical problems in geology such as magma flows and sedimentation, and perhaps future geologists will learn fluid dynamics as a standard part of their training.

This book describes seismology, the study of earthquakes and waves radiated by earthquakes and explosions, which is the major geophysical technique; its use in elucidating earth structure using simple principles of classical physics; and its use in developing and then verifying plate tectonic theory, the major advance in the earth sciences this century. Chapters 2–6 can be used for a basic grounding in seismology, but the book goes beyond what would normally be expected of a text on seismology. It also contains a significant amount of material on other geophysical results and about inferred properties of the earth's interior, and Chapter 7 contains a necessary background to quantitative plate tectonics.

The study of earthquakes themselves is an important branch of seismology; it has revealed a great deal of information about how fracturing occurs in the earth and about strains and short-term deformation processes. The study of seismic waves allows us to make inferences about certain properties of the parts of the earth through which the waves have travelled, as well as about the source of the waves. This book deals mainly with the former problem, the determination of structure within the earth from seismic waves, although there is one chapter (6) on earthquake mechanisms. There are other geophysical and geochemical techniques which "look" into the earth — electromagnetic methods for example — but none give so immediate or clear-cut results as seismology. Seismology provides the great bulk of geophysical data, but unfortunately it can only tell us directly about the elastic properties of the earth; it cannot, for example, tell us the temperature deep inside the earth, although we can try to infer the temperature indirectly by estimating its effect on elastic properties. We need to be more than just seismologists to understand the earth's constitution and dynamics.

1.1 Seismic waves

Earthquakes radiate waves with periods of tenths of seconds to several minutes. Rocks behave like elastic solids at these frequencies. Elastic solids allow a variety of wave types and this makes the ground motion after an earthquake or explosion (called an *event*) quite complex. There are two basic types of elastic wave: one involving compression and rarefaction of the elastic material in the direction of propagation of the wave, and one involving no compression but shear of the elastic material perpendicular to its direction of propagation. These are called P and S

waves respectively, for *primary* and *secondary* since the P wave travels fastest and arrives first. The combination of the two types of wave in the presence of a surface, like that of the earth, can lead to other types of waves, two of which are important for geophysics: the Rayleigh and Love waves. These are *surface waves*; the Rayleigh wave contains compressional motion and the Love waves do not. They travel slower than either P or S waves and arrive much later. They have large amplitude because they travel along the earth's surface, rather than through the main body of the earth, and the energy per unit area of wavefront diminishes only as $(\text{distance})^{-1}$ rather than $(\text{distance})^{-2}$.

There is another way to describe ground motion after an event: by its normal modes. After a large earthquake the earth "rings" like a bell; this motion can be observed on sensitive instruments up to a month after a large event. These oscillations have specific frequencies which are properties of the whole earth and which can now be measured very accurately indeed. The lowest frequency oscillation has a period of about one hour. Any combination of seismic waves can be represented as an equivalent combination of normal modes. In practice the mode representation is most useful at low frequency — for seismic waves with periods above about 40 s — since at higher frequencies the number of modes involved becomes prohibitively large. To interpret ground motion in terms of travelling waves we take displacement as a function of time in a *seismogram* and look for packets of incident energy (called *phases*); to interpret long period ground motion in terms of normal modes we take the Fourier transform of the seismogram and look for discrete peaks in the spectrum.

Seismic waves are reflected and refracted from discontinuities in the earth, which adds greatly to the complexity of an arriving wavetrain, with waves arriving at many different times and from different directions. The earth's surface is a particularly efficient reflector, and the earth is to some extent spherically layered and these layers produce still more reflections. Four examples of seismograms are shown in Figure 1.1. (a) shows the vertical ground motion recorded at a distance of 51.3°[1] from the event, which was in Alaska. Tick marks are every minute. P denotes the P wave arriving. The surface waves are the large long period oscillations arriving about 20 minutes after P. They have lower frequency than P. Other pulses of energy can be seen arriving after the P. These pulses could be identified with known reflections by a careful reading of the seismogram. (b) shows the northward component of ground motion for

[1]Distances are measured in degrees subtended at the earth's centre; $1° \approx 111$ km.

Figure 1.1: Seismograms after large earthquakes. (a) Vertical component of ground motion for an earthquake in Alaska recorded at a distance $\Delta = 51.3°$, redrawn with permission from SIMON, 1981. (b) As (a), horizontal component. (c) An earthquake in Mexico, $\Delta = 120.7°$. Phases are (1) P_{diff}, (2) *PKP*, (3) *PP*, (4) *SKP-PKS*, (5) *PPP*, (6) *SKKS*. (d) Earthquake in the Aleutian Islands recorded at State College in Pennsylvania.

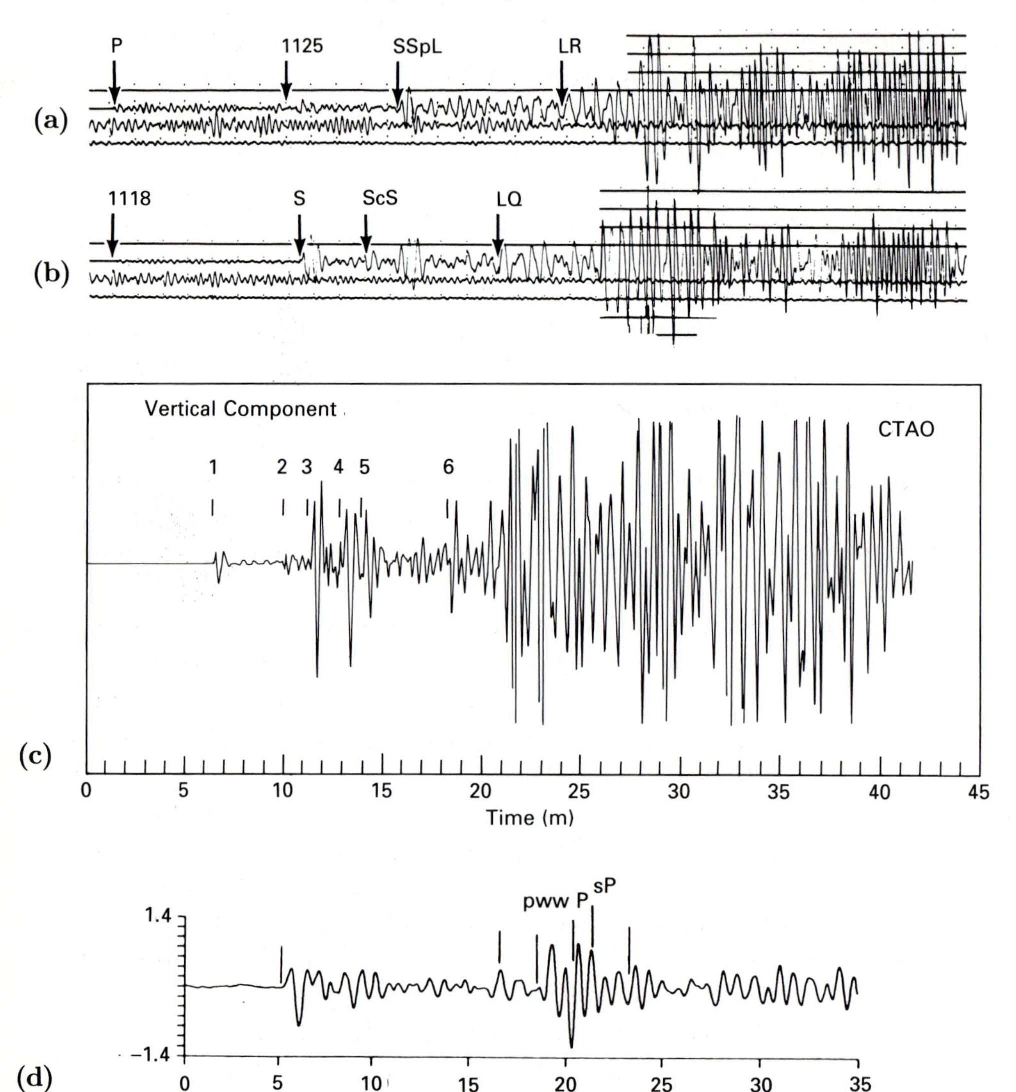

the same case as (a). The arrival of the S wave is marked. The S wave has small vertical displacement in this case because the particle motion of S waves is transverse and the wave is travelling almost vertically upwards; it is not visible in the vertical component seismogram (a). (c) shows the vertical component after an earthquake in Oaxaca, Mexico, recorded at Charters Towers in Australia at a distance of 120.5°. The phases marked have been identified by comparing with a *synthetic seismogram* calculated from a theoretical model of earth structure. Nomenclature is explained below. (d) shows the vertical component record at Jamestown, California, for a deep earthquake beneath the Aleutian Islands. The energy arriving about 15 s after P is caused by reflection from the earth's surface and reverberation between the surface and bottom of the Pacific Ocean near the source (Engdahl & Billington 1986).

The complexity of seismic waves is a great advantage because it helps in elucidating the structure inside the earth. The first achievement of seismology was to determine the main zones within the earth: the crust, mantle, outer core and inner core (Figure 1.2). Each are separated by an interface which reflects seismic waves; the reflections are obtained for events and observing points throughout the world and therefore they must be truly global features, although the reflections from the inner core are often difficult to see.

Within each main zone the speed of seismic waves generally (but not invariably) increases and the waves are refracted so that rays are concave

Figure 1.2: Major seismic zones in the earth: crust, mantle and cores. The outer core is liquid and does not transmit shear waves.

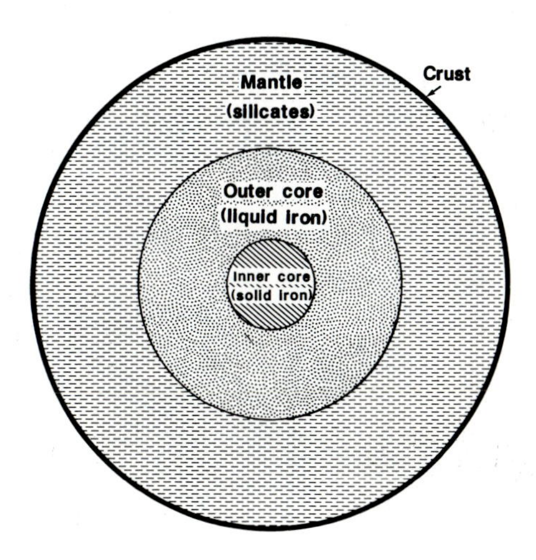

upwards, turned back towards the earth's surface. The outer core is a liquid with lower seismic velocity than the solid mantle and rays are refracted down into it, forming a shadow not reached by direct arrivals.

The various zones inside the earth lead to a very complex ray structure. Rays for compressional waves are shown in Figure 1.3. There is a naming convention for all these rays which takes into account the reflections and refractions at interfaces. Compressional waves in the mantle are denoted *P*, in the core *K*, and in the inner core *I*. Thus *PKP* is a compressional wave travelling through the mantle to the core and out via the mantle as a compressional wave; *PKIKP* is a compressional wave travelling into the inner core. Reflections from the earth's surface are denoted by repeated letters, so *PP* has suffered one reflection, and *PKKP* has reflected once off the lower side of the core-mantle interface before being refracted out into the mantle.

Shear waves in the mantle are denoted *S*. There are no shear waves in the core because it is a liquid. *J* has been reserved for shear waves in the inner core (although the existence of such a phase has never been confirmed). *P* can convert to *S* waves at sharp interfaces, and vice versa, through either reflection or refraction, and rays like *SP* or *PKS* are important in interpreting a seismogram. The efficiency of conversion from

Figure 1.3: Seismic waves travel along rays that are mainly refracted upwards back towards the earth's surface by the increase in wave speed with depth. An exception is refraction into the outer core which has lower wave speed. Tick marks are at 60 s intervals along the rays and represent wavefronts. Only compressional waves are shown. The velocity model is 1066B of Gilbert & Dziewonski (1975).

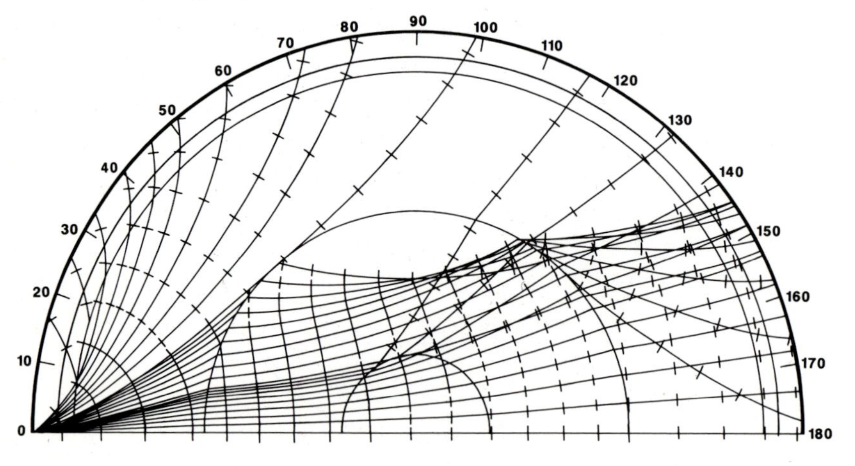

one wave type to another is determined from the reflection and refraction coefficients, which are derived in Chapter 3.

Reflections from the core surface are denoted by *c*, as *PcP* for example, and those from the inner core by *i*, as *PKiKP* for example. For deep earthquakes, waves which travel directly upwards and suffer a reflection from the surface before travelling off to be recorded at large distances are denoted *pP*, *sP*, etc. They are called *depth phases* and are important in determining the depth of an earthquake, as discussed in Chapter 4. At short distances the rays do not penetrate below the crust. P_g and S_g indicate a ray that has travelled in the upper part of the crust, P_* and S_* in the lower part, and P_n along the top of the mantle. The crust varies so much from place to place that it is now less common to distinguish the two sorts of crustal rays.

A complete list of phase names is given in AKI & RICHARDS, Appendix 1. Although the naming of these rays is important in understanding the literature and interpreting seismograms, we should bear in mind that the concept of a ray is only an approximation. The appearance of many records of ground motion suggests more complex behaviour such as diffraction and scattering. For example, the first arrival in Figure 1.1(c) has been diffracted around the core-mantle boundary.

1.2 Inference of earth structure

Once the basic rays have been identified their precise arrival times can be used to determine the seismic velocity in the earth, a model of which is shown in Figure 1.4. The letters indicate the major seismic zones. They were assigned by K.E. Bullen. Inevitably they have become somewhat out of date because of later refinements to earth structure, but they are still widely used in the literature. D' is rarely used now. We can then proceed to infer the composition. Different rays tell us information about different parts of the earth. The earth's crust is delineated by the *Moho*, a discontinuity in properties which reflects seismic waves and is present virtually everywhere in the world. The Moho is thought to represent a change in chemical composition, a change in atomic number which cannot be achieved by a mineral phase change. The crust is the only part of the earth that is accessible to drilling or explosion seismology, although some recent experiments have revealed reflections from within the mantle itself.

The crust beneath the deep oceans cannot be surveyed at all by ordinary geological methods; seismology provided the tool for its initial exploration. Very little was known about the ocean floor prior to seismic experiments, and it came as some surprise to find it so different from the continents. It is all young, thin (less than 10 km thick) and quite uniform. The main variation is with distance away from the mid-ocean ridges, the sediment layer becoming thicker with distance from the ridge. There are usually three discernible seismic layers, the sediment, basaltic upper crust, and lower crust. Continental crust is less homogeneous. Its thickness varies up to a maximum of about 50 km and it can contain several layers and faults separating blocks with different structures. It is composed mainly of lighter rocks than oceanic crust.

The mantle is the shell between the Moho and core-mantle boundary, another clear global seismic reflector. Its seismic structure is shown in Figure 1.4. There are weak seismic discontinuities within the mantle, notably at depths of 450 km and 650 km below the surface. These discontinuities are much more difficult to detect than the Moho, and in some localities are not seen at all. They are believed to be due to phase changes of the constituent minerals of the mantle, which are predominantly iron and magnesium silicates. These minerals have an olivine structure at ordinary temperatures and pressures, and this changes to a spinel structure at pressures corresponding to the 450 km discontinuity. The mixture of minerals, each of which undergoes a phase change at somewhat different pressure and therefore depth within the earth, suggests a diffuse transition spread over a range of depths of about 50–100 km. The phase change without change in atomic number, and the spread of the changes in physical properties over some distance, explains why the seismic reflections and refractions are sometimes difficult to detect. A change of some 5% in properties is required for the 450 km discontinuity.

The 650 km discontinuity is attributed to some other phase change

Figure 1.4: Seismic structure of the mantle

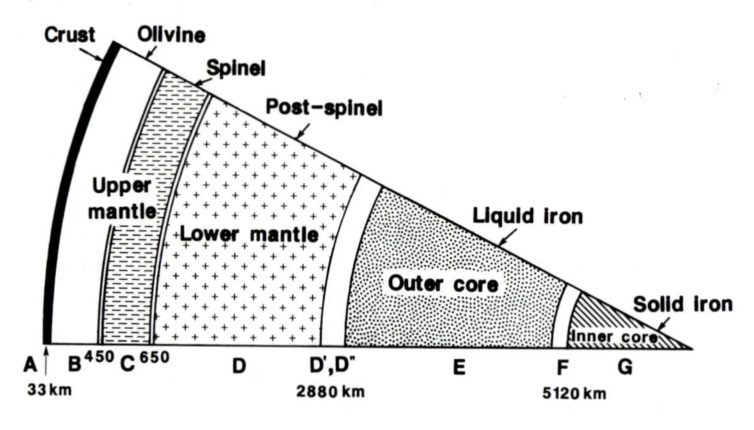

of the silicates from spinel to a more close-packed structure, possibly stishovite. The earthquakes stop near this boundary; there are no recorded earthquakes deeper than about 700 km. This absence of deeper earthquakes is a notable feature: the mantle is split into upper mantle (above the 650 km discontinuity) and the lower mantle (below the 650 km discontinuity). The mantle may well contain more discontinuities than these two major ones; as resolution of the seismology improves (mainly through the use of arrays of instruments, discussed in Chapter 4), and our understanding of the properties of silicates at high pressure improves, it may be possible to match more seismic discontinuities to the known behaviour of the minerals thought to be present.

The region at about 100–200 km depth often has low seismic wave speed. In Chapter 4 we show it is very difficult to obtain wave speeds within such low velocity zones. What information we have comes from studies of surface waves. Low seismic velocity could be associated with a low strength which could allow the observed movement of the plates (Chapter 7).

The main bulk of the lower mantle seems quite homogeneous with little evidence of scattering or anomalous refraction of seismic waves. The lowermost few hundred kilometres is much less homogeneous, and appears to scatter seismic waves considerably. It has been called the D'' region (Figure 1.4). This could be a thermal boundary layer at the interface with the core, a feature expected from simple ideas on convection in the mantle. The phase PcP is reflected from the core-mantle boundary. It can be read accurately and its arrival time is critically dependent on the core radius. It allows the core radius to be determined to within 10 km or so. The liquid core is a good transmitter of seismic waves. Attenuation of elastic waves arises mainly from shear motions, and so this attenuation largely disappears in a liquid. Seismic waves have been observed to bounce many times from the underside of the core-mantle boundary, and experiments on core waves designed to measure their attenuation have essentially failed to find any loss of energy.

The inner core does not give clear reflections. This led early seismologists to propose a layer, called the F region, of anomalous liquid or partially molten solid, around the solid inner core. Later work (Haddon & Cleary 1974), again with seismic arrays, indicated the scattered waves came from the D'' region, and evidence for the F region has now largely been removed.

Long wavelength waves are influenced mainly by large scale changes in the elastic properties of the earth. The low frequency normal modes have "wavelengths" comparable with the earth's radius, and their frequencies are gross properties of the whole earth. For this reason the normal modes are useful in establishing the average structure of the earth, particularly the deep interior. Surface waves, which are also long period and equivalent to fundamental normal modes, are confined to the earth's upper layers, and are dependent on the properties of the top few hundred kilometres of the upper mantle. Seismic waves have wavelengths of a few kilometres and sample elastic properties throughout the earth in some detail. They leave us the problem of deciding which part of the earth has produced a particular effect; unfortunately this problem is often insoluble.

1.3 Earthquakes

Earthquakes occur in narrow belts. Figure 1.5 shows the distribution of large and moderate sized earthquakes that occurred between 1977 and 1986. A very large number of small earthquakes are not shown on this map; they would be too small to be detected away from their immediate vicinity. Earthquakes generate elastic waves when one block of material slides against another; the break between the two blocks being called a *fault.* Explosions generate elastic waves by an impulsive change in volume in the material. Small explosive charges are used in controlled-source seismic experiments in which the waves penetrate only a few kilometres into the earth. Large nuclear explosions generate almost as much high frequency elastic radiation as large earthquakes, but much less low frequency radiation. This difference in radiation at high and low frequency forms a basis for discrimination between nuclear explosions and earthquakes.

The zones of earthquakes shown in Figure 1.5 are probably even thinner than appears on the map. The locations of deep earthquakes are quite difficult to obtain accurately, and the thickness of the zones in Figure 1.5 is probably determined by scatter rather than any real separation. Studies of deep seismic zones in which special techniques are used to refine the locations of the earthquakes relative to each other (these methods are described in Chapter 4) suggest the earthquakes are contained on fault systems less than 20 km thick.

As far as we can tell, changes in volume are rather unimportant in the way earthquakes radiate elastic energy. Large earthquakes rarely occur by a simple movement on a single surface; they are usually associated with a large number of smaller subsequent events, called foreshocks and aftershocks depending whether they occur before or after the main event. These associated events are usually on the same fault system as the main event.

Figure 1.6 shows two blocks of separated by a fault. By making accurate observations of this radiation pattern at large distances it is possible to deduce the nature of the faulting; in its simplest form this involves finding the orientation of the plane of the fault and the direction of slip. The radiation pattern for a simple type of faulting is derived in Chapter 6. The result can be understood by a heuristic argument. Consider the movement of one block against another in Figure 1.6. An observer watching the fault movement from a distance would see ground motion initially towards him if he were sited in either of the two quadrants marked +, away from him in either of the two quadrants marked −. The observer "looks" with seismic waves rather than with light, and uses a seismic recorder rather than his eyes. He must allow for refraction of the waves by changes of refractive index within the earth. Then if the first ground motion is upwards he must lie in one of the + quadrants; if downwards in one of the − quadrants. By making observations worldwide we can determine the planes separating + and − quadrants; it is then a matter of geometry to determine the direction normal to the fault and the direction of relative motion across it (in fact there remains an ambiguity as to which direction is which, to be resolved by other means). The full procedure is detailed in Chapter 6. It is one of the most useful techniques for investigating present-day earth motions and active tectonics.

1.4 Plate tectonics

The idea of continental drift was suggested to Wegener, who devised a theory for it in 1912, by the similarity of the coastlines of South America and Africa. It did not attract widespread attention in recent times until about 1960, and was only accepted by the scientific community at large some ten years later. Why did it take so long? There are a number of factors. First the geophysicists were concerned about the dynamics, the mechanism by which the plates were driven. This is still a contentious issue; McKenzie (1988) has asserted that progress was only made when the kinematics, or simple description of the past motion of the continents, was separated from the much more difficult dynamical problem, and attributes the idea to Bullard.

A second, probably larger, factor was the exploration of the ocean floors, which did not really get under way until after the second world war. Previous ideas about the deep ocean floor were based on a few dredged rocks. They were very primitive, less advanced than our present ideas on the geology of Mars for example. The oceans turned out to have a very young and very simple geology, and a simple theory evolved very quickly. There was also an element of luck in that the earth's magnetic field had left a record of the past motion of the oceanic crust by mag-

Figure 1.5: Locations of large earthquakes 1977–1986. Reproduced with permission from Susan K. Goter, United States Geological Survey.

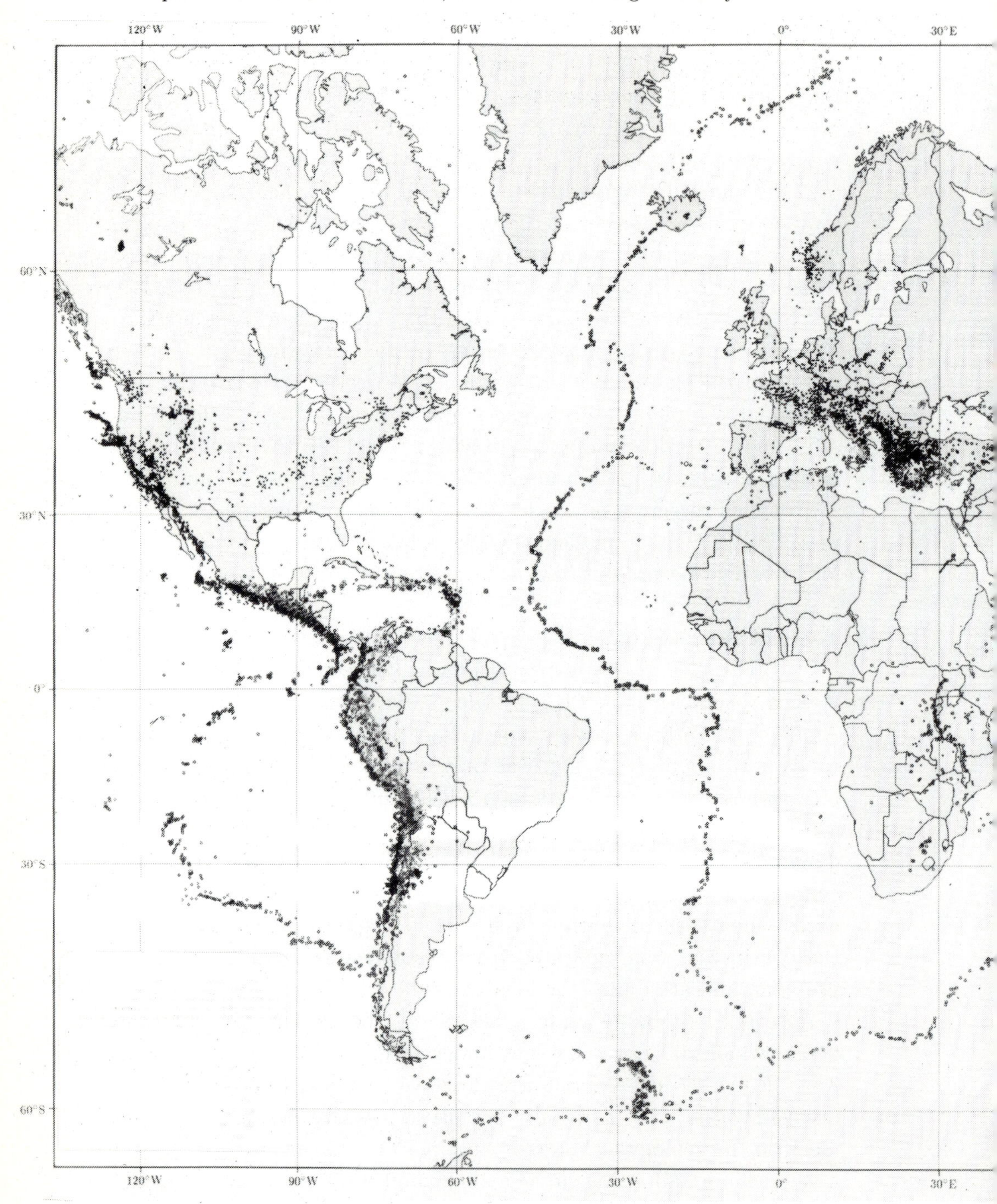

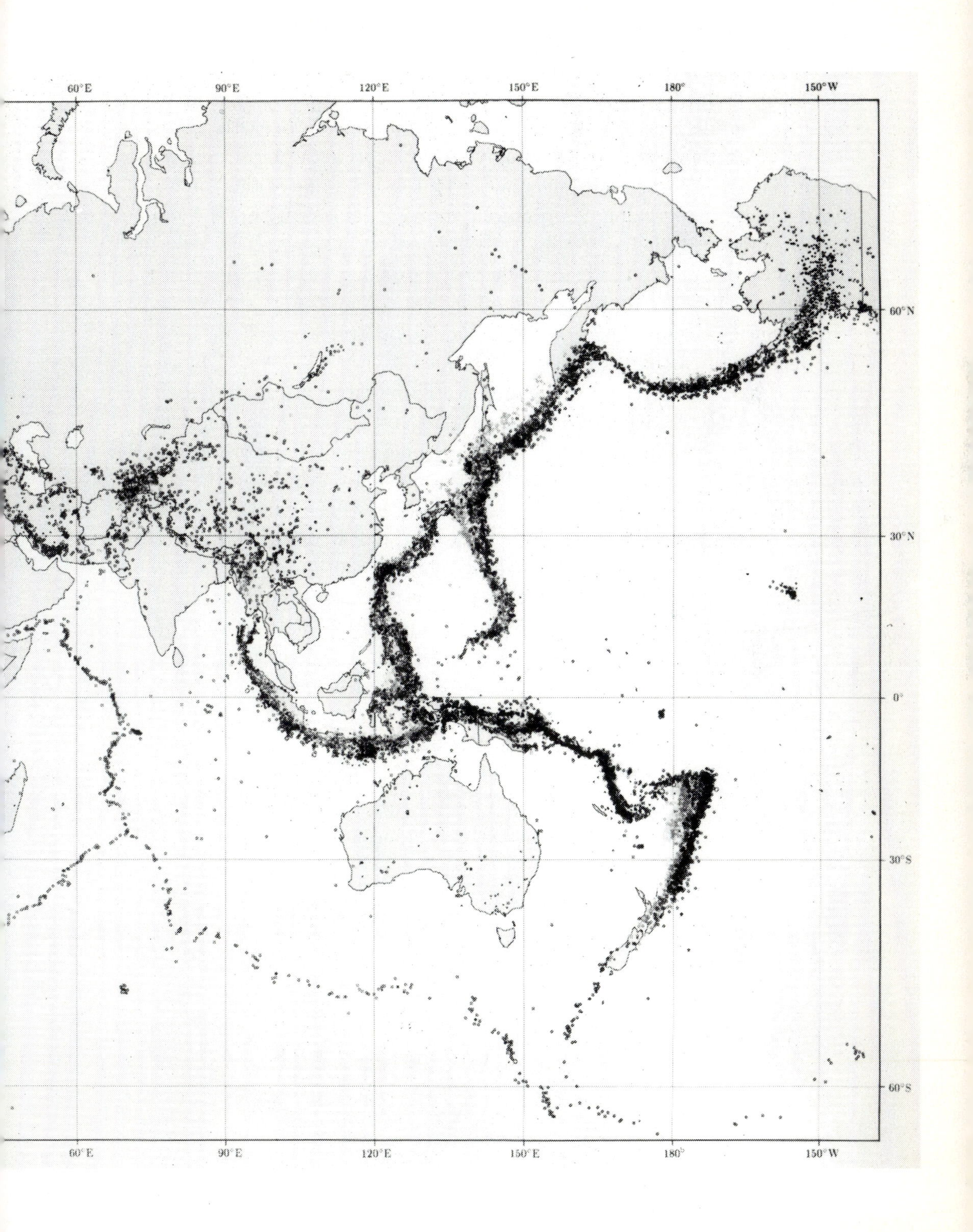

60°E
90°E
120°E
150°E
180°
150°W
60°N
30°N
0°
30°S
60°S

netising it in alternate directions as the main magnetic field reversed in polarity, leaving a permanent record of its age. A third factor was that improvements in earthquake locations showed the very narrow nature of seismic zones: maps prior to 1964 showed very diffuse belts of earthquakes because of inaccurate determinations of their locations. The seismicity map in Figure 1.5 suggests immediately that large areas of the earth's surface are not undergoing deformation, and that relative motion occurs only on boundaries delineated by the earthquakes. The areas where no deformation occurs are the *plates*, and the plate boundaries lie along the earthquake belts. The major plates are shown in Figure 1.7.

In 1968 the orientations of the fault planes and directions of slip for a large number of earthquakes were determined by Isacks, Oliver & Sykes using the method described in Chapter 6. These mechanisms were used to determine the relative motion across the narrow seismic zones between the plates. The results confirmed the developing calculations of relative motions of plates, which were based on the magnetic stripes on the sea floor. This was a most remarkable result: motions of the earth's surface estimated from the magnetisation of lavas erupted at mid-ocean ridges could be used to predict, albeit qualitatively, the motion of the earth's surface and the nature of the faulting and earthquake mechanisms half an ocean away. These areas of the earth's surface really do move huge distances without distortion; they were called plates and the theory, the central tenet of which is that plates move without deformation except at their edges, was called *plate tectonics*.

Plate tectonic theory has had an enormous effect in the earth sciences.

Figure 1.6: A simple type of faulting: thrust faulting. The first motion of the seismic waves observed at great distance will be upwards in the quadrants marked +, downwards in those marked −.

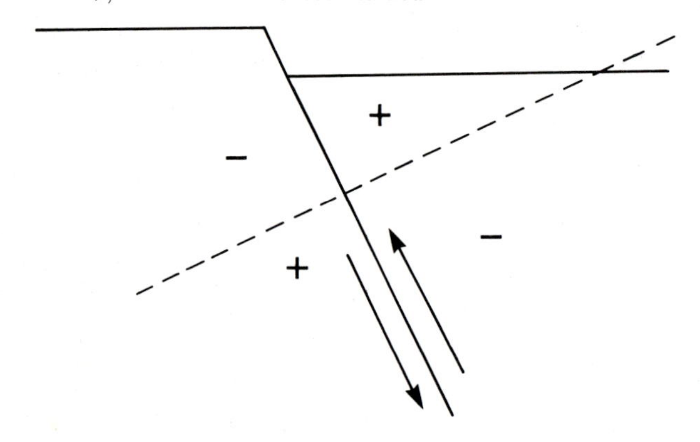

It gave geology a quantitative framework and confirmed the existence of convection inside the earth. Difficulties still remain with explaining the driving forces for the plates, but the observational evidence for plate motions is so overwhelmingly convincing that dynamical objections can no longer be used to oppose the idea of moving continents and plates.

The plates are best defined as the brittle, cold part of the mantle which is capable of sustaining earthquakes. It is not possible to give a definition in terms of seismic boundaries, as it is for the core and crust for example, as there appears to be no distinct lower boundary to the plates that can be detected with seismic waves. Probably there is no sharp lower boundary. The underlying flow of mantle material may be only indirectly related to the plate motions, making the study of deep mantle convection a very difficult task. The principles of plate tectonic theory are explained in Chapter 7. The discussion is there restricted to the kinematics, since the dynamics are still the subject of active research and many of the ideas are speculative.

1.5 Convection in the earth

With the acceptance of plate tectonics has come the acceptance of the concept of a convecting earth. Consider the idea, that was prevalent a few years ago, of a predominantly static earth in which heat flow was controlled by radiogenic heat generation and perhaps slow cooling of the surface layers by thermal conduction. Thermal conduction is slow in the

Figure 1.7: Major tectonic plates.

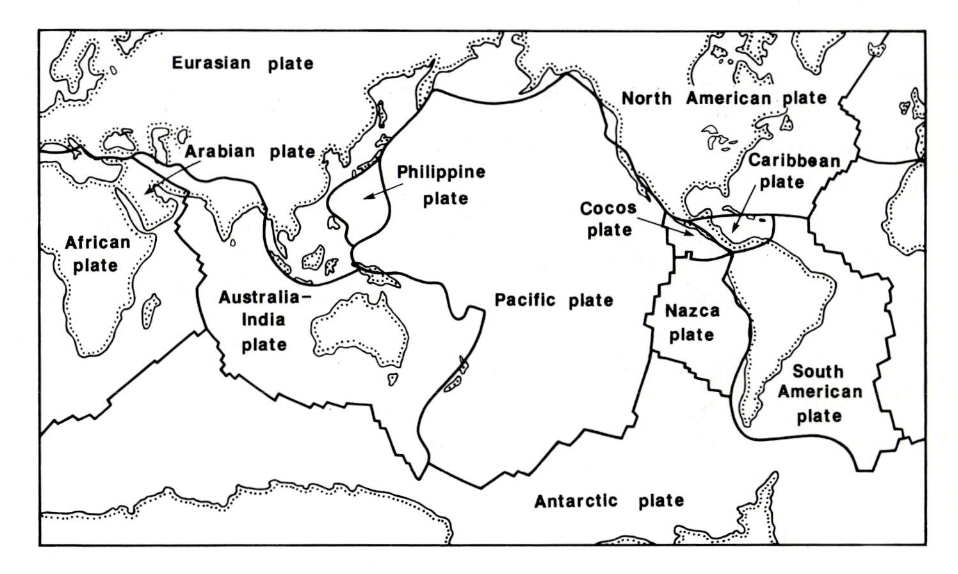

earth; in Chapter 2 it is shown that it takes several thousand years for heat to conduct a few kilometres; the time required to cool the whole earth by thermal conduction is several orders of magnitude longer than its age. In the absence of convection, the heat observed to flow from the earth's surface today must be due either to radioactive elements at shallow depth, or to cooling of near-surface layers. The latter formed the basis for Kelvin's classic calculation for the age of the earth. Helmholtz had already proposed a theory of gravitational collapse for the sun's heat (this was prior to any understanding of radioactive heating). According to this theory, the sun would have been so large as to engulf the earth 20 Myr ago [2], giving an upper limit to the age of the earth itself. Kelvin assumed an initially hot earth which cooled by conduction. Only the surface layers cooled, the interior remaining hot. By comparing the temperature gradient at the surface of a cooling sphere from model calculations with the measured temperature gradient at the earth's surface (the *geothermal gradient*) Kelvin was able to estimate the age of the earth independently, giving a figure of about 20 Ma[2], in satisfactory agreement with Helmholtz' estimate of the age of the sun.

This calculation was followed by a famous controversy lasting some forty years. The geologists, notably Lyell and Geike, asserted the earth must be much older than this because of the vast accumulation of sediments and the extent of evolutionary change. However, they came upon the problem which always accompanies such essentially qualitative arguments: they were unable to provide wholly convincing quantitative arguments. Kelvin's estimate for the age of the earth varied somewhat during this very long debate, but never reached anything like the high age we now know (and paleontologists then knew) the earth to have.

Kelvin's calculation, which is described fully in CARSLAW & JAEGER, requires estimates of the initial temperature of the earth and the geothermal gradient. He took 1200 K (the temperature of molten rock) and 37 K km^{-1} respectively, the best estimates he had, to obtain an age of 10 Ma. Nowadays we might choose an initial temperature of 5000 K (close to the current estimate of the centre of the earth) and a geothermal gradient of 10 K km^{-1}, giving the more reasonable figure of 1.6 Ga.

With hindsight we can say that Kelvin's calculation depended very much on uncertainties in numerical values chosen for the parameters, and that he should have paid more attention to the geological record, but the demise of his theory came first with the discovery of radioactivity,

[2]Myr denotes an interval of time of a million years; Ma will be used to denote an age of a million years before present.

which provided an explanation for both the high heat flow and the great age of the sun. There followed a long period in which it was thought that radioactive heating more-or-less balanced the heat flowing from the earth's surface; the earth was in steady state.

The earth's age is now determined from the rate of decay of certain radioactive isotopes. To make an important contribution to the earth's internal heat a radioactive isotope must have a half life comparable with the earth's age; short-lived isotopes will have decayed and will no longer be sufficiently abundant, while long-lived isotopes do not decay rapidly enough to contribute significantly to the heat flux. The main radioactive isotopes in the earth are U^{238}, U^{235}, Th^{232} and K^{40}; their abundances and contributions to the heat flux are given in Table 1.1. Values are given for three different compositions, appropriate to oceanic crust, continental crust, and mantle (chondritic meteorite model), respectively. Radioactive elements appear to be concentrated in the crust (probably in the upper crust) and more noticeably in continental crust. The total heat flux estimated from the figures in Table 1.1 is about 4×10^{13} W, about half coming from the mantle, most of the remainder from the continental crust, and only a few per cent from oceanic crust. This is comparable with current measurements of heat flow through the earth's surface. Thus it is quite possible to balance the heat budget with radioactive heating alone.

However, convection in the deep earth changes Kelvin's argument completely. Once we have convecting motions with an overturn time of a fraction of the earth's age (and plate motions of a few centimetres per year give an overturn time of about 100 Myr) then all of the material in the earth may be brought to the surface and may cool. The apparent heat capacity of the earth greatly exceeds that allowed by Kelvin, who permitted cooling of only the outermost layers, and a simple calculation allowing cooling of the whole earth suggests we could provide the heat

	Half life (Ga)	Continents (ppm granite)	Oceans (ppm basalt)	Chondrite (ppm)
U^{238} (99.3%)	4.50			
U^{235} (0.7%)	0.71	4.75	0.60	0.012
Th^{232} (100%)	13.9	18.5	2.70	0.0398
K^{40} (0.0118%)	1.3	379000	2400	845

Table 1.1: Radioactive heat-producing isotopes in the earth (after BOTT, 1971, Chapter 6).

flux without any assistance from radioactivity at all. The earth's internal temperature would have to fall a few tens of degrees in 1 Gyr, well within what might be expected.

These simple calculations are quite uncertain and could well be in error by a factor of two. Therefore we may have both primordial heat released by convective cooling and radiogenic heating contributing to today's heat flux. Whichever contribution dominates, and studies of the power supply to the earth's magnetic field suggest that convective cooling must be important; whole-earth convection on a time scale very much faster than that for thermal conduction has now become central to the modern view of the earth's evolution.

2

Mechanics of elastic media

Modern geophysics involves the study of deformation of a large range of different types of materials. Core, ocean, and some geological studies need fluid dynamics, some tectonic problems need visco-elastic solids, mantle convection needs a theory for solid state creep, and seismology needs elasticity theory. Texts on theoretical seismology usually begin with, or almost with, the equations governing infinitesimally small strains. The drawback of this approach is that one is ill-prepared for those rare occasions when the underlying assumptions of small strain and perfect elasticity break down. This treatment starts with a rather full description of continuum mechanics for geophysics, partly because such a treatment is lacking in books on seismology, and partly because it will be useful in other branches of geophysics.

2.1 The description of deformation and strain

The medium through which seismic waves travel will be treated as a *continuum*: we ignore its molecular structure and picture it as being without any gaps or empty spaces. A continuum may be subdivided to any desired extent, and we may therefore define an infinitesimally small element at position $\mathbf{x}$ and time t, and describe its properties point-wise by using the space coordinate $\mathbf{x}$. Thus $\mathbf{v}(\mathbf{x}, t)$ is the velocity of the material element that occupies position $\mathbf{x}$ at time t; and if dV is the volume of the element and $\rho(\mathbf{x}, t)dV$ its mass, then $\rho(\mathbf{x}, t)$ is the density of the material that occupies position $\mathbf{x}$ at time t.

This method of description, in which the material element is labelled according to its position $\mathbf{x}$ at a particular time, is termed *Eulerian.* The alternative *Lagrangian* description labels the material element by its position $\boldsymbol{\xi}$ at some reference time t_0. The current position of that material element which was initially at $\boldsymbol{\xi}$ is sufficient to specify the deformation of the medium. As a general rule we may say that the Eulerian formulation is convenient when deformations carry the material element a long way from its initial position, as in fluid flow for example, whereas the Lagrangian formulation is convenient when deformations are small. We shall be dealing with seismic waves which involve small oscillations about an equilibrium position, so the Lagrangian formulation, with $\boldsymbol{\xi}$ being the location of the element in its equilibrium position, will be particularly useful.

The displacement vector, $\mathbf{u}$, is defined as

$$\mathbf{u} = \mathbf{x} - \boldsymbol{\xi} \tag{2.1}$$

Consider the stretching of an infinitesimal material element of length $d\mathbf{x}$, deformed from an initial $d\boldsymbol{\xi}$. Let dL and dl be the lengths of the element before and after deformation. Then

$$\begin{aligned} dl^2 &= dx_i dx_i \\ &= (d\xi_i + du_i)(d\xi_i + du_i) \\ &= \left(d\xi_i + \frac{\partial u_i}{\partial \xi_j} d\xi_j\right)\left(d\xi_i + \frac{\partial u_i}{\partial \xi_k} d\xi_k\right) \\ &= dL^2 + 2\frac{\partial u_i}{\partial \xi_j} d\xi_i d\xi_j + \frac{\partial u_i}{\partial \xi_j}\frac{\partial u_i}{\partial \xi_k} d\xi_j d\xi_k \end{aligned} \tag{2.2}$$

The partial derivatives are deformation gradients; they determine the *strains.* Linear strain, of a rod extended along its length for example, is simply the extension divided by the original length of the rod, but deformation in three dimensions is more complex and we must specify changes in three components of vector displacement for initial lengths in each of three directions: these $3 \times 3 = 9$ scalar quantities are the partial derivatives.

We shall restrict ourselves to the case of small strains, i.e. when all elements of $\partial u_i/\partial \xi_j$ are small compared with unity, and neglect the last term on the right hand side of (2.2). Furthermore we define the symmetric strain tensor for small strains as

$$e_{ij} = \frac{1}{2}\left(\frac{\partial u_i}{\partial \xi_j} + \frac{\partial u_j}{\partial \xi_i}\right) \tag{2.3}$$

Noting that i, j in (2.2) are dummy subscripts because they are summed over, we can write (2.2) as

$$dl^2 = dL^2 + 2e_{ij} d\xi_i d\xi_j \tag{2.4}$$

There is an important distinction to be made between small strain and small displacement. It is a question of what we mean by small, and what the displacement or strain is small compared with. With strain it is easy: strain is dimensionless and therefore a small strain is one that is much less than unity. But what is a"small" displacement? Such small strains are typical of those set up at large (*teleseismic*) distances from an earthquake; we may expect strains near the earthquake to be governed by the strength of the material itself, which appears to fracture in a brittle fashion when subjected to strains that are too large. Measurements in the laboratory have shown that samples fracture at strains of about 10^{-3}; in the earth fracturing is believed to occur at strains of about 10^{-4}. The smaller value is thought to be due to the presence of pre-existing cracks and fissures in the natural rock which render it weaker than a small laboratory sample. Even the largest strains we shall have to deal with will be small compared with unity, and our small strain approximation should be a good one.

We shall find the symmetric strain (2.3) very useful in describing elastic media, and shall rarely be forced to make recourse to the full displacement derivative. For example, the volume change on deformation may be expressed in terms of the strain tensor e_{ij} alone provided the strains are small. Let (ξ_1, ξ_2, ξ_3) be rectangular Cartesian coordinates defining the position of a material element at time t_0 (Figure 2.1). The coordinate surfaces deform with the material until they form a curvilinear system. The ratio of deformed to original volumes is therefore that between volume elements in the rectangular system $\mathbf{x}$ and the curvilinear system $\boldsymbol{\xi}$. This is just the Jacobian, J, of the transformation from $\mathbf{x}$ to $\boldsymbol{\xi}$. Using (2.1) gives

Figure 2.1: Coordinates for describing deformation in two dimensions. Cartesian coordinates (x_1, x_2) are fixed in space; coordinates (ξ_1, ξ_2) are fixed in the material and deform to form a curvilinear coordinate system. The ratio of the shaded elemental areas is the Jacobian of the transformation.

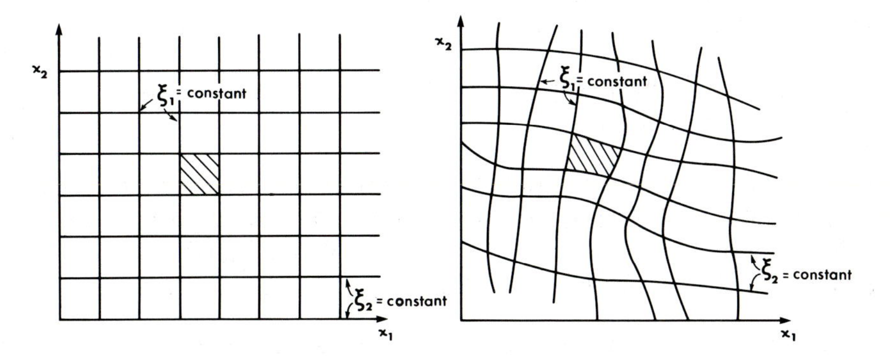

$$J = \det\left(\frac{\partial x_i}{\partial \xi_j}\right) = \det\left(\delta_{ij} + \frac{\partial u_i}{\partial \xi_j}\right) \tag{2.5}$$

For small strains we may expand the determinant and neglect terms smaller than those linear in the strains to give

$$J = 1 + \frac{\partial u_i}{\partial \xi_i} = 1 + e_{ii} \tag{2.6}$$

We may write the *dilatation* or fractional change in volume, θ, as

$$\theta = J - 1 \tag{2.7}$$

For small displacements we may replace the derivative with respect to $\boldsymbol{\xi}$ with one with respect to $\mathbf{x}$ and incur only a small error. The dilatation becomes

$$\theta = \nabla_x \cdot \mathbf{u} \tag{2.8}$$

This is an example of the lack of distinction between the Eulerian and Lagrangian formulations for small strains.

2.2 The continuity equation

Physical laws apply to finite amounts of material and are often in the form of conservation of a certain physical quantity. For example, conservation of mass requires that the integral

$$M_0 = \int_{V_0} \rho_0\,(\xi_1, \xi_2, \xi_3; t)\, dV_0 \tag{2.9}$$

remain invariant during the deformation provided V_0 is a *material volume*, i.e. one bounded by a surface that deforms with the material. Suppose the material in V_0 deforms to fill a volume V at time t; the density is then $\rho(x, y, z; t)$. The mass may be written as

$$\begin{aligned} M_0 &= \int_V \rho\,(x, y, z; t)\, dV \\ &= \int_{V_0} \rho |J| dV_0 \end{aligned} \tag{2.10}$$

where again we have used the Jacobian of the transformation defined in (2.5). Equating integrals in (2.9) and (2.10) gives a mathematical expression for the law of conservation of mass in the Lagrangian formulation:

$$\int_{V_0} (\rho_0 - \rho J)\, dV_0 = 0 \tag{2.11}$$

(we always have $J > 0$).

This expression is valid for any material volume V_0; therefore the integrand is zero. In making this deduction we have used the continuum hypothesis, that the material can be subdivided into arbitrarily small pieces. The continuum hypothesis enables us to replace an unwieldy integral expression (2.10) with a more useful differential equation that applies at a point:

$$\frac{d}{dt}(\rho J) = 0 \tag{2.12}$$

Lagrangian forms like (2.12) are typically easier to write down than their Eulerian counterparts, but unfortunately they are typically less useful. In the Eulerian formulation we consider the mass within a volume V that is fixed in space

$$M = \int_V \rho(x, y, z; t)\, dV \tag{2.13}$$

This mass will change with time as material crosses the boundary of V. Since V is fixed the rate of change of mass is simply

$$\frac{\partial M}{\partial t} = \int_V \frac{\partial \rho}{\partial t} dV \tag{2.14}$$

This must be equal to the rate of outflow of material as measured by $\mathbf{v}(\mathbf{x}, t)\cdot\hat{\mathbf{n}}$ where $\mathbf{v}$ is the velocity and $\hat{\mathbf{n}}$ the unit outward normal to V (Figure 2.2). Thus

$$\int_V \frac{\partial \rho}{\partial t} dV = -\oint_{\partial V} \rho\mathbf{v}\cdot\hat{\mathbf{n}} dS \tag{2.15}$$

Figure 2.2: A material volume V moves with the material. ∂V is the surface of V and $\hat{\mathbf{n}}$ the unit outward normal. As the material deforms V moves to the dashed volume enclosed by the surface $\partial V'$.

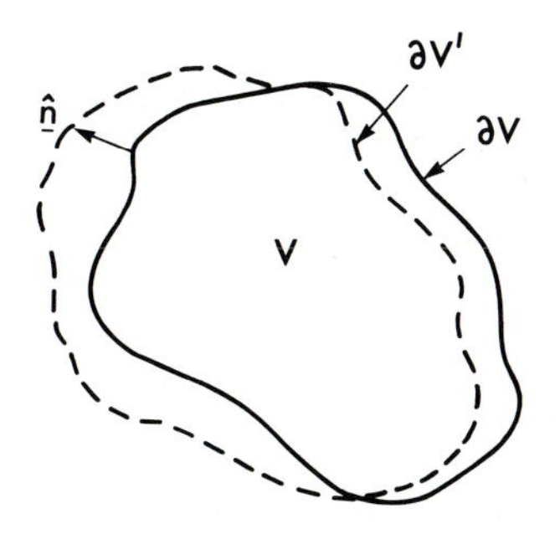

where ∂V is the surface of V. The divergence theorem may be used to transform the surface integral to give

$$\int_V \left\{ \frac{\partial \rho}{\partial t} + \nabla \cdot (\rho \mathbf{v}) \right\} dV = 0 \tag{2.16}$$

Again we invoke the continuum hypothesis and, because (2.16) holds for arbitrary V, we deduce the integrand to be zero everywhere. This gives the Eulerian form of the equation of mass conservation

$$\frac{\partial \rho}{\partial t} + \nabla \cdot (\rho \mathbf{v}) = 0 \tag{2.17}$$

(2.17) is often called the *continuity equation*; it is important in both fluid and solid mechanics. We can use it to obtain time derivatives of other material integrals. Let A be a property of the material defined per unit mass. A may be a scalar, vector, or tensor quantity. Consider the rate of change of A in volume V

$$\frac{d}{dt} \int_V \rho A dV \tag{2.18}$$

We must take account of the movement of the surface of V, with velocity $\mathbf{v}$, through space. (2.18) may therefore be written

$$\frac{d}{dt} \int_V \rho A dV = \int_V \frac{\partial}{\partial t} (\rho A) \, dV + \oint_{\partial V} \rho A \mathbf{v} \cdot \hat{\mathbf{n}} dS \tag{2.19}$$

and again using the divergence theorem and (2.17) we have

$$\frac{d}{dt} \int_V \rho A dV = \int \rho \left(\frac{\partial}{\partial t} + \mathbf{v} \cdot \nabla \right) A dV \tag{2.20}$$

The time derivative

$$\frac{D}{Dt} = \frac{\partial}{\partial t} + \mathbf{v} \cdot \nabla \tag{2.21}$$

is called the *Stokes derivative.* It is in fact the Lagrangian derivative since

$$\begin{aligned} \left(\frac{\partial A}{\partial t} \right)_\xi &= \left(\frac{\partial A}{\partial t} \right)_\mathbf{x} + \left(\frac{\partial x_i}{\partial t} \right)_\xi \\ &= \frac{\partial A}{\partial t} + \mathbf{v} \cdot \nabla A \end{aligned} \tag{2.22}$$

The Stokes derivative is also called the *material derivative* because it gives the rate of change of a quantity measured by an observer moving with the material. The physical meaning of (2.20) is now clear: ρdV is conserved and the material derivative can be taken under the integral sign.

2.3 Self-gravitation

Individual particles in the earth experience a gravitational force obeying Newton's law of gravity. The force of gravity is proportional to the mass it acts upon; in a continuum it can be expressed either as a force per unit mass or per unit volume. Such forces, which act on the main body of the continuum, are called *body forces.*

In the laboratory the gravitational force acting on any material is $\rho\mathbf{g}$ per unit volume, where $\mathbf{g}$ is the acceleration due to gravity; its magnitude is about 10 m s^{-2} and it is directed towards the earth's centre. In reality the situation is more complex because the gravitational acceleration is due to the earth's own mass, and in using constant $\mathbf{g}$ to calculate the fall of a mass under the action of gravity we are making an approximation: we ignore the effect of the moving mass on the value of $\mathbf{g}$ itself. Normally this is a very good approximation, but it is inadequate for calculations such as working out the pressure and density in the earth's interior. $\mathbf{g}$ is a function of the mass distribution within the earth itself, which is in turn dependent on the gravitational forces through the pressure. The earth creates its own gravitational field: it is said to be *self-gravitating.*

The gravitational force is a potential force. It may be written as

$$\mathbf{f}_g = \rho\nabla\psi \tag{2.23}$$

per unit volume, where the gravitational potential ψ satisfies a Poisson equation

$$\nabla^2\psi = 4\pi G\rho \tag{2.24}$$

and G is the gravitational constant: $G = 6.67 \times 10^{-11}$ $m^3s^{-2}kg^{-1}$.

The gravitational force is critical in determining density and pressure within the earth, as we shall see later in this chapter, but it plays only a minor role in seismic wave theory because fluctuations in the gravitational body force $\mathbf{f}_g$ are small compared with variations in elastic forces; for long period waves fluctuations in the gravitational force are significant but for short period waves they are negligible.

2.4 The stress tensor

Surface forces are reckoned per unit area and are usually called *tractions.* To specify them fully we must give not only the force per unit area but also the orientation of the surface on which it acts. This is done using the *stress tensor* $\boldsymbol{\sigma}$, the ji^{th} element of which gives the ith component of force per unit area (or stress) acting on a surface with normal in the jth direction. Thus for a surface with unit normal $\hat{\mathbf{n}}$

$$T_i = \sigma_{ji}\hat{n}_j \tag{2.25}$$

Within the material the short-range intermolecular forces cancel (by Newton's third law of motion, that reaction equals action) and therefore we may sum the tractions to a total force acting on the whole volume as

$$F_i = \oint_{\partial V} T_i dS = \oint_{\partial V} \sigma_{ji}\hat{n}_j dS \tag{2.26}$$

and transform it using the divergence theorem to give

$$F_i = \int_V \frac{\partial \sigma_{ji}}{\partial x_j} dV \tag{2.27}$$

which holds for arbitrary V. The divergence of the stress tensor has the form of a body force, and equation (2.27) shows that surface tractions may be converted to equivalent body forces by taking the divergence of the stress tensor.

Our surfaces here have been drawn in the deformed medium; $\boldsymbol{\sigma}$ is therefore an Eulerian quantity and is called the *Cauchy stress tensor*. We shall not need to resort to the Lagrangian formulation of stress.

According to Newton's second law of motion the rate of change of momentum is equal to the applied force. Consider an arbitrary material volume V. The momentum is $\int_V \rho \mathbf{v} dV$, and the applied force is the sum of body forces, $\mathbf{f}$, and surface tractions, $\mathbf{T}$. Newton's second law gives

$$\frac{d}{dt}\int_V \rho v_i dV = \oint_{\partial V} \sigma_{ji}\hat{n}_j dS + \int_V f_i dV \tag{2.28}$$

Transforming the surface integral by the divergence theorem and using (2.20) for the material time derivative we have

$$\int_V \left(\rho \left\{ \frac{\partial v_i}{\partial t} + v_j \frac{\partial v_i}{\partial x_j} \right\} - \frac{\partial \sigma_{ji}}{\partial x_j} - f_i \right) dV = 0 \tag{2.29}$$

and since the integral applies to arbitrary volume V the integrand is zero

$$\rho \left\{ \frac{\partial v_i}{\partial t} + v_j \frac{\partial v_i}{\partial x_j} \right\} = \frac{\partial \sigma_{ji}}{\partial x_j} + f_i \tag{2.30}$$

(2.30) is called *Cauchy's equation of motion.*

2.5 Angular momentum and symmetry of the stress tensor

The rate of change of angular momentum of an arbitrary volume of material is equal to the applied couple. Thus

$$\frac{d}{dt}\int_V \rho\,(\mathbf{r}\times\mathbf{v})\,dV = \mathbf{C} \tag{2.31}$$

where $\mathbf{C}$ is the couple due to body and surface forces. Note that (2.31) ignores the angular momentum of the material at rest and couples other than those produced by forces (for instance couples produced by the magnetic field). Write (2.31) as

$$\frac{d}{dt}\int_V \rho\epsilon_{ijk}x_j v_k dV = \int_V \epsilon_{ijk}x_j f_k dV + \oint_{\partial V} \epsilon_{ijk}x_j\sigma_{lk}\hat{n}_l dS \tag{2.32}$$

and use (2.20) to evaluate the left hand side as

$$\int_V \rho\epsilon_{ijk}x_j\frac{Dv_k}{Dt}dV + \int_V \rho\epsilon_{ijk}\left(v_j v_k + v_l\frac{\partial x_j}{\partial x_l}v_k\right)dV \tag{2.33}$$

The second integral in (2.33) is zero because it is proportional to $\mathbf{v}\times\mathbf{v}$.

The second integral on the right hand side of (2.32) may be transformed to a volume integral using the divergence theorem:

$$\begin{aligned}\int_V \epsilon_{ijk}\frac{\partial}{\partial x_l}(x_j\sigma_{lk})\,dV &= \int_V \epsilon_{ijk}\left(\delta_{jl}\sigma_{lk} + x_j\frac{\partial\sigma_{lk}}{\partial x_l}\right)dV \\ &= \int_V \left(\epsilon_{ijk}\sigma_{jk} + \epsilon_{ijk}x_j\frac{\partial\sigma_{lk}}{\partial x_l}\right)dV\end{aligned} \tag{2.34}$$

Combining (2.33) and (2.34) in (2.32) gives

$$\int_V \epsilon_{ijk}x_j\left(\rho\frac{Dv_k}{Dt} - f_k - \frac{\partial\sigma_{lk}}{\partial x_l}\right)dV = \int_V \epsilon_{ijk}\sigma_{jk}dV \tag{2.35}$$

but the terms in parentheses on the left hand side are zero by the equation of motion (2.30). This leaves

$$\int_V \epsilon_{ijk}\sigma_{jk}dV = 0 \tag{2.36}$$

which is true for arbitrary volume, and therefore we may set the integrand to zero at every point. An equivalent condition is

$$\sigma_{kj} = \sigma_{jk} \tag{2.37}$$

This result shows the stress tensor to be symmetric. There are only six independent components of stress rather than nine; σ_{21}, σ_{32}, and σ_{31} are equal to σ_{12}, σ_{23}, and σ_{13} respectively. It is helpful to separate the stress tensor into *isotropic* and *deviatoric* parts and define the hýdrostatic pressure, p, by

$$\sigma_{ij} = -p\delta_{ij} + \sigma'_{ij} \tag{2.38}$$

where σ'_{ij} is called the deviatoric stress tensor or stress deviator. It has zero trace

$$\sigma'_{kk} = 0 \tag{2.39}$$

Setting $i = j$ in (2.38) and summing gives a relationship between the pressure and the stress tensor

$$p = -\frac{1}{3}\sigma_{kk} \tag{2.40}$$

2.6 Hydrostatic pressure

A fluid at rest cannot support shear stresses; consequently the stress tensor is isotropic. The momentum equation for a fluid at rest is obtained from (2.30) and (2.38) with $\boldsymbol{\sigma}' = 0$

$$\nabla p = \mathbf{f} \tag{2.41}$$

Taking the body force to be the gravitational force $\rho\mathbf{g}$ gives

$$\nabla p = \rho\mathbf{g} \tag{2.42}$$

The earth is mainly solid and therefore capable of supporting shear stresses, but they are very much less than the huge hydrostatic pressures pertaining deep within the earth. They are usually neglected in a first approximation to the basic state of stress. Furthermore all materials creep over a sufficiently long period of time when under sufficiently high stress, and on the geological time scale the earth is thought to behave as a fluid. Thus (2.42) will be a good first approximation to the force balance.

For a spherically symmetric earth (2.43) may be written as

$$\frac{dp}{dr} = -\rho(r)g(r) = -\rho(r)\frac{4\pi}{r^2}\int_0^r G\rho(r')r'^2 dr' \tag{2.43}$$

where we have used the simple result from Newton's law of gravity, that the gravitational acceleration due to mass within radius r is equal to that of the same mass concentrated on a point at the origin, provided the density is spherically symmetric. (2.43) is integrated to give the pressure in terms of the density

$$p(r) = -\int_a^r \rho(r')g(r')dr' \tag{2.44}$$

where a is the earth's radius and we have taken $p(a)$, the pressure at the earth's surface, to be zero. (2.44) allows calculation of the pressure at any depth in a self-gravitating liquid body from the density function.

A rough idea of the pressure at the centre of the earth is obtained by assuming uniform density. The variation with radius of the acceleration due to gravity is

$$g(r) = \frac{4}{3}\pi r \rho_0 G \tag{2.45}$$

Equation (2.44) then gives

$$\begin{aligned} p(0) &= \int_0^a \frac{4}{3}\pi r' \rho_0^2 G dr' \\ &= \frac{2}{3}\pi a^2 \rho_0^2 G \end{aligned} \tag{2.46}$$

Taking $a = 6371$ km, $\rho_0 = 5 \times 10^3$ kg m^{-3}, and $G = 6.67 \times 10^{-11}$ m^3 s^{-2} kg^{-1} gives $p(0) = 140$ GPa, or about a million atmospheres. Pressures in the earth's core are indeed measured in millions of atmospheres, although the pressure at the centre of the earth is rather more than this value because of the compression with depth: it is about 360 GPa.

Equation (2.42) is simply the differential vector form of the familiar equation for hydrostatic pressure in a liquid: $p = \rho g h$, where h is the head of fluid. The spherical geometry gives some interesting effects, one of which is illustrated by the following simple problem due to Professor F.C. Frank. A milk bottle is left to stand until the cream separates (Figure 2.3); it is then inverted and left for the cream to separate once more. Has the pressure at the bottom of the milk increased or decreased? Or perhaps it remains the same?

A sector through the earth's centre, defined in terms of spherical coordinates (r, θ, ϕ) by $\theta_0 < \theta < \theta_1, \phi_0 < \phi < \phi_1$, has the essential feature of an inverted milk bottle: it is wider at the top than at the bottom. Rearrangement of the earth's mass leads to a change in the hydrostatic pressure according to (2.42). If the reader finds the answer to Professor Frank's problem easy he can try the more geophysical example in problem 8 below.

2.7 The energy equation

It is not possible to discuss the purely mechanical aspects of deformation separately from thermal and other changes; the discussion must be based on thermodynamics. In seismology the deformations are often small and, for reasons discussed below, thermal changes can often be ignored. This is a rather dangerous approach if our understanding of continuum mechanics is to embrace other aspects of geophysics, which require a more general understanding. Here we develop a definition of an elastic solid which is based on thermodynamic principles.

The rate of working by body and surface forces is

$$\frac{dW}{dt} = \int_V v_i f_i dV + \oint_{\partial V} v_i \sigma_{ji} \hat{n}_j dS \qquad (2.47)$$

which can be converted to a volume integral by the divergence theorem

$$\frac{dW}{dt} = \int_V v_i f_i dV + \int_V \frac{\partial}{\partial x_j} (v_i \sigma_{ji})\, dV \qquad (2.48)$$

The rate of increase of thermal energy of a material volume is the sum of internal heat sources, h per unit volume, and the heat conducted across its boundary

$$\frac{dH}{dt} = \int_V h dV - \oint_{\partial V} \mathbf{q} \cdot \hat{\mathbf{n}} dS \qquad (2.49)$$

where $\mathbf{q}$ is the heat flux vector, the heat conducted across unit area in unit time. Fourier's law of heat conduction relates $\mathbf{q}$ to the temperature gradient

Figure 2.3: The milk bottle problem. The cream has separated in each case. Which bottle has the higher pressure on its base?

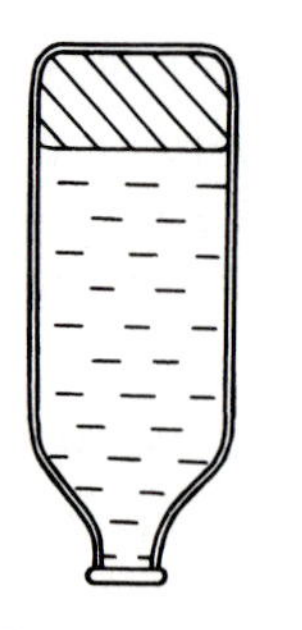

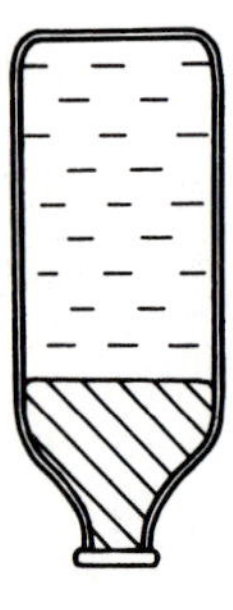

$$\mathbf{q} = -k\nabla T \tag{2.50}$$

where k is the thermal conductivity of the material. Writing (2.49) as a volume integral using the divergence theorem and combining with (2.50) gives

$$\frac{dH}{dt} = \int_V \left\{ h + \frac{\partial}{\partial x_j}\left(k\frac{\partial T}{\partial x_j} \right) \right\} dV \tag{2.51}$$

Conservation of energy requires the gain in internal and kinetic energy to be equal to the increase in thermal energy plus the rate of working by external forces. Thus

$$\frac{d}{dt}\int_V \left(\rho U + \frac{1}{2}\rho \mathbf{v}^2 \right) dV = \frac{d}{dt}(H + W) \tag{2.52}$$

where U is the internal energy per unit volume. Using (2.20) for the material derivative the left hand side of (2.52) becomes

$$\int_V \left(\rho\frac{DU}{Dt} + \rho\mathbf{v}\cdot\frac{D\mathbf{v}}{Dt} \right) dV \tag{2.53}$$

Substituting for $D\mathbf{v}/Dt$ from the equation of motion (2.30) gives

$$\rho\mathbf{v}\cdot\frac{D\mathbf{v}}{Dt} = v_i f_i + v_i\frac{\partial\sigma_{ji}}{\partial x_j} \tag{2.54}$$

Combining (2.48), (2.51), and (2.54) in (2.52) and cancelling terms leaves

$$\int_V \rho\frac{DU}{Dt}dV = \int_V \left(h + \frac{\partial}{\partial x_i}\left(k\frac{\partial T}{\partial x_i} \right) + \sigma_{ji}\frac{\partial v_i}{\partial x_j} \right) dV \tag{2.55}$$

This equation holds for arbitrary volume and we may therefore equate the integrands. First note that, as a consequence of symmetry of the stress tensor,

$$\sigma_{ji}\frac{\partial v_i}{\partial x_j} = \sigma_{ji}\frac{\partial v_j}{\partial x_i} = \sigma_{ji}\frac{De_{ij}}{Dt} = \sigma_{ij}\frac{De_{ij}}{Dt} \tag{2.56}$$

and therefore

$$\rho\frac{DU}{Dt} = h + \frac{\partial}{\partial x_i}\left(k\frac{\partial T}{\partial x_i} \right) + \sigma_{ij}\frac{De_{ij}}{Dt} \tag{2.57}$$

(2.57) is a local expression of the *principle of conservation of energy.*

2.8 Heat conduction

The differential expression for internal energy of a gas is probably more familiar to the reader than (2.57):

$$dU = TdS - pdV \tag{2.58}$$

where S is the specific entropy of the material. In continuum mechanics we deal with specific thermodynamic variables, and it is often more convenient to deal with the density of the material rather than the volume of the system. Density is the inverse of volume per unit mass, and in thermodynamic formulae we may write $\rho = V^{-1}$. Multiplying (2.58) by ρ gives

$$\rho dU = \rho TdS - p\frac{dV}{V} = \rho TdS + p\frac{d\rho}{\rho} \tag{2.59}$$

where dV/V is an infinitesimal volumetric strain.

From the definition of entropy, $dS = dq_{\mathrm{rev}}/T$, we may write

$$\rho TdS = h + \frac{\partial}{\partial x_i}\left(k\frac{\partial T}{\partial x_i}\right) \tag{2.60}$$

Substituting (2.60) into (2.58) gives

$$\rho\frac{DU}{Dt} = \rho T\frac{DS}{Dt} + \sigma_{ij}\frac{De_{ij}}{Dt} \tag{2.61}$$

In differential form this equation is

$$\rho dU = \rho TdS + \sigma_{ij}de_{ij} \tag{2.62}$$

Equation (2.62) shows that in a solid the internal energy depends on all components of the strain, not only the volumetric strain, and comparing (2.62) and (2.58) shows the strain energy density is $\sigma_{ij}de_{ij}$ rather than $-pdV/V$. (2.62) allows a thermodynamic interpretation of the stress tensor

$$\sigma_{ij} = \rho\left(\frac{\partial U}{\partial e_{ij}}\right)_S \tag{2.63}$$

Other thermodynamic relations generalise readily to the form for a deforming solid. Maxwell relations involving the strain and stress tensors replace the more familiar expressions containing the pressure and volume. For example, the *free energy* generalises to

$$\rho dF = -\rho S dT + \sigma_{ij} de_{ij} \tag{2.64}$$

and from the two expressions for the second derivative $\partial^2 F/\partial T \partial e_{ij}$ there follows the Maxwell relation

$$\left(\frac{\partial S}{\partial e_{ij}}\right)_T = -\left(\frac{\partial}{\partial T}\left\{\frac{\sigma_{ij}}{\rho}\right\}\right)_{e_{ij}} \tag{2.65}$$

We are now in a position to derive the differential equation governing the temperature. Regarding the internal energy as a function of temperature and elements of the strain tensor we have

$$\frac{DU}{Dt} = \left(\frac{\partial U}{\partial T}\right)_{e_{ij}} \frac{DT}{Dt} + \left(\frac{\partial U}{\partial e_{ij}}\right)_T \frac{De_{ij}}{Dt} \tag{2.66}$$

The first partial derivative is, by definition, the specific heat at constant strain (volume for a perfect gas)

$$C_V = \left(\frac{\partial U}{\partial T}\right)_{e_{ij}} \tag{2.67}$$

and from (2.62) we have

$$\left(\frac{\partial U}{\partial e_{ij}}\right)_T = T\left(\frac{\partial S}{\partial e_{ij}}\right)_T + \frac{\sigma_{ij}}{\rho} \tag{2.68}$$

and from the Maxwell relation (2.65)

$$\left(\frac{\partial U}{\partial e_{ij}}\right)_T = -T\left\{\frac{\partial}{\partial T}\left(\frac{\sigma_{ij}}{\rho}\right)\right\}_{e_{ij}} + \frac{\sigma_{ij}}{\rho} \tag{2.69}$$

Substituting (2.66), (2.67), and (2.69) into (2.57) gives

$$\rho C_V \frac{DT}{Dt} = \rho T\left\{\frac{\partial}{\partial T}\left(\frac{\sigma_{ij}}{\rho}\right)\right\}_{e_{ij}} \frac{De_{ij}}{Dt} + \frac{\partial}{\partial x_i}\left(k\frac{\partial T}{\partial x_i}\right) + h \tag{2.70}$$

The more familiar form of the equation of heat conduction is obtained by neglecting stress effects in (2.70) and taking constant thermal conductivity

$$\frac{DT}{Dt} - \kappa \nabla^2 T = h \tag{2.71}$$

where

$$\kappa = \frac{k}{\rho C_p} \tag{2.72}$$

is the *thermal diffusivity*. Now in (2.71) DT/Dt will typically be of order the temperature divided by the time scale of the changes, τ_k, and $\kappa\nabla^2 T$ will be of order $\kappa T/l^2$ where l is a typical dimension. Equating the two quantities and in the absence of heat sources gives the estimate

$$\tau_k = l^2/\kappa \tag{2.73}$$

τ_k is the time taken for thermal conduction to be effective; it is the *thermal diffusion time*.

Taking numerical values for olivine ($k = 4$ J K^{-1} m^{-1} s^{-1}, $\rho = 3.3\times 10^3$ kg m^{-3}, $C_p = 10^3$ J kg^{-1} K^{-1}) gives $\kappa = 1.2 \times 10^{-6}$m^{2}s^{-1}. Substituting into (2.73) with

(1) $l = 3$ m gives the conduction time $\tau_k = 4$ months
(2) $l = 1$ km gives $\tau_k = 26$ Kyr
(3) $l = 6000$ km gives $\tau_k = 950$ Gyr

We deduce from (1) that a wine cellar 3 m deep is reasonably well shielded from seasonal temperature variations; from (2) that tapping geothermal energy from a reservoir of linear dimension 1 km is in effect "mining" the heat because it takes a very long time to replenish the reservoir by thermal conduction through the rock; and from (3) that heat has not yet had time to have been conducted away from the centre of the earth because τ_k is several hundred times longer than the earth's age of 4.5 Ga.

We shall be concerned with deformations associated with seismic waves of period a second or so, and wavelengths of a few kilometres. The heat conducted over a kilometre in one second is quite negligible because from (2) above the conduction time is very long compared with the period of the wave; therefore these deformations may be regarded as adiabatic. It is commonly said that the period of the wave is too short for heat conduction to be important. In fact this is misleading because adiabaticity holds with even greater force at longer period: a wave with period a hundred times as great will have a conduction time 10^4 times as long for the same phase speed. The factor 10^4 increase in l^2 in (2.73) far exceeds the factor 10^2 increase in the time available for heat conduction.

2.9 Perfect elasticity

We all know that an elastic material is one that springs back to its original shape when the stresses are removed. An elastic body is more precisely defined as one which, having undergone deformation, returns to its original configuration when stresses are removed. Even this definition is somewhat loose because we have not specified what we mean by "configuration", which should include the full thermodynamic description of the body. It is better to work with the *strain energy function* which we can define through the differential relation (cf. equation (2.62))

$$\rho dR = \sigma_{ij} de_{ij} \tag{2.74}$$

An *elastic solid* is defined as one for which dR is a function only of the strain at the current time. A plastic body which can be permanently deformed by the applied stress will have a strain energy that depends on its deformation history (or strains up to the present time) as well as the current deformation. An elastic body has no memory of its deformation history.

We consider only adiabatic changes, and set $dS = 0$ in equation (2.62) so that $dU = dR$. We could equally deal with isothermal or more complicated changes, but the adiabatic condition is relevant to seismology because heat conduction is so slow. The resulting elastic parameters we derive will be the adiabatic ones, which are the pertinent material properties for application to most problems in geophysics. We write the internal energy as a power series in the strains and write the first two terms in the series as

$$U = \alpha_{ij} e_{ij} + \frac{1}{2} C_{ijkl} e_{ij} e_{kl} + \ldots \tag{2.75}$$

where the tensor C is a property of the material. From (2.63) and (2.75) we have

$$\sigma_{ij} = \alpha_{ij} + \frac{1}{2} \left(C_{ijkl} + C_{klij} \right) e_{kl} + \ldots \tag{2.76}$$

and without loss of generality we can choose C to satisfy

$$C_{ijkl} = C_{klij} \tag{2.77}$$

Differentiating (2.76) and using (2.63) gives

$$C_{ijkl} = \left(\frac{\partial^2 U}{\partial e_{ij} \partial e_{kl}} \right)_S \tag{2.78}$$

It is clear from (2.76) that α is the stress present in the unstrained medium. It is called the pre-stress; it has some relevance to seismology, for example in the study of waves in media under permanent stress, but we do not consider it in this book.

When strains are small we neglect higher order terms in (2.76) to give the linear relationship

$$\sigma_{ij} = C_{ijkl}e_{kl} \tag{2.79}$$

which is a generalised form of *Hooke's law* for linearised elasticity.

The fourth order tensor C is subject to some restrictions from the symmetry of the stress and strain tensors. These conditions require that

$$C_{ijkl} = C_{jikl} = C_{jilk} \tag{2.80}$$

This leaves the elasticity tensor with 21 independent components. The strain energy is given by

$$\rho R = \frac{1}{2}\sigma_{ij}e_{ij} \tag{2.81}$$

Hooke's law (2.83) allows us to estimate the stress release during an earthquake. Taking shear modulus $\mu = 5 \times 10^{10}$ Pa and a strain of 10^{-4} gives a stress of 5 MPa, or about 50 bar. Compare this with the internal hydrostatic pressures of some 100 GPa. The stresses generated by propagating seismic waves are smaller than those released by earthquakes by several orders of magnitude.

2.10 The elastic moduli

An elastic solid is said to be *homogeneous* if the elements of the elasticity tensor are independent of position; *isotropic* if they are independent of direction. Homogeneous media have constant C. Isotropic media have elasticity tensors that are invariant under rotation, the most general form of which is

$$C_{ijkl} = \lambda\delta_{ij}\delta_{kl} + \mu\left(\delta_{ik}\delta_{jl} + \delta_{il}\delta_{jk}\right) \tag{2.82}$$

(This is not the most general fourth order isotropic tensor — we have used the symmetry properties of C — see problem 2.) λ and μ are scalar functions of position; they are called the *Lamé parameters.*

Substituting (2.82) into (2.77) gives Hooke's law for an isotropic solid

$$\sigma_{ij} = \lambda e_{kk}\delta_{ij} + 2\mu e_{ij} \tag{2.83}$$

Many materials display anisotropy but few require the full 21 independent components of C. For example, the elastic behaviour of crystals that exhibit high symmetry may be described with only a few components, the interdependence of the elements of C being determined by the crystal symmetry (see problem 12).

The variation of the Lamé constants with depth within the earth is an important inhomogeneity; we shall consider it in detail in the subsequent chapters. The earth is to a good approximation spherically symmetric, but departures from spherical symmetry, or *lateral inhomogeneities*, are an important area of current geophysical research. We shall mention it only in passing. Likewise the earth is to a good approximation isotropic, although again anisotropy is significant and an active area of research.

In seismology anisotropy manifests itself most straightforwardly by a variation of the phase speed of seismic waves with their direction of propagation. The upper mantle beneath the ocean basins, at a depth of about 5 km beneath the sea bottom, is anisotropic, with the faster wave speed perpendicular to the mid-ocean ridges. This was shown directly in a carefully designed experiment carried out in the north-east Pacific Ocean by Raitt, Shor, Francis & Morris (1969).

Seismic velocity in the uppermost mantle is measured by recording the time of arrival of seismic waves generated by explosions. Recording along a line of receivers to give a range of distances up to 100 km or more allows removal of the effects of propagation through the crust. Early experiments in the Pacific between California and Hawaii had yielded seismic wave speeds ranging from 8.6 km s^{-1}, in a roughly east west direction, down to 8.0 km s^{-1}, in a roughly north south direction, suggesting anisotropy. No experiments had been repeated at the same site however, so the results were ambiguous and could be interpreted as either inhomogeneity or anisotropy.

An experiment was designed to measure seismic velocity in several directions at the same site; consequently the shot and receiver patterns shown in Figure 2.4 were devised. *Flip* was a receiving ship; *Baird* was capable of both firing shots and receiving arrivals. For the *Quartet* configuration (Figure 2.4(a)) a third ship fired shots along the line joining *Flip* and *Baird*, and its perpendicular bisector. By recording arrivals at both ships it was possible to eliminate time delays due to effects near the source (similar techniques will be discussed in Chapter 4). Only two ships were available for the *Flora* operation (Figure 2.4(b)); *Baird* fired shots on circular and in-and-out tracks for the receiving ship.

The results for the wave speed are shown in Figure 2.5. There is clear evidence of anisotropy in excess of 0.2 km s^{-1} between the E W and N S directions.

For isotropic media it is convenient to separate the strain tensor into isotropic and deviatoric parts, in the same way in which we separated the stress tensor in (2.38)

$$e_{ij} = \frac{1}{3} e_{kk} \delta_{ij} + e'_{ij} \tag{2.84}$$

Figure 2.4: (a) Shot and receiver pattern for the *Quartet* experiment. *Flip* and *Baird* were ships that received signals from shots fired from a third ship on the two lines shown. (b) Shot and receiver pattern for the *Flora* experiment. Only two ships were available for this experiment. *Baird* fired shots along circular and in-and-out tracks, with *Flip* receiving. From Raitt *et al.* (1969).

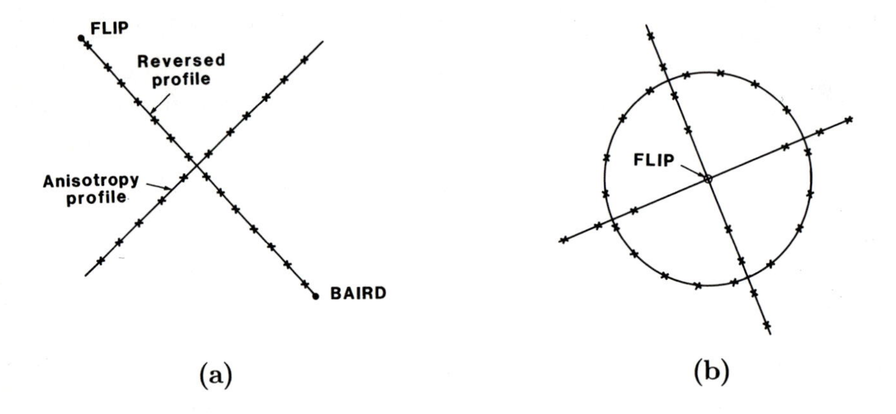

Figure 2.5: Estimated seismic velocities of the uppermost mantle from the *Quartet* experiment as a function of direction showing variation of about 0.2 km s^{-1}, with the fast direction roughly east-west. From Raitt *et al.* (1969).

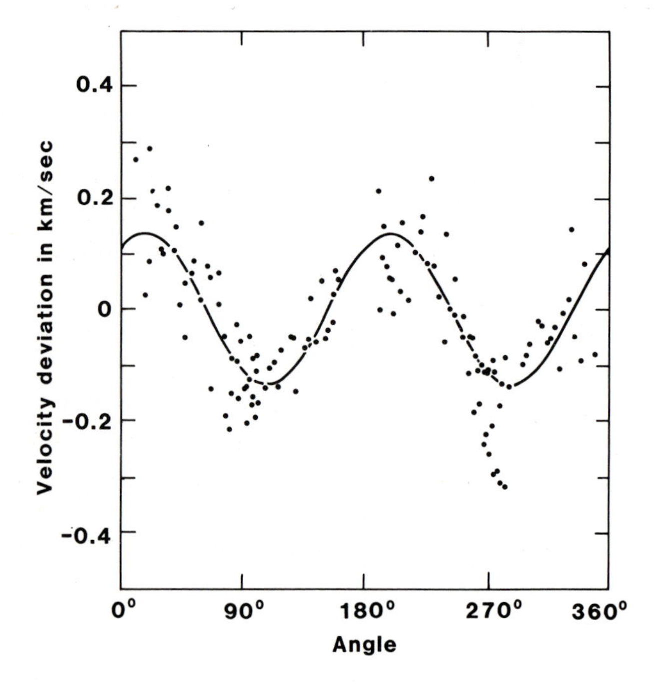

where the deviatoric strain

$$e'_{ij} = e_{ij} - \frac{1}{3} e_{kk} \delta_{ij} \tag{2.85}$$

represents *pure shear* (no change in volume), and the remaining part of e_{ij} represents the volume change. Then Hooke's law (2.83) may be written as

$$\sigma_{ij} = \left(\lambda + \frac{2}{3} \mu \right) e_{kk} \delta_{ij} + 2\mu e'_{ij} \tag{2.86}$$

Separating the stress into isotropic and deviatoric parts according to (2.38) gives

$$p = -\left(\lambda + \frac{2}{3} \mu \right) e_{kk} \tag{2.87}$$

$$\sigma'_{ij} = 2\mu e'_{ij} \tag{2.88}$$

Here μ relates the stress to pure shear; it is called the *shear modulus.* $(\lambda + 2\mu/3)$ relates changes in pressure to changes in volume; it is called the *adiabatic bulk modulus*

$$\kappa_S = \left(\lambda + \frac{2}{3} \mu \right) \tag{2.89}$$

From (2.87) we see that

$$\kappa_S = -V \left(\frac{\partial p}{\partial V} \right)_S \tag{2.90}$$

and (2.89) is consistent with the thermodynamic definition of the adiabatic bulk modulus.

2.11 The Navier equation

We are now in a position to specialise the equation of motion (2.30) to small displacements of a linear elastic solid. If velocities are small we may neglect the term in $(\mathbf{v} \cdot \nabla)\mathbf{v}$, which is quadratic in the small quantity, compared with $\partial \mathbf{v}/\partial t$. Furthermore the difference between Lagrangian and Eulerian derivatives is also small (see problem 2) and we may replace $\partial \mathbf{v}/\partial t$ with $\partial^2 \mathbf{u}/\partial t^2$.

Using Hooke's law in the form (2.83) gives

$$\begin{aligned} \frac{\partial \sigma_{ij}}{\partial x_j} &= \frac{\partial}{\partial x_j} (\lambda e_{kk} \delta_{ij} + 2\mu e_{ij}) \qquad (2.91) \\ &= \frac{\partial}{\partial x_i} (\lambda \nabla \cdot \mathbf{u}) + \mu \left(\nabla^2 u_i + \frac{\partial}{\partial x_i} \nabla \cdot \mathbf{u} \right) + 2 \frac{\partial \mu}{\partial x_j} e_{ij} \end{aligned}$$

We may use the vector notation $\nabla\mu \cdot \mathrm{e}$ to denote the contraction of e with $\nabla\mu$ without ambiguity because $\mathbf{e}$ is symmetric. The divergence of the stress tensor is then

$$\frac{\partial \sigma_{ij}}{\partial x_j} = \left[\nabla\left(\lambda\nabla\cdot\mathbf{u}\right) + \mu\nabla^2\mathbf{u} + \mu\nabla\left(\nabla\cdot\mathbf{u}\right) + 2\nabla\mu\cdot\mathrm{e}\right]_i \tag{2.92}$$

The only body forces we shall include are those due to gravity: $\mathbf{f}_g = \rho\nabla\psi$. Then (2.30) becomes

$$\rho\frac{\partial^2\mathbf{u}}{\partial t^2} = \nabla\left(\lambda\nabla\cdot\mathbf{u}\right) + 2\nabla\mu\cdot\mathrm{e} + \mu\nabla^2\mathbf{u} + \mu\nabla\left(\nabla\cdot\mathbf{u}\right) + \rho\nabla\psi \tag{2.93}$$

This is called the *Navier equation*; it governs small displacements from an equilibrium state of zero elastic stress.

The very large hydrostatic pressure within the earth appears in (2.93) along with its deformation. Our primary aim is to investigate the displacements associated with seismic disturbances and this large static pressure need not concern us provided there are no shear stresses associated with it. There will always be some shear stresses sustained by an elastic solid after compression, and in the earth they can be significant for seismology. However, the material cannot sustain shear stresses indefinitely and they relax by plastic flow after a sufficiently long period of time. The ice sheets which formed in northern Canada and Fennoscandia during the last glaciation caused substantial deformation, and as they melt the ice load is removed, setting up shear stresses. From the present day uplift in these regions we can estimate that mantle rocks must flow to remove these applied stresses over a period of some tens of thousands of years. This is a very short time compared with most geological processes, and we can justify ignoring these shear stresses when considering the gross properties of the earth and the static state at depth.

The hydrostatic stress and its accompanying deformation may therefore be subtracted out of (2.93); the hydrostatic balance (2.42) may be written in terms of displacements as

$$\rho_0\nabla\psi_0 + \nabla\left\{\left(\lambda + \frac{2}{3}\mu\right)\nabla\cdot\mathbf{u}_0\right\} = 0 \tag{2.94}$$

where subscript zero denotes the value in the static equilibrium. Subtracting (2.94) from (2.93) leaves the Navier equation unchanged provided we replace $\rho\nabla\psi$ with the departure of the gravitational force from the equilibrium value, $\rho\nabla\psi - \rho_0\nabla\psi_0$.

2.12 Determination of density within the earth

We end this chapter with an illustration of how simple assumptions about the thermodynamic state in the interior of the earth have given a remarkably accurate picture of the density deep within the earth, a calculation that was subsequently used to determine the probable composition. The calculation is probably the major achievement of seismology in the first half of the twentieth century. We show in the next three chapters that seismology can be used to determine the speeds of elastic waves rather well, but these are a combination of the density of the earth and the elastic moduli. We assume for the present that we have estimated the ratio κ_S/ρ throughout the earth. An assumption of adiabaticity, coupled with the assumption of chemical homogeneity, is now used to calculate the density. The earth is probably very close to adiabatic in most regions. It is not chemically homogeneous, but the relationship derived in this section can be used to relate densities throughout homogeneous layers of the earth.

We first need to derive an expression relating density to pressure. Consider the density of a chemically homogeneous region as a function of pressure and entropy. Elemental changes of density are given by

$$d\rho = \left(\frac{\partial\rho}{\partial p}\right)_S dp + \left(\frac{\partial\rho}{\partial S}\right)_p dS \tag{2.95}$$

The total radial derivative of the density is therefore given by

$$\frac{d\rho}{dr} = \left(\frac{\partial\rho}{\partial p}\right)_S \frac{dp}{dr} + \left(\frac{\partial\rho}{\partial S}\right)_p \frac{dS}{dr} \tag{2.96}$$

Assume the pressure is hydrostatic and given by (2.43), then (2.96) becomes

$$\frac{dp}{dr} = -\frac{\rho^2 g}{\kappa_S} + \rho\alpha\tau \tag{2.97}$$

where

$$\alpha = -\frac{1}{\rho}\left(\frac{\partial\rho}{\partial T}\right)_p = \frac{1}{V}\left(\frac{\partial V}{\partial T}\right)_p \tag{2.98}$$

is the thermal expansion coefficient and

$$\tau = -\left(\frac{\partial T}{\partial S}\right)_p \frac{dS}{dr} \tag{2.99}$$

τ represents a departure from the adiabatic state. To see this, consider the temperature a function of radius and entropy (since pressure must increase monotonically with decreasing radius, the radius may be thought of as playing the part of the thermodynamic variable $-p$)

$$\frac{dT}{dr} = \left(\frac{\partial T}{\partial r}\right)_S + \left(\frac{\partial T}{\partial S}\right)_p \frac{dS}{dr} \tag{2.100}$$

The last term is just τ, so

$$\tau = \left(\frac{\partial T}{\partial r}\right)_S - \frac{dT}{dr} \tag{2.101}$$

which is just the difference in temperature gradients (adiabatic minus actual).

In an adiabatic region we take $\tau = 0$, in which case (2.97) becomes the *Adams-Williamson* equation, after the first two people to use it to determine a density model of the earth in 1923. Rewriting (2.97) in the adiabatic case as

$$\frac{d\rho}{dr} = \frac{G\rho^2}{\kappa_S r^2}\left\{\int_r^a 4\pi\rho(x)x^2 dx - M\right\} \tag{2.102}$$

where M is the mass of the earth. Knowing M from astronomical measurements, κ_S/ρ from seismology (Chapter 4), and having some estimate of the density at the earth's surface, $\rho(a)$, we can integrate (2.102) downwards from the earth's surface into the interior until some depth is reached where the assumption of either adiabaticity or chemical homogeneity is violated. For example, we could divide the earth into uniform layers of thickness Δr, choose a value of ρ in the top layer, evaluate the integral on the right hand side to give a value of $d\rho/dr$ at $r = a$, determine $\rho(a - \Delta r) = \rho(a) - \Delta r d\rho/dr$, repeat the process for the next layer down, and so on.

Bullen used this process with a model of the κ/ρ to obtain an early but definitive model for the earth's density. He first assumed $\rho(a) = 3.3 \times 10^3$ kg m^{-3} and obtained a density for the entire mantle. He then calculated the mass and moment of inertia of his model and subtracted the results from the measured mass and moment of inertia of the whole earth (the latter can be deduced from the earth's precession frequency) to obtain an estimate for the mass and moment of inertia of the core, M_C and I_C. This estimate gave $I_C = 0.57 M_C a^2$. The constant is much larger than the figure of 0.4 for a uniform sphere, and if correct it would imply the outer layers of the core were more dense than the deep interior. This situation is quite unstable, and if it were ever to be imposed it would immediately collapse to a state with heavier material beneath lighter material: we must have $I_C < 0.4 M_C a^2$ for gravitational stability.

Bullen deduced that his density model was in error because of chemical inhomogeneity, and introduced discontinuities into his next calculation. Each discontinuity requires a new value of the density at one point to start the calculation off; he therefore evaluated the density for the whole earth and used the moment of inertia and some conditions for matching the densities and density gradients in the chemically inhomogeneous regions. His model of the earth is still grossly correct today; a modern estimate of density and the corresponding regions within the earth are given in Figure 2.6. The crustal layers vary laterally to such an extent that it is not possible to categorise them in any simple way. The Moho is a major seismic discontinuity that occurs everywhere on the earth and marks the interface between crust and mantle. It is almost certainly a boundary between materials of different compositions because the change in seismic parameters is so sharp. Bullen began his density calculation at the top of the mantle, at the Moho.

The first discontinuity is at 450 km; this is almost certainly a phase change of the minerals of the mantle from an olivine structure to a high pressure form called spinel. The next major discontinuity is at about 650 km, again believed to be a phase change to post-spinel, probably a

Figure 2.6: A model of density in the earth showing the major discontinuities, which are interpreted as changes in composition. The smooth variations within each homogeneous zone are attributable to changing pressure and adiabatic conditions. (Model A, from BULLEN (1975).)

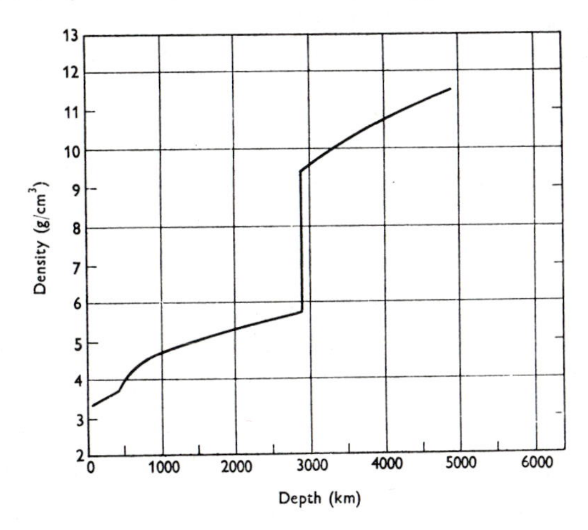

stishovite structure. Bullen had classified the whole region between these discontinuities as a transition zone of gradual change in properties; the discontinuities were detected subsequently by the use of seismic arrays as described in Chapter 4. The whole region down to the 650 km discontinuity is known as the *upper mantle*; there are no earthquakes below this depth and all mantle motions as mapped by earthquakes are restricted above this discontinuity.

The region below the 650 km discontinuity is called the *lower mantle*. It is very uniform and appears to be chemically homogeneous, except for a layer a few hundred kilometres thick at the boundary with the core, which appears to scatter seismic waves and may therefore be laterally inhomogeneous. The core is a very good transmitter of seismic waves, and appears very homogeneous. Bullen placed a layer around the inner core because reflections from the surface of the inner core were not clear, and sharp refractions were sometimes seen. This layer is no longer believed by many seismologists because the energy once thought to be refracted from the inner core surface is now known to originate at the base of the mantle, the D'' region (another success for seismic array studies, discussed in Chapter 4).

There is now a method of estimating density directly, using free oscillation frequencies, which are described in Chapter 5. This modern work has produced improvements in our knowledge of the density of the deep interior of the earth, but nowhere has it substantially changed the picture derived from the method given here.

Exercises

1 Use the continuum hypothesis to argue that two elements of fluid that are initially adjacent will always remain so, and that any fluid element initially adjacent to an interface with a solid will always remain on the interface.

2 Show that there is no difference between the time derivatives in the two descriptions for small displacements, and that D/Dt may be replaced with $\partial/\partial t$. Deduce that there is no difference between Eulerian and Lagrangian representations of the rate of change of dilatation when strains are small. Hence show, starting from the Lagrangian form, the Eulerian form of the equation of continuity for small strains.

3 Any tensor may be separated into symmetric and antisymmetric parts. Give expressions for θ and e$'$ in terms of elements of the strain tensor if

$$e_{ij} = \theta\delta_{ij} + e'_{ij} \tag{2.103}$$

where $e'_{kk} = 0$. Find the vector **a** such that

$$e'_{ij} = \epsilon_{ijk}a_k \tag{2.104}$$

This strain tensor is said to have been separated into dilatation and *deviatoric strain*, e′. The latter describes pure rotation. Give a geometrical interpretation of the vector **a**.

4 A strainmeter is simply a device for measuring changes in distance between two points fixed on the ground. It may consist of a wire stretched between two points, or use some form of interferometer and a laser beam. Strainmeter records show large regular oscillations with tidal frequencies: they are called solid earth tides and are driven by the varying gravitational attraction of the sun and moon, in similar fashion to the driving of ocean tides. The strainmeters show typical tidal amplitudes of 3×10^{-8}.
   a Estimate the ground displacement for a typical earth tide (assume the wavelength of an earth tide to be approximately the earth's radius, 6371 km).
   b Surface waves are studied in the next chapter. They typically have periods of 20 s and wavelengths of 100 km; the strains measured teleseismically (for example in Cambridge for a large earthquake in Alaska) are typically 10^{-7}. Estimate the ground displacement. You should find it surprisingly large. (There is no cause for concern because the associated strain is so small!)
   c Estimate also the ground displacement of higher frequency waves with period 1 s and wavelength 10 km (typical of body waves which we shall also study in the next chapter), with strain 10^{-10}.

5 Consider a seismic wave with period 20 s, wavelength 100 km, and ground displacement 10^{-2} m. Estimate the gravitational restoring force caused by the displacement of the equal density surfaces (assuming g remains constant) and compare the result with the rate of change of momentum. Repeat the calculation for 1 s period waves with wavelength 10 km. Deduce the importance (or lack of it) for these waves of self-gravitation.

6 Estimate the acceleration of the ground during passage of each of the wave types in exercise 4 and compare the result with the value of g. How large is the acceleration for an earth tide? Could you detect these oscillations on a gravimeter with a sensitivity of 10^{-11} m s^{-2}?

7 Use Hooke's law and a bulk modulus of 100 GPa to estimate the compression of the iron in the earth's core caused by the hydrostatic pressure. The density of iron at standard temperature and pressure is 7.5×10^3 kg m^{-3}; estimate the density of iron in the core.

8 A 3×3 matrix contains nine elements; a symmetric matrix will have only six independent elements — the three diagonal elements and three of the remaining six. Show that symmetry of the stress tensor reduces the number of independent ij combinations in the elasticity tensor C_{ijkl} from 9 to 6; symmetry of the strain tensor in addition reduces the number of independent elements in all from 81 to 36. Show the condition of differentiability of the strain energy function (2.77) reduces the number of independent components from 36 to 21. Find the number of independent components when the material structure is invariant under any

rotation of the coordinate axis that leaves each of the new axes parallel or antiparallel to one of the old axes.

9 The most general form of the fourth-order isotropic tensor is

$$a\delta_{ij}\delta_{kl} + b\delta_{ik}\delta_{jl} + c\delta_{il}\delta_{jk} \tag{2.105}$$

Show that the elasticity tensor for an isotropic medium has the form (2.82).

10 Taking strains of 10^{-8} as typical of seismic waves, estimate the pressure and strain energy for the waves listed below. From the pressure changes use (2.70) to estimate the temperature change and compare TdS with the strain energy. Hence justify the assumption of adiabaticity.

a Period 1 s, wavelength 5 km.

b Period 1 hr, wavelength 1000 km.

11 Using the thermodynamic relation for a fluid

$$\kappa_S = \kappa_T(1 + T\alpha\Gamma); \; \Gamma = \kappa_S\alpha/\rho C_p \tag{2.106}$$

where κ_T is the isothermal bulk modulus, estimate the percentage difference between adiabatic and isothermal bulk moduli for olivine. The following numerical values are typical: $\alpha = 4 \times 10^{-4}$ K^{-1}, $k = 4$ J K^{-1} m^{-1} s^{-1}, $\rho = 3.3$ Mg m^{-3}, $C_p = 10^{-3}$ J kg K^{-1}.

12 The student should familiarise himself with the solution of the linear flow of heat in a rod. It is described in detail in CARSLAW & JAEGER, p. 133ff. In particular solve the problem of the temperature $T(x,t)$ of a semi-infinite rod initially at temperature T_0 with boundary condition $T(0,t) = 0$ where $x = 0$ is the end of the rod. By linear we mean that heat is conducted only along the rod, and no heat is lost from the sides of the rod. The temperature satisfies a one-dimensional form of equation (2.71), which can be solved by a variety of methods. This simple problem is a model for some of the calculations that are done in Chapter 7 on plate tectonics; it also gives a quantitative value for the diffusion time for this particular geometry.

13 Calculate the pressure at the centre of the earth assuming a constant density as in the text, but using only the value of g at the surface as data.

14 Calculate g at the core–mantle boundary ($r = 3485$ km) and at the earth's surface ($r = 6371$ km) for an earth model with uniform density 10^3 kg m^{-3} in the core and uniform density 4×10^3 kg m^{-3} in the mantle and crust. Compare the results with the earth model in Table 5.1.

15 Calculate the change in hydrostatic pressure within the earth caused by each of the following:

a An ice age producing a uniform 100 m thick layer of ice over the earth's surface.

b Formation of the upper mantle from undifferentiated mantle.

Take the density of sea water to be 1.1×10^3 kg m^{-3}, of ice 10^3 kg m^{-3}, and the mean depth of the oceans 2 km. Take the densities of upper mantle material to be 4.1×10^3 kg m^{-3}, for that of undifferentiated mantle 4.53×10^3 kg m^{-3} and the depth of the upper mantle 700 km. The depth of the whole mantle can be found from the data given for exercise 14.

Summary

We have derived basic equations that govern the elastic deformation in the earth, emphasising those basic concepts of continuum mechanics and thermodynamics on which the equations governing seismic waves rest.

The earth is treated as a CONTINUUM; its molecular structure is of no consequence and we may talk of an ELEMENT of material that is infinitesimally small. The continuum hypothesis allows us to express physical laws, which are formulated in terms of bodies of finite size, as differential equations which apply to an element of the continuum, effectively at a point. The movement and deformation of the continuum may be described mathematically in two ways: we may label an element of material by its position at some reference time and express its position in space at some later time as a function of that initial position (LAGRANGIAN); or we may specify the deformation of the element in terms of its current position in space (EULERIAN). As a general rule the Lagrangian formulation is the more useful in elasticity theory and the Eulerian formulation is used widely in fluid dynamics. For small deformations the two formulations are in many respects indistinct, but we make use of both.

Displacement of the material is less important than STRAIN. Linear strain is simply the extension per unit length (of a rod for example) but our medium is three-dimensional and there are nine components of strain corresponding to the three directions of displacement and the three directions of "unit length". The strain must be represented as a tensor, but only the six independent components of the symmetric strain tensor are needed for elasticity theory.

Forces acting on the medium are of two types: BODY FORCES that act on the main body of the material and may be expressed per unit volume; and surface forces (more commonly called TRACTIONS) that act on surfaces in the material and are expressed per unit area. Gravitational attraction is an example of a body force. The earth generates its own gravitational force, through the mass of material itself, according to Newton's law of gravity. The earth is said to be SELF-GRAVITATING. The gravitational forces are crucial in determining the pressure within the earth and the equilibrium density of the material. We can neglect its effect on short-period seismic waves; its effect on the very longest period waves is still quite small but measureable and important in determining the density of the earth's interior.

Surface tractions are represented by the STRESS TENSOR, an element of which gives the force in a given direction per unit area aligned in a second given direction. There are nine elements of the stress tensor in general, corresponding to the 3×3 directions involved, but conservation of angular momentum requires that only six of these be independent: the stress tensor is symmetric. Three physical laws were formulated as differential equations using the continuum hypothesis: conservation of mass, momentum, and energy. Momentum conservation gives us a general equation of motion that is further specialised to elastic solids. Energy considerations and basic concepts of thermodynamics tell us how work done by the various forces interacts with the thermal state of the material. The conclusion, even at this early stage, is that thermal conduction is very slow in the earth and virtually all of the dynamic behaviour we are likely to study will be adiabatic.

Having dispensed with the preliminary fundamentals we define an ELASTIC MATERIAL: a material for which the deformational energy depends only upon the current strain (and not the strain history). A further specialisation to small strains allows us to develop linear elasticity and use HOOKE'S LAW to relate the stress and strain tensors. For a rod these are simply related through Young's modulus, but in our three-dimensional medium we need a fourth-order tensor for the relationship. Symmetry of the stress tensor reduces the number of components from 81 to 21.

The earth is both INHOMOGENEOUS (its properties vary from place to place) and ANISOTROPIC (its properties at any one place vary with direction); there is plenty of evidence for both of these complications from seismology but the major effect is the variation of properties with depth and we specialise to an isotropic solid. This leaves only two independent components of the elasticity tensor which can be grouped in various ways. We use the Lamé constants and relate them to the bulk and shear moduli. The simplified version of Hooke's law is put into the equation of motion to leave the NAVIER EQUATION for an isotropic elastic solid. It is a differential equation for the displacement of the material.

The conditions of hydrostatic equilibrium and self-gravitation were combined to give the ADAMS WILLIAMSON equation, which is used to infer density within the earth from estimates of the adiabatic bulk modulus.

Further reading

There are a great many standard works on continuum mechanics: MALVERN deals comprehensively and rigorously with continua of all types and is particularly strong on plasticity and all kinds of stress-strain relations; LANDAU & LIFSHITZ (vol. 7) is specific to elastic media and is strong on thermodynamics; BATCHELOR is specific to fluid dynamics. Generally speaking texts on theoretical seismology tend to start with the elasticity equations and do not delve deeply into their underlying physical basis as we have done here. The author knows of no good discussion of self-gravitation. CARSLAW & JAEGER gives a comprehensive discussion of heat conduction. Tensors can be brushed up with JEFFREYS(1975). BULLEN describes determination of density within the earth.

3

Elastic waves in simple media

We are concerned with observations of ground motion at long distances from the earthquake source–*teleseismic* distances. The displacement travels from the source by radiation of elastic waves. Some of these waves are attenuated only very slightly in their passage through the earth: indeed, some wavetrains have been observed to traverse the earth several times, and long period disturbances from great earthquakes can be observed for several days after the event. The ground strains are very small at teleseismic distances, but the record of ground motion contains a great deal of information about the source and, more importantly for our purposes, about the propagation of elastic waves within the earth.

The simplest measurements to make are the time of onset of a particular wave packet and perhaps its dominant frequency and amplitude. Very considerable progress has made recently in matching the wave form by some theoretical model, particularly at long period, but only brief mention is made of this in the book. It is a complicated subject requiring computer matching of observations and theory. We can gain considerable insight into earth structure by simply observing the first onset of packets of energy – *phases* – and identifying them with elastic waves that are known to travel in simple media. By "simple media" we mean ones in which the elastic parameters are homogeneous and isotropic, or consist of a few homogeneous layers.

3.1 The seismometer

A *seismometer* is any instrument which measures the time dependence of the ground displacement. The measurement is not straightforward because the ground on which the instrument is placed moves. We use inertia: for sufficiently rapid ground motion a suspended mass will remain at rest, providing a reference point against which we can measure movement of the ground. The *pendulum seismometer* consists of a mass on a spring with some form of damping; the principle is shown schematically in Figure 3.1 where the damping is provided by a dashpot in parallel with the spring. The spring provides a restoring force proportional to its extension, the dashpot a force proportional to the velocity of the mass. Let the constants of proportionality be k and D respectively. Assume all displacements are in the x direction. Let u be the ground displacement and ξ the displacement of the mass from its equilibrium position, ξ_0. $\xi + \xi_0$ is the displacement of the mass from the ground. The equation of motion is

$$M(\ddot{\xi} + \ddot{u}) + D\dot{\xi} + k\xi = 0 \tag{3.1}$$

which on rearrangement gives

$$\ddot{\xi} + 2\epsilon\dot{\xi} + \omega_s^2\xi = -\ddot{u}. \tag{3.2}$$

where $\epsilon = D/2M$ and $\omega_s^2 = k/M$. Clearly we can use (3.2) to find the ground motion, u, by measuring ξ and its time derivatives.

When ground acceleration is large equation (3.2) reduces to $\xi = -u$ and the output of the instrument is interpreted directly as ground displacement. When ground acceleration and velocity are small, that is the frequency of ground motion is low, equation (3.2) reduces to $\xi = -\ddot{u}/\omega_s^2$ and the output is interpreted in terms of ground acceleration. In this case if the ground exhibits simple harmonic motion with frequency ω the

Figure 3.1: A schematic representation of the pendulum seismometer. From Aki & Richards, copyright ©1980 W.H.Freeman & Co. Reprinted with permission.

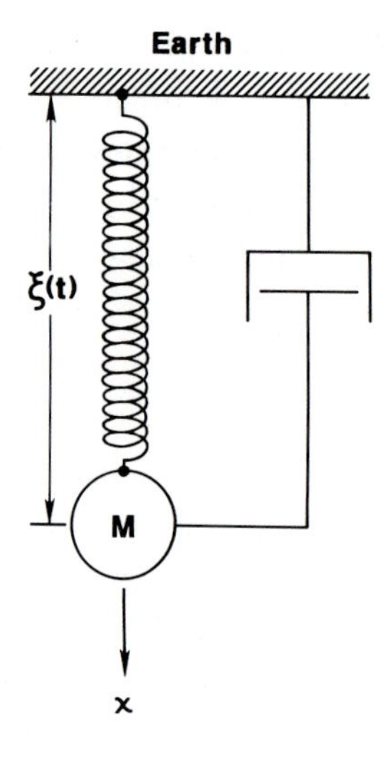

response of the instrument is $\xi = (\omega/\omega_s)^2 u$. This is small if $\omega << \omega_s$. Instruments sensitive to low frequencies, *long period seismometers*, are designed by decreasing the frequency ω_s, either by increasing the mass M (20 ton masses have been used in the past) or by using a horizontal beam supported by a spring. In principle the period of such a pendulum can be made infinite, in which case the instrument is called a *gravimeter* because it measures changes in the vertical gravitational force. Two other types of instrument for measuring long period displacements are strainmeters, which measure the distance between two points fixed in the ground by, for example, setting up an optical interference pattern and counting the movement of fringes; and tiltmeters, which measure tilt by, for example, measuring the difference in water level heights at different points.

The responses of the instrument to different frequency inputs from the ground are described by the *frequency response*. For the simple pendulum seismometer it is obtained from (3.2) by applying a single frequency input of unit amplitude: $u = \exp -i\omega t$. The frequency response is

$$X(\omega) = \frac{-\omega^2}{\omega^2 + 2i\epsilon\omega - \omega_s^2} \tag{3.3}$$

which can be separated into an *amplitude response* $|X(\omega)|$ and a *phase response* $\phi(\omega)$ where

$$X(\omega) = |X(\omega)| \exp i\phi(\omega) \tag{3.4}$$

An equivalent description is via the *impulse response* of the instrument, which gives the output ξ for a delta-function input; the frequency response is the Fourier transform of the impulse response.

Ocean waves produce a large amount of energy at periods around 6 s. When measured on a seismometer these are called *microseisms*; they dominate the ground motion at these frequencies. Instruments for seismology must therefore be designed to be unresponsive in the microseism frequency band. This is accomplished in the conventional pendulum seismometer by using two separate instruments, one with a period $(2\pi/\omega_s)$ less than 6 s which responds to short period ground motion, and one with a period longer than 6 s which responds to long period ground motion, both instruments having low response in the microseismic band. In this book we shall use measurements from instruments with periods of 1 s and 20 s respectively. Their frequency responses are shown in Figure 3.2.

Modern electronic filtering means that we can use a single mechanical sensor with a high response over a broad range of frequencies, a *broad-band* instrument, and subsequently shape the response into one appropriate for either a short- or long-period instrument. The frequency response is usually specified in terms of the Laplace transform of the impulse response as the ratio of two polynomials; the positions of the poles and zeroes of the Laplace transform in the complex plane are sufficient to describe the entire response. The *SRO* (*Seismological Research Observatory*) instrument responses are specified in this way; we study records from these instruments in Chapter 5. Digital recording gives greater dynamic range, which allows recording of both small and large displacements in the same record. We can therefore dispense with separate short and long period instruments and record all of the information given by a broad-band sensor and process the microseismic band subsequently. The main limitation at present is the sheer volume of data that must be collected if the signal is sampled sufficiently often to provide a valuable record for short-period studies and for a sufficiently long period of time for long-period studies. Broad-band instruments are likely to replace conventional seismometers as the technology for handling large volumes of data improves.

The Wood-Anderson seismometer is particularly important because it was used in 1935 by C. F. Richter to define *earthquake magnitude.* This lightweight, portable instrument uses a 5 gram mass suspended on tor-

Figure 3.2: Instrument responses for *WWSSN*, from Peterson & Orsini (1976).

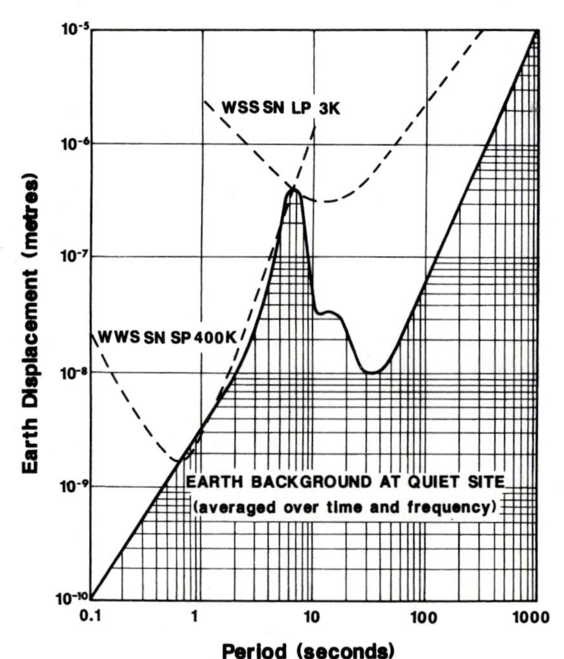

sion wire with optical recording and electromagnetic damping. It has a period of 0.8 s and magnification of the ground motion of 3000. The local magnitude was defined as $\log_{10}(\Delta)$, where Δ is the maximum recorded amplitude in μm at a distance of 100 km from the earthquake. Gutenberg and Richter worked in California to establish a relationship between earthquake magnitude and energy released by the earthquake. At teleseismic distances the short period waves measured by the Wood-Anderson seismometer are scattered, giving an unreliable estimate of the size of the earthquake, and longer period waves (surface waves, discussed later in this chapter) were found to give a more reliable estimate of size. We should note that magnitude bears only an empirical relation to earthquake size; in Chapter 6 we discuss earthquake sources in more detail and find their size is properly measured by the seismic moment, not the magnitude.

3.2 *WWSSN* recordings

A record of one component of ground motion is called a *seismogram.* Its appearance depends very much on the type of instrument used. In this book we shall study three types of instrument: the older *WWSSN* in this and the next two chapters, and more recent digital instruments in Chapter 5. *WWSSN* stands for *World Wide Standardised Seismograph Network.* Prior to 1960 the seismologist who made use of global, rather than his own local, data had to rely on a collection of instruments of widely different types that were run by separate organisations without a common agreed practice. For example, some instruments were wired the wrong way round, making it virtually impossible to obtain source mechanisms for distant earthquakes (see Chapter 6). A ban on the testing of nuclear weapons required a means of monitoring explosions at great distances and of discriminating between the explosions and natural phenomena. The *WWSSN* was deployed with funding that was far in excess of anything received by seismology before, and by 1964 a network of several hundred standard instruments had been installed. Time and again data from the *WWSSN* has played a critical role in establishing new results in geophysics, and the importance of this network cannot be overstated. The *WWSSN* is now being replaced by digital instruments in a new network called the *GDSN* (*Global Digital Seismometer Network*). The *WWSSN* has three short period (SP) and three long period (LP) instruments, covering the three directions. These are oriented upwards, north, and east. The six instruments are numbered 1–6 in the order SP (Z,N,E) and LP (Z,N,E). The output from all six instruments is recorded on paper on a drum which in most cases rotates once every hour (LP) or fifteen minutes (SP) and translates along its axis so that the trace describes a spiral on the drum. When the paper is removed from the drum and laid flat the trace appears as a set of lines which go off the end of the paper to the right and reappear on the left side. An example is given in Figure 3.3. Every minute a mark is made just above the trace; the minute starts at the beginning of this mark. Thus on the LP record of Figure 3.3, 08:43 is at the beginning of the mark indicated, the mark on the trace underneath (offset slightly to the right) is one hour later corresponding to one complete revolution of the drum. For the SP record the minute mark on the trace beneath the 08:43 mark is fifteen minutes later. The marks on the LP records are offset slightly; this makes it easier to tell which trace is which when long period noise makes the traces cross each other.

The records in Figure 3.3 were made in Golden, Colorado, for an earthquake in the Gulf of Alaska at a distance of 34°and depth of 31 km. The arrivals of several different types of wave are marked. P denotes the arrival of the P wave; it is the earliest arrival because it has the fastest group velocity. S denotes the S wave, which is small on the vertical instrument because of the polarisation of the motion – the wave travels

Figure 3.3: An example of a *WWSSN* seismogram. It was written at Golden, Colorado, for an event in the Gulf of Alaska, from SIMON.

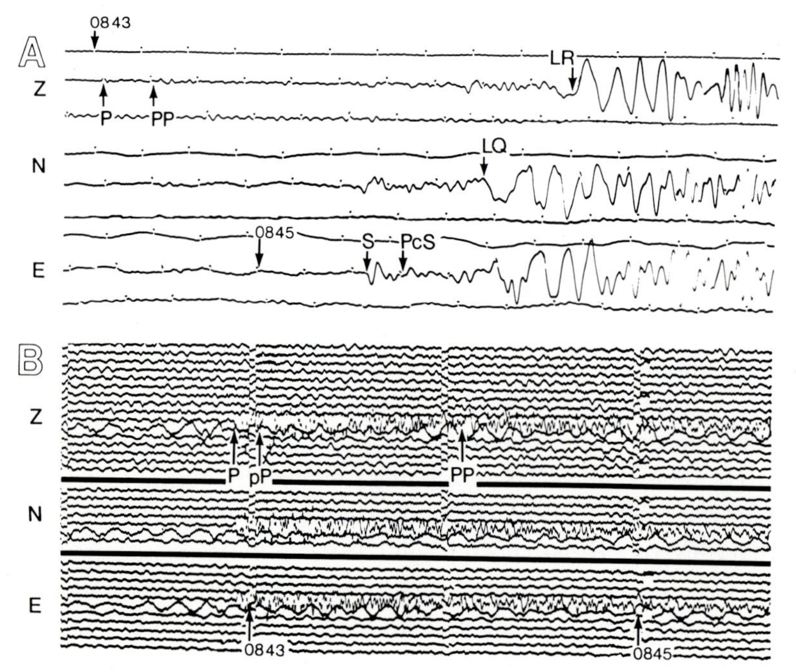

quite steeply up to the earth's surface even at this short distance from the source. These arrivals are much clearer and more accurately read on the SP instruments because of the faster speed of the recording drum. Other phases are marked on Figure 3.3. These will be discussed in this chapter as the theory for them is developed.

3.3 *P* and *S* waves

Consider the simplest case of wave propagation in a homogeneous medium. We take ρ, λ, and μ to be constant in the Navier equation (2.93) and ignore self-gravitation to give

$$\rho\ddot{\mathbf{u}} = (\lambda + \mu)\nabla(\nabla \cdot \mathbf{u}) + \mu\nabla^2\mathbf{u} \tag{3.5}$$

Using the identity, valid for any vector,

$$\nabla^2\mathbf{u} = \nabla(\nabla \cdot \mathbf{u}) - \nabla \times \nabla \times \mathbf{u} \tag{3.6}$$

we transform (3.5) into the useful form

$$\rho\ddot{\mathbf{u}} = (\lambda + 2\mu)\nabla(\nabla \cdot \mathbf{u}) - \mu\nabla \times \nabla \times \mathbf{u} \tag{3.7}$$

We can now effect a separation of (3.7) into two equations, one governing the dilatation (volume changes) and the other the rotation. We make use of the following vector identities

$$\nabla \cdot (\nabla \times \mathbf{u}) = 0 \tag{3.8}$$

$$\nabla \times \nabla\theta = 0 \tag{3.9}$$

for any $\mathbf{u}$ and θ. The divergence of (3.7) yields

$$\rho\ddot{\theta} = (\lambda + 2\mu)\nabla^2\theta \tag{3.10}$$

where we have used (2.8), and

$$\theta = \nabla \cdot \mathbf{u} \tag{3.11}$$

is the dilatation. The curl of (3.7) yields

$$\rho\ddot{\boldsymbol{\omega}} = \mu\nabla^2\boldsymbol{\omega} \tag{3.12}$$

where

$$\boldsymbol{\omega} = \nabla \times \mathbf{u} \tag{3.13}$$

is the rotation associated with the vorticity. The general equation (3.7) therefore contains two separate wave equations, (3.10) and (3.12).

Equation (3.10) governs P waves. The phase speed, α, is given by

$$\alpha^2 = \frac{(\lambda + 2\mu)}{\rho} = \frac{\left(\kappa_s + \frac{4}{3}\mu\right)}{\rho} \tag{3.14}$$

where κ_s is the adiabatic bulk modulus given by (2.89). Change in volume is an essential feature of the P wave. Equation (3.12) governs S waves. The phase speed, β, is given by

$$\beta^2 = \frac{\mu}{\rho} \tag{3.15}$$

Rotation is an essential feature of S waves. For normal materials $\kappa_s > 0$ and so (3.14) and (3.15) give $\alpha > \beta$; the P waves are the faster of the two and will therefore arrive first.

3.4 Elastodynamic potentials

Any vector may be separated into scalar and vector potentials. We write the displacement as

$$\mathbf{u} = \nabla\Phi + \nabla \times \mathbf{\Psi} \tag{3.16}$$

and choose the gauge condition

$$\nabla \cdot \mathbf{\Psi} = 0 \tag{3.17}$$

Φ and $\mathbf{\Psi}$ are called the *elastodynamic potentials*. It follows from (3.11) and (3.16) that

$$\theta = \nabla^2\Phi \tag{3.18}$$

Φ is the potential for the P waves. From (3.13) and (3.16)

$$\omega = \nabla \times \mathbf{u} = \nabla \times \nabla \times \mathbf{\Psi} = -\nabla^2\mathbf{\Psi} \tag{3.19}$$

$\mathbf{\Psi}$ is the potential for the S waves. The displacement may be separated into a P wave displacement, $\mathbf{P}$, and an S wave displacement, $\mathbf{S}$,

$$\mathbf{u} = \mathbf{P} + \mathbf{S} \tag{3.20}$$

where

$$\mathbf{P} = \nabla\Phi \tag{3.21}$$

and

$$\mathbf{S} = \nabla \times \mathbf{\Psi} \tag{3.22}$$

With this definition of P and S motion we see that P waves have no rotation while S waves have no change in volume — their motion is pure shear.

3.5 Plane waves

For plane waves with frequency ω and wavevector $\mathbf{k}$ we take both potentials to be proportional to $\exp i\,(\mathbf{k}\cdot\mathbf{x} - \omega t)$. Then from (3.21)

$$\mathbf{P} = A\mathbf{k}\exp i\,(\mathbf{k}\cdot\mathbf{x} - \omega t) \tag{3.23}$$

where A is a complex amplitude and we have used the useful result that the vector operator ∇ can be replaced by the vector $i\mathbf{k}$ when operating on the exponential function. Equation (3.21) shows that P waves are *longitudinal* because $\mathbf{P}$ is parallel to $\mathbf{k}$, which is normal to the wavefronts.

From (3.22)

$$\mathbf{S} = \mathbf{k} \times \mathbf{B}\exp i\,(\mathbf{k}\cdot\mathbf{x} - \omega t) \tag{3.24}$$

where $\mathbf{B}$ is a complex vector amplitude. S waves are transverse because the displacement $\mathbf{S}$ is normal to the wavevector $\mathbf{k}$. These waves are polarised, the direction of polarisation depending on the direction of $\mathbf{B}$ within the wavefront.

Note that both P and S waves are non-dispersive in a uniform medium. The group and phase velocities are equal in a uniform medium for both P and S waves. The group velocity is the relevant speed of transmission of the wavepacket from source to receiver, the phase velocity can be estimated from oscillations measured at the receiver.

3.6 Boundary conditions

Consider now the conditions prevailing on an interface separating two uniform media. At a *welded interface* we require continuity of displacement and traction. Continuity of $\mathbf{u}$ is essential if the interface is to remain welded. Traction must also be continuous, otherwise accelerations will develop to separate the interface. These conditions may be applied at the undeformed surface with an accuracy consistent with our small displacement approximation — the deformation of the surface away from its equilibrium position produces only a small change in the evaluation of the stress components.

The interface with a vacuum is an important special case called a *free surface*; we expect surface tractions to be zero there. The earth's surface is treated as a free surface. If the surface has normal in the z direction the boundary condition is

$$\sigma_{zx} = \sigma_{zy} = \sigma_{zz} = 0 \tag{3.25}$$

which may be expressed in terms of strains by using Hooke's law (2.83).

3.7 Reflection of plane waves at a free surface

Consider a uniform half space with free surface at $z = 0$ as shown in Figure 3.4 and consider reflection of an incident P wave. The axes are chosen so that the incident wave travels in the xz plane. When the wave impinges on the free surface then in general both P and S motions will be generated in the reflected response. The incident P motion is longitudinal and confined to the xz plane. No y displacement can be set up in the reflected waves because such a response would imply stress on the free surface. The reflected S wave will therefore be polarised in the xz plane. Such a wave is called the SV wave where the V stands for vertical polarisation (as distinct from the SH wave which has horizontal polarisation).

In Figure 3.4 $\hat{\boldsymbol{\ell}}$, $\hat{\mathbf{m}}$, and $\hat{\mathbf{n}}$ are unit vectors in the direction of propagation of incident P, reflected P, and reflected S waves, respectively.

$$\hat{\boldsymbol{\ell}} = (\sin I, 0, \cos I)$$

Figure 3.4: Geometry for the reflection of an incident P wave from a free surface. Arrows indicate particle motion. The angle of reflection of the S wave is smaller than the angle of incidence because the velocity is slower. y points into the diagram.

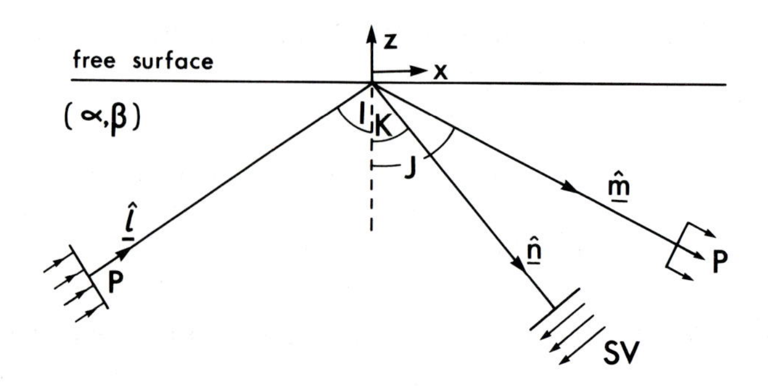

$$\begin{aligned}\hat{\mathbf{m}} &= (\sin J, 0, -\cos J) \\ \hat{\mathbf{n}} &= (\sin K, 0, -\cos K)\end{aligned} \tag{3.26}$$

The wavevectors themselves are

$$\begin{aligned}\boldsymbol{\ell} &= \hat{\boldsymbol{\ell}}\omega/\alpha \\ \mathbf{m} &= \hat{\mathbf{m}}\omega/\alpha \\ \mathbf{n} &= \hat{\mathbf{n}}\omega/\beta\end{aligned} \tag{3.27}$$

and the displacement for an incident P wave of unit amplitude is

$$\begin{aligned}\hat{\boldsymbol{\ell}} \exp i\,(\boldsymbol{\ell}\cdot\mathbf{x} - \omega t) \;&+\; B\hat{\mathbf{m}} \exp i\,(\mathbf{m}\cdot\mathbf{x} - \omega t) \\ &+\; C\,(\hat{\mathbf{n}}\times\hat{\mathbf{y}}) \exp i\,(\mathbf{n}\cdot\mathbf{x} - \omega t)\end{aligned} \tag{3.28}$$

where $\hat{\mathbf{y}}$ is a unit vector in the y direction. The total displacements in the x and z directions are then

$$\begin{aligned}u_x = \hat{\ell}_x \exp i\,(\boldsymbol{\ell}\cdot\mathbf{x} - \omega t) \;&+\; B\hat{m}_x \exp i\,(\mathbf{m}\cdot\mathbf{x} - \omega t) \\ &-\; C\hat{n}_z \exp i\,(\mathbf{n}\cdot\mathbf{x} - \omega t)\end{aligned} \tag{3.29}$$

$$\begin{aligned}u_z = \hat{\ell}_z \exp i\,(\boldsymbol{\ell}\cdot\mathbf{x} - \omega t) \;&+\; B\hat{m}_z \exp i\,(\mathbf{m}\cdot\mathbf{x} - \omega t) \\ &+\; C\hat{n}_x \exp i\,(\mathbf{n}\cdot\mathbf{x} - \omega t)\end{aligned} \tag{3.30}$$

At the free surface $z = 0$ the tractions must be zero. $\sigma_{zy} = 0$ automatically and (2.83) gives

$$\sigma_{zx} = \mu\left(\frac{\partial u_x}{\partial z} + \frac{\partial u_z}{\partial x}\right) = 0 \tag{3.31}$$

$$\sigma_{zz} = \lambda\left(\frac{\partial u_x}{\partial x} + \frac{\partial u_z}{\partial z}\right) + 2\mu\frac{\partial u_z}{\partial z} = 0 \tag{3.32}$$

$\hat{\mathbf{m}}, \hat{\mathbf{n}}$, B and C must be chosen to satisfy (3.31) and (3.32). Differentiating (3.29) and (3.30) and setting $z = 0$ gives

$$\begin{aligned}\frac{\partial u_x}{\partial z} &= i\frac{\omega}{\alpha}\hat{\ell}_x\hat{\ell}_z \exp i\,(\ell_x x - \omega t) + iB\frac{\omega}{\alpha}\hat{m}_x\hat{m}_z \exp i\,(m_x x - \omega t) \\ &\quad -iC\frac{\omega}{\beta}\hat{n}_z^2 \exp i\,(n_x x - \omega t)\end{aligned} \tag{3.33}$$

$$\begin{aligned}\frac{\partial u_z}{\partial x} &= i\frac{\omega}{\alpha}\hat{\ell}_x\hat{\ell}_z \exp i\,(\ell_x x - \omega t) + iB\frac{\omega}{\alpha}\hat{m}_x\hat{m}_z \exp i\,(m_x x - \omega t) \\ &\quad +iC\frac{\omega}{\beta}\hat{n}_x^2 \exp i\,(n_x x - \omega t)\end{aligned} \tag{3.34}$$

Substituting (3.33) and (3.34) into (3.31) gives

$$2\frac{\omega}{\alpha}\hat{\ell}_x\hat{\ell}_z \exp i\left(\ell_x x - \omega t\right) + 2B\frac{\omega}{\alpha}\hat{m}_x\hat{m}_z \exp i\left(m_x x - \omega t\right) \\ + C\frac{\omega}{\beta}\left(\hat{n}_x^2 - \hat{n}_z^2\right)\exp i\left(n_x x - \omega t\right) = 0 \qquad (3.35)$$

This equation can only be satisfied for all x and t if the functional dependence is the same in all three terms, i.e. $\ell_x = m_x = n_x$. The wavespeeds of both P waves are the same and $\mathbf{m}$ represents a downgoing wave, hence

$$\begin{aligned} \hat{m}_x &= \hat{\ell}_x \\ \hat{m}_z &= -\hat{\ell}_z \end{aligned} \qquad (3.36)$$

Also from (3.28)

$$\hat{n}_x = \frac{\beta\hat{\ell}_x}{\alpha} \qquad (3.37)$$

(3.36) and (3.37) may be written in terms of angles as the laws of reflection

$$J = I \qquad (3.38)$$

$$\sin K = \frac{\beta}{\alpha}\sin I \qquad (3.39)$$

The first of these laws was expected: the angle of reflection is equal to the angle of incidence, but (3.40) shows the SV wave to be reflected with a smaller angle of reflection, according to the ratio of S to P velocities in the medium.

Using (3.27) and (3.39), (3.35) becomes

$$\beta\sin 2I\ B + \alpha\cos 2K\ C = \beta\sin 2I \qquad (3.40)$$

Differentiating (3.29) and (3.30), setting $z = 0$, and using (3.36) and (3.39), we obtain

$$\frac{\partial u_x}{\partial x} = i\left\{(1+B)\frac{\omega}{\alpha}\hat{\ell}_x^2 - C\frac{\omega}{\beta}\hat{n}_x\hat{n}_z\right\}\exp i\left(\ell_x x - \omega t\right) \qquad (3.41)$$

$$\frac{\partial u_z}{\partial z} = i\left\{(1+B)\frac{\omega}{\alpha}\hat{\ell}_z^2 + C\frac{\omega}{\beta}\hat{n}_x\hat{n}_z\right\}\exp i\left(\ell_x x - \omega t\right) \qquad (3.42)$$

and the boundary condition (3.32) gives

$$(1+B)\,\beta\,(\lambda + 2\mu)\cos^2 I - C\mu\alpha\sin 2K = 0 \qquad (3.43)$$

Using the wave speeds (3.14) and (3.15): $\rho\alpha^2 = \lambda + 2\mu$; $\rho\beta^2 = \mu$; (3.43) becomes

$$(\alpha^2 - 2\beta^2 \sin^2 I)\, B - \alpha\beta \sin 2K\, C = -(\alpha^2 - 2\beta^2 \sin^2 I) \qquad (3.44)$$

(3.40) and (3.44) are two simultaneous equations to be solved for the *reflection coefficients* B and C.

We adopt the notation of AKI & RICHARDS and set $B = \acute{P}\grave{P}$ (amplitude coefficient for the downward propagating P wave, $\grave{P}$, reflected from upward propagating P wave, $\acute{P}$) and $A = \acute{P}\grave{S}$ (for amplitude of downward propagating S wave, $\grave{S}$, reflected from upward propagating P). Then (3.40) and (3.44) give

$$\acute{P}\grave{P} = \frac{\sigma - q^2}{\sigma + q^2} \qquad (3.45)$$

$$\acute{P}\grave{S} = \frac{4pq\cos I}{\beta\,(\sigma + q^2)} \qquad (3.46)$$

where

$$p = \frac{\sin I}{\alpha} \qquad (3.47)$$

$$q = \frac{1}{\beta^2} - 2p^2 \qquad (3.48)$$

$$\sigma = \frac{4p^2 \cos I \cos K}{\alpha\beta} \qquad (3.49)$$

The reflection coefficients in Figure 3.5 are plotted as a function of the angle of incidence, I, for $\alpha = 5$ km s^{-1} and β=3 km s^{-1}. For normal incidence ($I = 0$) the P wave is reflected with no conversion to SV motion. P to S conversion increases with I until total conversion is achieved near $I = 52°$, where $\acute{P}\grave{P} = 0$.

Exercise 5 asks you to consider reflection of an incident SV wave. It is readily shown that the same law of reflection holds (in fact the rays are reversible — see Chapter 6) and

$$\sin r_P = \frac{\alpha}{\beta} \sin i_S \qquad (3.50)$$

where i_S is the angle of incidence of the S wave and r_P is the angle of reflection of the P wave. The four reflection coefficients may be gathered into a reflection or *scattering matrix*

$$\mathrm{S} = \begin{pmatrix} \acute{P}\grave{P} & \acute{S}\grave{P} \\ \acute{P}\grave{S} & \acute{S}\grave{S} \end{pmatrix} \qquad (3.51)$$

S operates on a 2-vector of P and S wave incident motions to give the reflected motions. Scattering matrices are fundamental building blocks for the construction of the response of plane layered media to a source of elastic waves.

Most seismograms show some evidence of waves that have been reflected from the earth's surface or at some internal discontinuity inside the earth. In Figure 3.3 the phase PP has suffered one reflection at the earth's surface and arrives later than P because of its longer path. The depth phase pP has also been identified in Figure 3.3. We shall discuss uses of depth phases in detail in Chapter 4. Finally, PcS has been reflected from the core-mantle boundary and converted to an S wave by the reflection.

3.8 Reflection and refraction of *SH* waves

Coupling of P and SV waves is a feature of all laterally homogeneous media, but what of the other S motions? They are horizontally polarised and are labelled *SH*. An incident *SH* wave on a free surface is reflected with angle of reflection equal to the angle of incidence. The behaviour of the *SH* system of waves is therefore much simpler than that of the *P-SV* system. It is easy to show that *SH* waves are reflected perfectly at a free surface, with reflection coefficient unity. We use it here to study the reflection and transmission of waves at a plane welded interface

Figure 3.5: Reflection coefficients for a P wave reflected at a free surface as a function of the angle of incidence. Note the lack of any P to S conversion at normal incidence and the total conversion $\acute{P}\grave{P} = 0$ near an angle of incidence of 52°. From AKI & RICHARDS, copyright ©1980 W.H.Freeman & Co. Reprinted with permission.

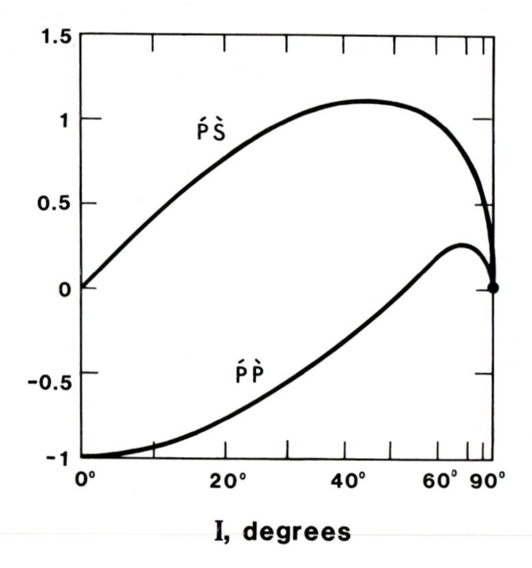

between two uniform media. The two uniform media (1 and 2) are shown in Figure 3.6; they have densities and shear velocities ρ_1, β_1 and ρ_2, β_2 respectively. Consider an *SH* wave incident on the interface from below as shown in Figure 3.6. *SH* motion is transmitted into the upper medium, unlike the free surface case, as well as being reflected downwards into the lower medium.

The wavevectors $\boldsymbol{\ell}, \mathbf{m}, \mathbf{n}$ for incident, reflected and transmitted waves are

$$\hat{\boldsymbol{\ell}} = \omega\boldsymbol{\ell}/\beta_2; \quad \hat{\mathbf{m}} = \omega\mathbf{m}/\beta_2; \quad \hat{\mathbf{n}} = \omega\mathbf{n}/\beta_1 \tag{3.52}$$

where the unit vectors $\hat{\boldsymbol{\ell}}$ and $\hat{\mathbf{m}}$ are given by (3.27a,b) and

$$\hat{\mathbf{n}} = (\sin K, 0, \cos K) \tag{3.53}$$

The displacements associated with each wave are

$$\begin{aligned} \text{incident} \quad & S: \quad \hat{\mathbf{y}} \exp i\,(\boldsymbol{\ell}\cdot\mathbf{x} - \omega t) \\ \text{reflected} \quad & S: \quad R\hat{\mathbf{y}} \exp i\,(\mathbf{m}\cdot\mathbf{x} - \omega t) \\ \text{transmitted} \quad & S: \quad T\hat{\mathbf{y}} \exp i\,(\mathbf{n}\cdot\mathbf{x} - \omega t) \end{aligned} \tag{3.54}$$

where R and T are the reflection and transmission coefficients.

The response in medium 2 is the sum of incident and reflected waves. Continuity of displacement across the interface $z = 0$ gives

$$\exp(i\ell_x x) + R\exp(im_x x) = T\exp(in_x x) \tag{3.55}$$

Figure 3.6: Reflection and transmission of *SH* waves at an interface between two different homogeneous media. Particle motion is in the y direction and the waves propagate in the xz plane. Snell's law of refraction and the usual reflection law apply. ⊗ indicates motion into the paper.

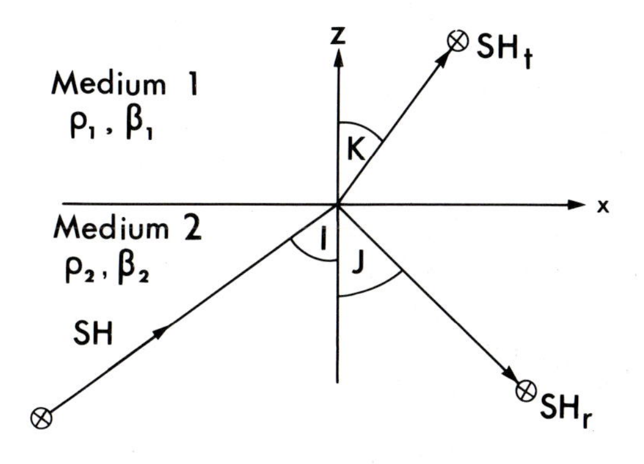

(3.55) must be valid for all x. This is only possible if the functional form is the same for each term, therefore $\ell_x = m_x = n_x$ or

$$\frac{\sin I}{\beta_2} = \frac{\sin J}{\beta_2} = \frac{\sin K}{\beta_1} \tag{3.56}$$

The first equality is the simple law of reflection: $I = J$. The second gives Snell's law of refraction. (3.55) then gives

$$1 + R = T \tag{3.57}$$

The stress must be continuous across the interface. The elements of the stress tensor σ_{zx} and σ_{zz} are zero for these *SH* waves. The element σ_{zy} is given by (2.83) as

$$\sigma_{zy} = \mu \frac{\partial u_y}{\partial z} = \rho\beta^2 \frac{\partial u_y}{\partial z} \tag{3.58}$$

Differentiating the three expressions (3.54), substituting into (3.58), and equating σ_{zy} in the two media at $z = 0$ gives

$$\rho_2\beta_2 \left(\hat{\ell}_z + R\hat{m}_z\right) = \rho_1\beta_1 T\hat{n}_z \tag{3.59}$$

Substituting $\hat{\ell}_z = \cos I$ and applying the law of reflection gives

$$\rho_2\beta_2 (1 - R)\cos I = \rho_1\beta_1 T \cos K \tag{3.60}$$

Equations (3.57) and (3.60) must be solved for the reflection and transmission coefficients R and T. The solution is readily obtained as

$$T = \acute{S}\acute{S} = \frac{2\rho_2\beta_2 \cos I}{\rho_1\beta_1 \cos K + \rho_2\beta_2 \cos I} \tag{3.61}$$

$$R = \acute{S}\grave{S} = T - 1 \tag{3.62}$$

It is left as an (elementary) exercise to the reader to show the responses to a downward propagating *SH* wave in the upper medium (1) are

$$\grave{S}\acute{S} = -\acute{S}\grave{S} = \frac{\rho_1\beta_1 \cos K - \rho_2\beta_2 \cos I}{\rho_1\beta_1 \cos K + \rho_2\beta_2 \cos I} \tag{3.63}$$

$$\grave{S}\grave{S} = \frac{2\rho_1\beta_1 \cos K}{\rho_1\beta_1 \cos K - \rho_2\beta_2 \cos I} \tag{3.64}$$

The 2×2 scattering matrix

$$\begin{pmatrix} \grave{S}\acute{S} & \acute{S}\acute{S} \\ \grave{S}\grave{S} & \grave{S}\acute{S} \end{pmatrix} \tag{3.65}$$

then gives the full response for *SH* motion for this interface. We shall make use of equations (3.63)–(3.64) in discussing surface waves later in the chapter. R and T depend on the product $\rho\beta \cos I$ for each medium. This quantity is the *impedance* of the *SH* waves.

The scattering matrix for the *P-SV* system from an interface between two different uniform media is evaluated in similar fashion but there are four waves in each medium (two upward propagating and two downward propagating); the scattering matrix is therefore fourth order and there are 16 coefficients to be calculated. The calculation is more complicated than the *SH* case but no more difficult in principle. The reflection and transmission coefficients are given in AKI & RICHARDS (pp. 150–151).

3.9 Rayleigh waves

Interconversion of P and S waves is not the only way the stress-free conditions can be met on a free surface. We now study a second way these boundary conditions can be satisfied, and find it leads to a quite different type of wave that travels along the free surface. Such waves are called *surface waves*; the particular wave containing P and SV motions is called the *Rayleigh wave*, after Lord Rayleigh who first studied it.

Consider again the system illustrated in Figure 3.5. We seek a surface wave that propagates in the x direction, at some phase speed c_R, with an amplitude function that falls away with z, the distance from the surface, thus being a wave that is confined to the vicinity of the surface. Any vector displacement may be separated into P and S motion (3.20). P and S motions each satisfy a wave equation with wave speeds α and β respectively. For plane waves travelling in the x direction we write

$$u(x, z) = f(z) \exp i\,(kx - \omega t) \tag{3.66}$$

Substituting into the wave equation gives a differential equation for the depth dependence $f(z)$

$$\frac{d^2 f}{dz^2} = \left(k^2 - \frac{\omega^2}{c^2}\right) f \tag{3.67}$$

where c is the phase speed (α or β). It follows that

$$f(z) = f_0 \exp\left(\kappa_c z\right) \tag{3.68}$$

where

$$\kappa_c^2 = k^2 - \frac{\omega^2}{c^2} \tag{3.69}$$

We note for future reference that if $\omega/k < c$ then κ_c is real, and the amplitude function will fall off exponentially with depth. This is the characteristic feature of a surface wave; waves with $\omega/k > c$ would propagate into the body of the medium. The surface waves contain both P and S motion, and therefore their phase speed must be slower than both the P and S velocities.

Take P and S motions of the form (3.20)–(3.22) with potentials Φ and $\boldsymbol{\Psi}$ proportional to

$$\exp\left[\kappa_\alpha z + i\left(kx - \omega t\right)\right]$$

and

$$\exp\left[\kappa_\beta z + i\left(kx - \omega t\right)\right]$$

respectively giving

$$P_x = ka \exp\left[\kappa_\alpha z + i\left(kx - \omega t\right)\right] \tag{3.70}$$
$$P_z = -i\kappa_\alpha a \exp\left[\kappa_\alpha z + i\left(kx - \omega t\right)\right] \tag{3.71}$$
$$S_x = -\kappa_\beta b \exp\left[\kappa_\beta z + i\left(kx - \omega t\right)\right] \tag{3.72}$$
$$S_z = ikb \exp\left[\kappa_\beta z + i\left(kx - \omega t\right)\right] \tag{3.73}$$

where a and b are amplitudes which must be chosen, along with the phase speed $c_R = \omega/k$, to make the free surface tractions zero. We have $\sigma_{zy} = 0$ always. From Hooke's law (2.83) we have

$$\sigma_{zx} = \mu\left(\frac{\partial u_x}{\partial z} + \frac{\partial u_z}{\partial x}\right) \tag{3.74}$$

Substituting (3.70)–(3.73) into (3.20) for $\mathbf{u}$, differentiating and setting $z = 0$, and substituting into (3.74) gives

$$2\kappa_\alpha ka - \left(k^2 + \kappa_\beta^2\right) b = 0 \tag{3.75}$$

Again from Hooke's law (2.83) we have

$$\sigma_{zz} = \lambda\left(\frac{\partial u_x}{\partial x} + \frac{\partial u_z}{\partial z}\right) + 2\mu\frac{\partial u_z}{\partial z} \tag{3.76}$$
$$= \rho\left[\left(\alpha^2 - 2\beta^2\right)\frac{\partial u_x}{\partial x} + \alpha^2\frac{\partial u_z}{\partial z}\right] = 0 \tag{3.77}$$

Substituting for the derivatives of $\mathbf{u}$ and setting $z = 0$ gives

$$\left(\alpha^2 - 2\beta^2\right)\left(ik^2 a - ik\kappa_\beta b\right) + \alpha^2\left(-i\kappa_\alpha^2 a + ik\kappa_\beta b\right) = 0 \tag{3.78}$$

This expression can be simplified with the identity

$$\left(\alpha^2 - 2\beta^2\right) k^2 - \alpha^2\kappa_\alpha^2 = -\beta^2\left(k^2 + \kappa_\beta^2\right) \tag{3.79}$$

which follows from (3.69), the definition of κ_α and κ_β. (3.78) becomes

$$\left(k^2 + \kappa_\beta^2\right) a - 2k\kappa_\beta b = 0 \tag{3.80}$$

Equations (3.75) and (3.80) are two simultaneous equations for a and b, but they are homogeneous and the only solution will be the trivial one $a = b = 0$ unless the determinant of the coefficients vanishes. The condition of the vanishing determinant gives us the *dispersion relation* which determines the allowed frequencies of the wave. The condition is

$$\left(k^2 + \kappa_\beta^2\right)^2 = 4k^2\kappa_\alpha\kappa_\beta \tag{3.81}$$

Substituting for κ_α and κ_β and squaring gives the more explicit form

$$\left(2k^2 - \frac{\omega^2}{\beta^2}\right)^4 = 16k^4\left(k^2 - \frac{\omega^2}{\alpha^2}\right)\left(k^2 - \frac{\omega^2}{\beta^2}\right) \tag{3.82}$$

If we regard this as an equation for wavenumber k at given frequency ω we factor out roots at $k = 0$ leaving a cubic in k^2. Writing

$$\zeta = \omega^2/k^2\beta^2; \quad \eta = \beta^2/\alpha^2 = \mu/\left(\lambda + 2\mu\right) \tag{3.83}$$

gives

$$R(\zeta) = \zeta^3 - 8\zeta^2 + 8\zeta(3 - 2\eta) - 16(1 - \eta) = 0 \tag{3.84}$$

Since κ_c must be real, (3.69) gives

$$c_R = \omega/k < \beta \tag{3.85}$$

and $\zeta < 1$. The solution of this cubic equation is left as an exercise in the problems at the end of this chapter, but it is clear from the second equation of (3.83) that $0 < \eta < \frac{1}{2}$ and therefore that $R(0) < 0$. Also $R(1) = 1$, and therefore there must be at least one positive root less than unity. In fact there is only one such real root, which corresponds to the Rayleigh wave. The root for a Poisson solid, which has $\eta = \frac{1}{3}$, is $\zeta_P = 0.8453$. c_R is independent of frequency and therefore Rayleigh waves are non-dispersive in a uniform medium: their group velocity is equal to the phase velocity.

Solving for a and b gives the particle motion for the Rayleigh wave. We may arbitrarily set $a = 1$ and obtain from (3.75)

$$b = \frac{2k\kappa_\alpha}{\left(k^2 + \kappa_\beta^2\right)} = \frac{\left(k^2 + \kappa_\beta^2\right)}{2k\kappa_\beta} \tag{3.86}$$

where the second form has been derived using (3.81). Substituting $a = 1$ and b from (3.86) into (3.70)–(3.74) setting $z = 0$, and taking the real part gives the horizontal motion at the surface

$$(ka - \kappa_\beta b)\cos(kx - \omega t) = \frac{k\left(k^2 + \kappa_\beta^2\right) - 2k\kappa_\alpha\kappa_\beta}{\left(k^2 + \kappa_\beta^2\right)}\cos(kx - \omega t) \tag{3.87}$$

Similarly the vertical motion at the surface $z = 0$ is

$$(\kappa_\alpha a - kb)\sin(kx - \omega t) = \frac{2k\kappa_\alpha\kappa_\beta - k\left(k^2 + \kappa_\beta^2\right)}{2k\kappa_\beta}\sin(kx - \omega t) \tag{3.88}$$

Comparing (3.87) and (3.88) shows the particle motion to be elliptically polarised; the amplitude factors are of opposite sign and so the particle motion must be retrograde, as the following illustration shows. Suppose the amplitude coefficient for horizontal motion is positive; that for vertical motion must then be negative. A particle at time $t = 0$ will be at the right side of the ellipse and moving towards the top. The motion is like that of a ball rolling on the underside of the free surface. The quantities κ_α and κ_β determine the attenuation with depth of the P and S components of the motion. Since $\alpha > \beta$, from (3.69) we find $\kappa_\alpha > \kappa_\beta$; therefore the P motion attenuates faster with depth than the SV motion, and the latter ultimately dominates. This leads to prograde particle motion at depth,

Figure 3.7: Polarisation of Rayleigh waves.

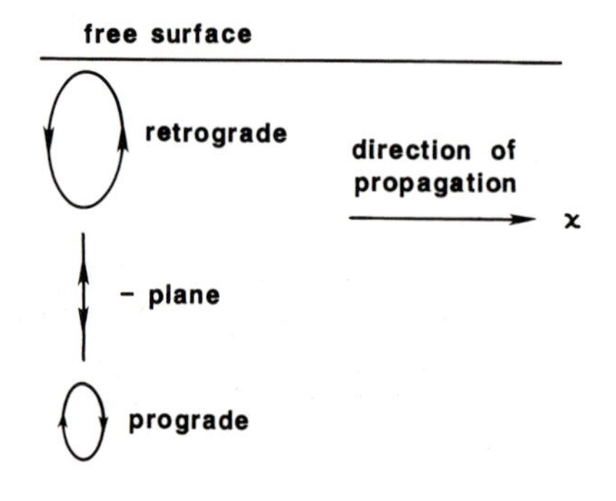

and vertically polarised motion at just one critical depth, as shown in Figure 3.7.

Surface waves are often the largest amplitude arrival on the seismogram. The arrival marked *LR* in Figure 3.3 is the Rayleigh wave; the wavetrain has a period of about 20 s, a wavelength of about 100 km, and arrives after both the *P* and *S* wave.

3.10 Dispersion of Rayleigh waves

Solving (3.84) yields a phase speed for Rayleigh waves that is independent of frequency; Rayleigh waves in a uniform half space are non-dispersive. However, observation shows Rayleigh waves in the earth to be highly dispersed. The dispersion is due to variation with depth of the elastic moduli and density. The solution of the Navier equation for Rayleigh waves in inhomogeneous media is a formidable undertaking in terms of the algebra involved, if not the general principles. We therefore restrict ourselves to the problem of a fluid layer overlying a solid half space. This configuration has the dual virtues of mathematical simplicity and geophysical relevance: the water in the oceans causes considerable dispersion of Rayleigh waves.

The parameters are set out in Figure 3.8. The water layer is of thickness H. There are no shear waves in the fluid and displacement must satisfy the equation for P waves

$$\frac{\partial^2 u}{\partial t^2} = \alpha_1^2 \nabla^2 u \tag{3.89}$$

The P motion is written as

$$P_x \quad = \quad k\left[A\exp\left(i\gamma_L z\right) - B\exp\left(-i\gamma_L z\right)\right]\exp i\left(kx - \omega t\right)$$

Figure 3.8: The model for studying dispersive surface waves. The surface $z = H$ is stress-free, that at $z = 0$ is an interface between the solid half space and the upper fluid layer.

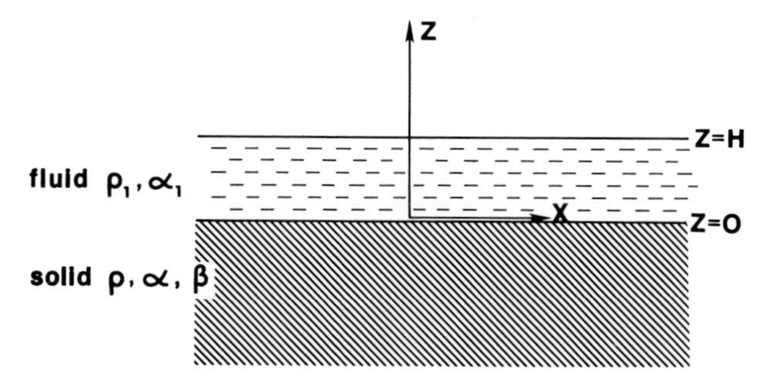

$$P_z = \gamma_L \left[A \exp\left(i\gamma_L z\right) + B \exp\left(-i\gamma_L z\right)\right] \exp i\left(kx - \omega t\right) \quad (3.90)$$

where

$$\gamma_L^2 = \omega^2/\alpha_1^2 - k^2 \quad (3.91)$$

In (3.91) we have defined γ_L to be real for oscillatory solutions, in contrast to (3.69) where we required κ_c to be real for surface-wave type solutions that decayed exponentially with depth. In this case we still expect solutions in which the amplitude dies away with distance from the water surface, but there are also important modes that oscillate with depth (γ_L real), or even die away in both directions from the fluid-solid interface (γ_L imaginary).

Vertical displacement and traction are continuous across the fluid-solid interface at $z = 0$ in Figure 3.8. We must relax the condition of continuous horizontal displacement because we have neglected viscosity in the fluid. The resulting solutions for the horizontal displacement are discontinuous (see Figure 3.10). The only boundary condition that arises at the water surface $z = H$ is $\sigma_{zz} = 0$, which leads to

$$B = A \exp 2i\gamma_L H \quad (3.92)$$

Within the solid the same expressions apply as did for Rayleigh waves in a uniform half space (3.70)–(3.73). The shear stress σ_{zx} on the interface between the solid and fluid must vanish because the fluid cannot support shear stresses; this condition is therefore the same as before (3.75). Continuity of u_z gives

$$-\kappa_\alpha a + kb = -2i\gamma_L A \exp\left(i\gamma_L H\right) \cos\gamma_L H \quad (3.93)$$

and continuity of σ_{zz}

$$a(\omega^2 - 2k^2\beta^2) + 2bk\kappa_\beta\beta^2 = -2i(\rho_1/\rho)\omega^2 A \exp i\gamma_L H \sin\gamma_L H \quad (3.94)$$

(3.75), (3.93) and (3.94) form three homogeneous simultaneous equations to be solved for a, b and A. There will be a non-trivial solution only when the determinant of the coefficients vanishes; this determinant gives the period equation for the allowed frequencies of the modes just as it gave (3.82) for the uniform half space. To obtain an equation for the phase speed of the Rayleigh waves we write $c_R = \omega/k$ and use (3.69) and (3.91) in the determinant. The resulting expression is:

$$\tan \gamma_L H = \frac{(c_R^2/\alpha_1^2 - 1)^{\frac{1}{2}}\, \rho\beta^4}{(1 - c_R^2/\alpha^2)^{\frac{1}{2}}\, \alpha\rho_1 c_R^4} \left\{ 4\left(1 - \frac{c_R^2}{\alpha^2}\right)^{\frac{1}{2}} \left(1 - \frac{c_R^2}{\beta^2}\right)^{\frac{1}{2}} - \left(2 - \frac{c_R^2}{\beta^2}\right)^2 \right\} \tag{3.95}$$

As $kH \to 0$ (3.95) reduces to (3.75), the period equation for Rayleigh waves in a uniform half space. This limit corresponds to the wavelength of the Rayleigh waves being much greater than the fluid depth; the waves do not "see" the fluid layer. The group velocity $c_g = \partial\omega/\partial k$ can also be found from (3.95).

A useful dimensionless measure of the frequency is obtained by scaling with the time taken for P waves to cross the fluid layer.

$$\mathcal{N} = H\omega/\alpha_1 \tag{3.96}$$

For any particular choice of $\mathcal{N}$ there are discrete solutions to the period equation (3.95). Phase and group velocities for the first two modes are shown in Figure 3.9. The eigenfunctions for the two components of displacement u_x and u_z are shown in Figure 3.10. When $\alpha, \beta > c_R > \alpha_1$ then (3.91) shows the solutions to be waves that propagate within the water layer, but decay with depth in the solid. Some higher modes have nodal planes within the fluid (the second mode (b) in Figure 3.10 for example). This behaviour is quite unlike that of Rayleigh waves on a half space. When $\mathcal{N}$ is small all the modes except the first have complex solutions to the period equation (3.95). These modes propagate into the half space and attenuate; they are *leaky modes.* When $\mathcal{N}$ is large there is a quite different type of solution with $c_R < \alpha_1, \alpha, \beta$ (and hence imaginary γ_L from (3.91)). For this case the displacement decays exponentially in both directions away from the fluid-solid interface; it is called a *Stoneley wave* and an example is shown in (d) of Figure 3.10. When $\mathcal{N} \to \infty$ its phase velocity approaches $0.998\alpha_1$.

The displacement u_x is discontinuous at the fluid-solid interface (see Figure 3.10). This is physically unrealistic; it is a consequence of neglecting viscosity in the fluid, which if it were present could provide shear stresses at the interface and damp the waves. When a small viscosity is included in the calculations it can be shown that a thin viscous boundary layer forms within the fluid which accomodates the discontinuity in displacement.

A much nastier example of the role of viscosity arose when Smylie & Mansinha (1971) were calculating the response of the earth's rotation to internal deformation (and consequent changes to its inertia tensor) using the equations of seismology in a low frequency limit. These Lagrangian equations are quite satisfactory in the solid mantle but the earth's liquid core can flow, and in the low frequency limit the displacements become arbitrarily large. In fact the "small strain" assumption breaks down and the distinction between Eulerian and Lagrangian formulations, so often neglected in seismology, becomes crucial. Smylie and Mansinha deduced (correctly) that they needed to relax the condition of continuity of *normal* as well as tangential displacement across the boundary in order to patch up the Lagrangian equations. This discontinuity in normal displacement, which implies cavitation (or overlap!), caused some alarm, and great controversy ensued. The issue was settled in no fewer than three ways: by a mathematical argument in which the Lagrangian equations were assumed valid in both media away from the solid-fluid interface (Crossley & Gubbins 1975), a fluid dynamical approach to the low frequency limit which

Figure 3.9: Phase (c_R) and group (c_g) velocities as functions of the dimensionless frequency $\mathcal{N}$ (equation (3.95)), after EWING, JARDETSKY & PRESS, for Rayleigh waves in a liquid layer over a solid half space. Parameter values $\rho/\rho_1 = 2.5$; $\alpha/\beta = 3$; $\beta/\alpha_1 = 2$. Reprinted with permission.

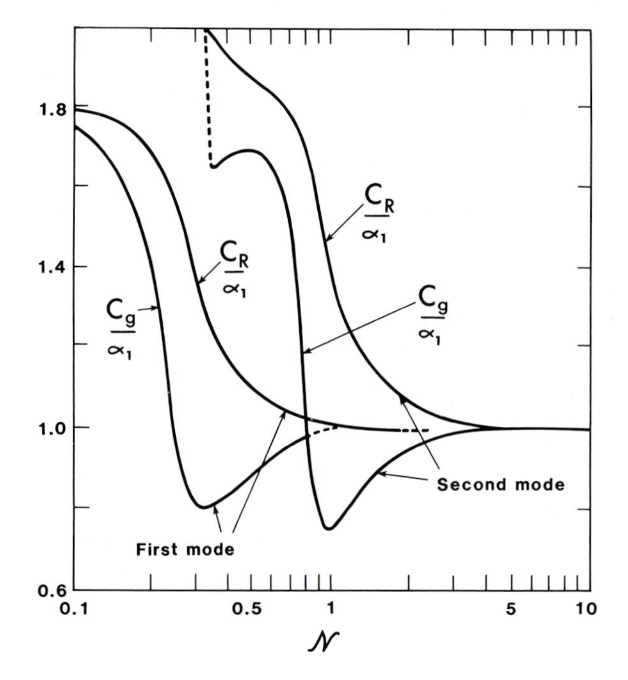

Figure 3.10: Eigenfunctions for various solutions of the period equation (3.95). (a), (b) and (c) are dispersive Rayleigh waves and (d) a Stoneley wave that travels along the fluid-solid interface. From EWING, JARDETSKY & PRESS. Reprinted with permission.

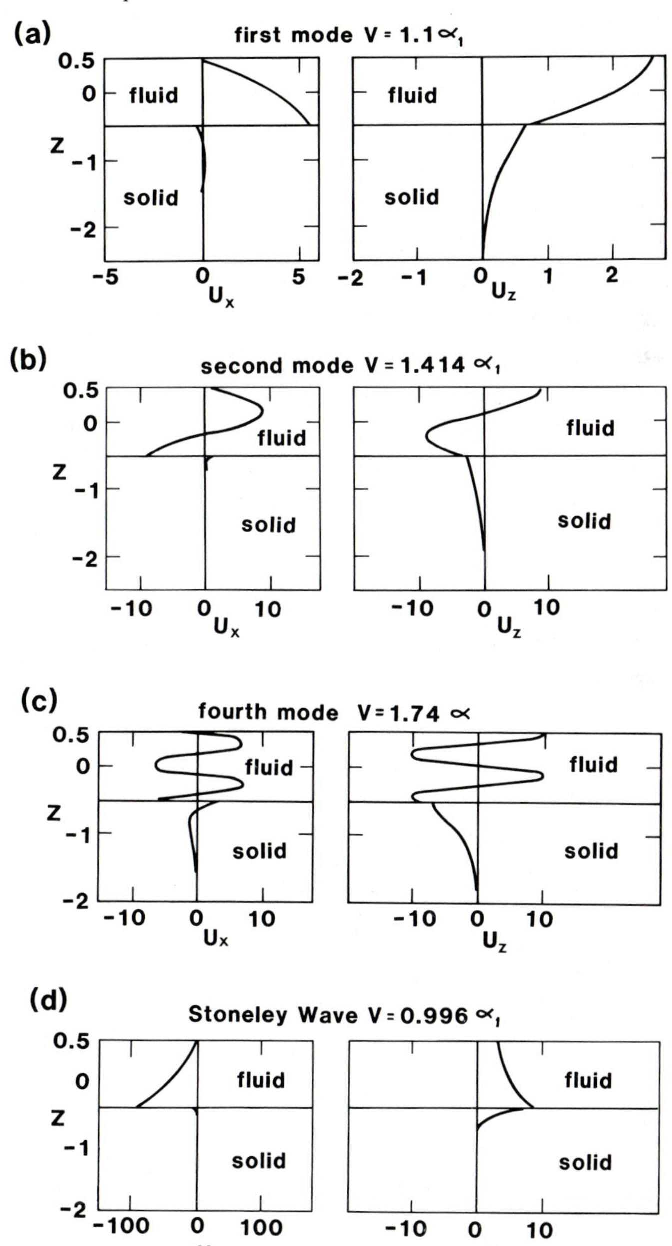

involved solving explicitly for the large displacements in the fluid (Wunsch 1974), and a full treatment using the Eulerian equations (Dahlen 1974).

3.11 Love waves

There are no surface waves with purely *SH* motion in uniform half space because it is not possible to satisfy the stress-free conditions with purely *SH* motion. Such surface waves can occur, however, if the half space is overlain by a uniform layer of different material; they are seen regularly on seismograms and are direct evidence of the earth's vertical inhomogeneity. They are called *Love waves.* They can be analysed in the same way as we analysed Rayleigh waves, by combining all possible up- and down-going waves in the layered medium (all of which will be of *SH* type) and satisfying the boundary conditions. The Love waves then emerge as solutions in addition to the usual combination of body waves. The reader is referred to AKI & RICHARDS (p.262) for the standard derivation of the dispersion relation. Here we present a simple argument due to KENNETT which uses the reflection coefficients for body waves derived in Section 3.8.

Consider reflection and transmission of body *SH* waves in a medium with one layer over a half space and an upper free surface, as shown in Figure 3.11, where $\beta_1 < \beta_2$. The reflection coefficient for *SH* waves at a free surface is unity. A wave refracted at A will reflect at the free surface and then reflect again at B, at the interface between the two media, but this time with the reflection coefficient given by $\grave{S}\acute{S}$ in equation (3.63). The condition for the propagation of surface waves is given when this reflected wave interferes constructively with the upcoming wave in medium 2.

Let c_L be the horizontal phase speed of the disturbance; it is the same in both media by the law of refraction. It will also be the phase speed of the surface wave. The vertical wavenumber in medium 1 is then

$$n_z = \omega \left(\frac{1}{\beta_1^2} - \frac{1}{c_L^2} \right)^{\frac{1}{2}} \tag{3.97}$$

The phase lag suffered by the wave reflected from the free surface relative to the upcoming wave in medium 2 is then

$$2H\omega \left(\frac{1}{\beta_1^2} - \frac{1}{c_L^2} \right)^{\frac{1}{2}} \tag{3.98}$$

Constructive interference occurs between this reflected wave and the upcoming wave if

$$\grave{S}\acute{S} \exp\left[2iH\omega\left(\frac{1}{\beta_1^2} - \frac{1}{c_L^2}\right)^{\frac{1}{2}}\right] = 1 \tag{3.99}$$

The angles of refraction I and K can be expressed as

$$\sin K = \frac{\beta_1}{c_L}; \quad \cos K = \beta_1\left(\frac{1}{\beta_1^2} - \frac{1}{c_L^2}\right)^{\frac{1}{2}} \tag{3.100}$$

$$\sin I = \frac{\beta_2}{c_L}; \quad \cos I = \beta_2\left(\frac{1}{\beta_2^2} - \frac{1}{c_L^2}\right)^{\frac{1}{2}} \tag{3.101}$$

Substituting (3.100)–(3.101) into (3.63) gives $\grave{S}\acute{S}$, and (3.99) becomes

$$\exp\left[-2iH\omega\left(\frac{1}{\beta_1^2} - \frac{1}{c_L^2}\right)^{\frac{1}{2}}\right] = \frac{\mu_1\left(\frac{1}{\beta_1^2} - \frac{1}{c_L^2}\right)^{\frac{1}{2}} - \mu_2\left(\frac{1}{\beta_2^2} - \frac{1}{c_L^2}\right)^{\frac{1}{2}}}{\mu_1\left(\frac{1}{\beta_1^2} - \frac{1}{c_L^2}\right)^{\frac{1}{2}} + \mu_2\left(\frac{1}{\beta_2^2} - \frac{1}{c_L^2}\right)^{\frac{1}{2}}} \tag{3.102}$$

We need the mathematical result, that if

$$\exp(-2i\theta) = \frac{a-b}{a+b} \tag{3.103}$$

then $\tan\theta = -ib/a$. Hence (3.102) gives

Figure 3.11: Two layer medium supporting surface waves with SH motion only

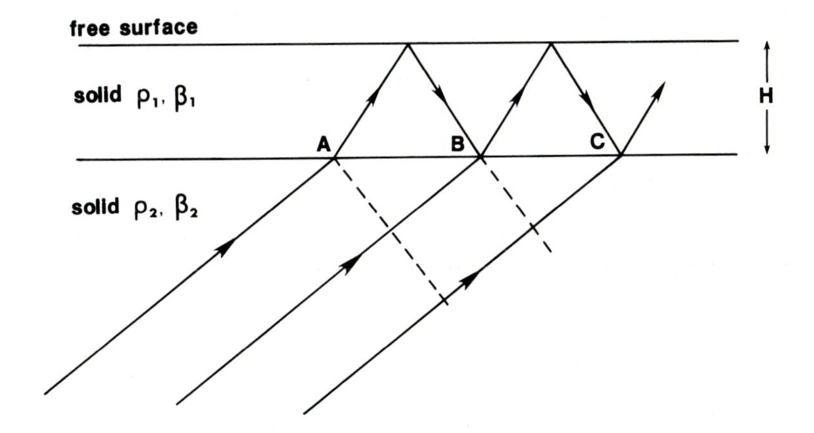

$$\tan\left[\omega H\left(\frac{1}{\beta_1^2}-\frac{1}{c_L^2}\right)^{\frac{1}{2}}\right]=\frac{\mu_2\left(\frac{1}{c_L^2}-\frac{1}{\beta_2^2}\right)^{\frac{1}{2}}}{\mu_1\left(\frac{1}{\beta_1^2}-\frac{1}{c_L^2}\right)^{\frac{1}{2}}} \tag{3.104}$$

This is the dispersion relation for Love waves. Equation (3.104) possesses a finite number of solutions, giving a finite number of modes for a given layer, with phase velocity lying between the two shear velocities β_1 and β_2. Section 5.4 of BULLEN & BOLT (p. 114) describes other solutions to this dispersion relation, and the implications for Love waves. In the earth Love waves travel faster than Rayleigh waves. They are marked LQ in Figure 3.3. Being horizontally polarised they do not appear on the Z instrument.

In fact Love waves do exist on the surface of a uniform solid sphere, even though they do not exist in a plane half space, as a limit of body SH waves multiply reflected from the spherical free surface. We shall return to them under the guise of free oscillations in Chapter 5, where their analysis will be found to be simpler than the corresponding P-SV case.

3.12 Propagation in inhomogeneous media

The earth is strongly heterogeneous in the radial direction and we must address the difficult question of wave propagation in heterogeneous media before progressing any further. We have seen that displacement separates into P and S motion in a uniform medium, and that the two types of waves are coupled at a sharp interface. In a generally inhomogeneous medium we therefore expect coupling between P and S motions everywhere, and perhaps the whole concept of separation into P and S waves will lose its meaning. We show here that one can usually ignore P to S conversion away from sharp interfaces and strong gradient zones: the coupling is weak provided the properties of the medium do not change too much over distances comparable with one wavelength. The approximations involved are related to those of geometrical optics.

Consider P waves as a simple illustration. To derive the wave equation (3.10) for the dilatation for a uniform medium we took the divergence of the Navier equation of motion (2.93) and neglected gravitational forces. When the Lamé parameters λ and μ are functions of position the divergence of the Navier equation gives

$$\begin{aligned}\rho\ddot{\theta}-(\lambda+2\mu)\nabla^2\theta = & -\nabla\rho\cdot\ddot{\mathbf{u}}+\nabla\lambda\cdot\nabla\theta-\nabla\mu\cdot\nabla\times\nabla\times\mathbf{u} \\ & -\nabla\mu\cdot\nabla\theta+2\nabla\cdot(\nabla\mu\cdot\mathrm{e})+\theta\nabla^2\lambda\end{aligned} \tag{3.105}$$

The right hand side of this equation contains terms that vanish for homogeneous media, in which case it reduces to the wave equation for the dilatation. The displacement associated with P waves has $\nabla \times \mathbf{u} = 0$: therefore the third term on the right involves coupling of compressional motion with rotational motion, so also do the terms in $\nabla\rho \cdot \mathbf{u}$ and $2\nabla \cdot (\nabla\mu \cdot \mathrm{e})$. The separation into P and S motion has broken down.

The magnitudes of all the terms on the right hand side of (3.105) depend on the gradients of the material properties. Suppose they vary appreciably over some distance L. Then a typical order of magnitude of $\nabla\rho$, for example, would be ρ/L; that of $\rho/|\nabla\rho|$, L. Suppose the displacement itself has wavelength ℓ. A typical magnitude of θ is then $|u|/\ell$, and $\nabla^2\theta = O[|u|/\ell^3]$. If the wavelength of the wave is very short compared with the distance L we can estimate the relative sizes of terms in equation (3.105), for example $\rho\theta$ and $\nabla\rho \cdot \mathbf{u}$ are in the ratio $L/\ell \gg 1$. Other terms are tabulated in Table 3.1. Here we have set ω to be of the same order as α/ℓ, making a high frequency as well as a short wavelength approximation; all orders of magnitude have been expressed relative to the terms on the left hand side of (3.105).

This simple argument shows that, away from sharp interfaces, P waves will propagate without coupling to S waves, according to the wave equation, provided $\ell \ll L$. A similar result can be demonstrated for S waves. In the earth we take take a "sharp gradient zone" to be one thinner than a wavelength, or about 10 km.

3.13 Geometrical optics

If P to S conversion can be ignored we can employ the usual laws of geometrical optics in a medium in which the phase velocity varies slowly with position. This follows simply because the dilatation (for P waves) and rotation (for S waves) satisfy the wave equation. The laws of geo-

term	approximate value	order of magnitude
$\rho\ddot{\theta}$	$\omega^2\rho\theta$	$O(1)$
$(\lambda + 2\mu)\,\nabla^2\theta$	$\rho\alpha^2\theta/\ell^2$	$O(1)$
$\nabla\rho \cdot \ddot{\mathbf{u}}$	$\omega^2\rho\theta\ell/L$	$O(\ell/L)$
$\nabla\lambda \cdot \nabla\theta$	$\rho\alpha^2\theta/\ell L$	$O(\ell/L)$
$\theta\nabla^2\lambda$	$\rho\alpha^2\theta/L^2$	$O\left[(\ell/L)^2\right]$
$\nabla\mu \cdot \nabla \times \nabla \times \mathbf{u}$	$\rho\beta^2\theta/\ell L$	$O(\ell/L)$
$\nabla\mu \cdot \nabla\theta$	$\rho\beta^2\theta/\ell L$	$O(\ell/L)$
$\nabla \cdot (\nabla\mu \cdot \mathrm{e})$	$\rho\beta^2\theta/\ell L$	$O(\ell/L)$

Table 3.1: Relative sizes of terms in equation (3.105)

metrical optics are well known from our experience with optics: opaque objects cast a sharp shadow, a vanishingly small opening defines an infinitesimally thin pencil (the ray), and the ray obeys the laws of refraction and reflection. All these laws carry over from optics to seismology. We shall review some of the results here: a more complete discussion is to be found in Chapter 3 of BORN & WOLF. In seismology we must use ray theory with caution. While light has a typical wavelength of 10^{-10} m, many orders of magnitude smaller than everyday objects, seismic waves have wavelengths of some tens of kilometres, which may be comparable with the linear dimensions of the variations in seismic velocity.

We represent the disturbance in dilatation as a sinusoidal oscillation of the form

$$\theta(\mathbf{r}, t) = \hat{\theta}(\mathbf{r}) \exp i\omega t \tag{3.106}$$

and

$$\hat{\theta}(\mathbf{r}) = \hat{\theta}_0(\mathbf{r}) \exp\left[ik_0 S(\mathbf{r})\right] \tag{3.107}$$

where $k_0 = \omega/\alpha_o$ and α_o is some reference wave speed. There is no loss of generality in choosing this form of variation because any disturbance can be decomposed into sinusoidal waves, and any complex number can be written in the form (3.106). The form (3.107) suggests an approximation to plane waves provided the amplitude $\hat{\theta}_0(\mathbf{r})$ and phase $S(\mathbf{r})$ do not vary too rapidly in space, and $k_0 \gg 1$.

Substituting (3.106) into the wave equation gives Helmholtz' equation

$$\nabla^2\hat{\theta} + k^2\hat{\theta} = 0 \tag{3.108}$$

where $k = \omega/\alpha$. From (3.107) we have

$$\nabla^2\hat{\theta} = \left[\nabla^2\hat{\theta}_0 + 2ik_0\nabla\hat{\theta}_0 \cdot \nabla S + ik_0\nabla^2 S\hat{\theta}_0 - k_0^2(\nabla S)^2\hat{\theta}_0\right] \exp ik_0 S(r) \tag{3.109}$$

If S and $\hat{\theta}_0$ vary slowly in space, or more precisely

$$|\nabla S|/S, |\nabla\hat{\theta}_0|/\hat{\theta}_0 \ll k_0$$

then the last term dominates (3.109) and it becomes

$$(\nabla S)^2 = \frac{\alpha^2}{\alpha_0^2} \tag{3.110}$$

The function $S(\mathbf{r})$ is called the *Eikonal* and (3.110) the *Eikonal equation.* When $S(\mathbf{r})$ is constant then (3.107) shows $\hat{\theta}$ to have the same phase: it defines the wavefront. The unit vector

$$\mathbf{s} = \frac{\alpha_0}{\alpha} \nabla S \tag{3.111}$$

is always normal to the wavefront, in the direction of maximum rate of change of phase. The ray paths are defined as trajectories orthogonal to the wave fronts. The wave propagates normal to the wavefront and parallel to the ray path.

The most important results of geometrical ray theory are Snell's law of refraction and Fermat's principle, that the time of travel along a ray path is stationary. We shall make use of them in discussing rays within the earth in the next chapter, and ignore P to S conversion except at sharp interfaces.

3.14 Energy density

Consider a plane P wave: $\mathbf{s}$ will be parallel to $\mathbf{u}$ because P waves are longitudinal, and

$$\mathbf{u} = \mathbf{u}(t - \mathbf{s} \cdot \mathbf{x}/\alpha) \tag{3.112}$$

The strain tensor is

$$e_{ij} = -\frac{1}{2\alpha}\left(\dot{u}_i s_j + \dot{u}_j s_i\right) \tag{3.113}$$

where we have made use of

$$\frac{\partial u_i}{\partial x_j} = \frac{\partial u_i}{\partial q}\frac{\partial q}{\partial x_j} \tag{3.114}$$

where

$$q = t - \mathbf{s} \cdot \mathbf{x}/\alpha \tag{3.115}$$

The strain energy density is, from (2.81) and (2.83),

$$\begin{aligned} \rho R = \frac{1}{2}\sigma_{ij}e_{ij} &= \frac{1}{2}\lambda e_{ii}e_{jj} + \mu e_{ij}e_{ij} \\ &= \frac{1}{2\alpha^2}\left\{\lambda\left(\dot{\mathbf{u}} \cdot \mathbf{s}\right)^2 + \mu\left[\dot{\mathbf{u}}^2\mathbf{s}^2 + \left(\dot{\mathbf{u}} \cdot \mathbf{s}\right)^2\right]\right\} \end{aligned} \tag{3.116}$$

but $\dot{\mathbf{u}}$ and $\mathbf{s}$ are parallel and $\mathbf{s}$ is a unit vector, so

$$\rho R = \frac{(\lambda + 2\mu)}{2\alpha^2}\dot{\mathbf{u}}^2 = \frac{1}{2}\rho\dot{\mathbf{u}}^2 \tag{3.117}$$

which demonstrates equipartition between strain energy and kinetic energy.

Consider a *ray tube* as shown in Figure 3.12. The surface of a ray tube is formed by individual ray paths. The wave propagates along the ray tube, and so also does the energy. No energy propagates through the surface of the tube. The intensity of the wave on a surface is defined in terms of the incident energy per unit area of surface. The energy flow remains constant along the tube, and so the intensity is determined solely by the area normal to the ray tube. Body waves in a uniform medium have spherical wavefronts; the energy spreads out over an increasingly large area as the wave propagates and the energy per unit area of wavefront falls off as the inverse square of the distance. The observed amplitude of the wave therefore falls off as the inverse of the distance. In an inhomogeneous medium we must calculate the decrease in amplitude due to geometrical spreading from the change in area of a ray tube. We shall make use of this result in Chapter 4 when calculating amplitudes of seismic rays in the earth.

Practical 1. Surface wave dispersion

We are now in a position to read some seismograms ourselves. This first practical is very short and deals only with the surface waves,

Figure 3.12: A ray tube. The surface of the tube is formed by rays. No energy propagates through the surface of the ray tube. conservation of energy applied to the volume V shows that energy flows across surface S_1 at the same average rate as across S_2. The intensity is inversely proportional to the area normal to the ray tube.

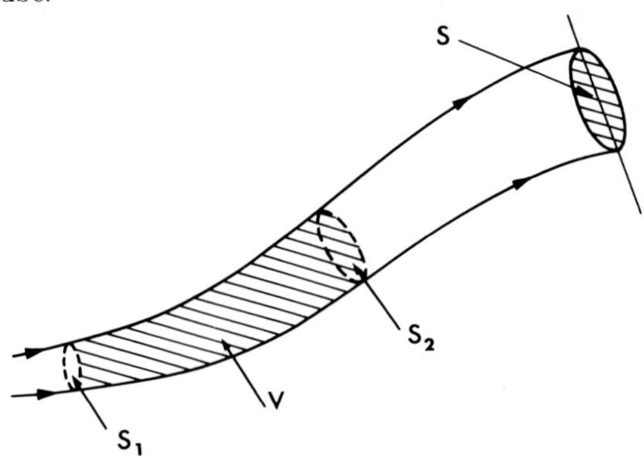

which are usually the largest arrivals on the seismogram. We shall tackle the identification of the earlier phases in the next practical in Chapter 4.

The LP seismograms in Figures 3.14–3.16 were written at Dugway in Utah by an earthquake in the Solomon Islands which occurred at 3:11:0.2 on 22/APR/77 at a distance of about 10440 km. The onset of the Rayleigh waves are indicated on the trace. The full-size seismogram is too large to display in this book, and only a part is reproduced in Figures 3.14–3.16; the Z and E records are given in two overlapping parts.

Look at the Z instrument first. Dispersion can be seen as a progressive change in frequency of the waves with time. Find the period of the Rayleigh waves as a function of time by measuring the spacing between the peaks. Use all three traces. Find a time scale by measuring the distance between the nearest minute marks; do not use just one scale for the whole seismogram because the rotation speed of the drum will not be reliably uniform and the paper may have stretched.

Find the group velocity for a particular part of the wavetrain by calculating the total travel time and dividing it into the distance. Tabulate the group velocities against your measured frequencies and plot the results on the graph of group velocities in Figure 3.13, which were derived from equation (3.95) for a variety of water depths. Choose the best-fitting theoretical curve and hence estimate the mean water depth along the path.

The epicentre of this earthquake was south of Guadalcanal at 10.2°N, 160.7°E, and the great circle path to Dugway in southern Utah lies almost east-west. The signal on instrument 5 (oriented north-south) should therefore be very small. This is true until about 4:06 but not later. Beats also appear on instrument 4 (vertical) at this time. Study the great circle path on a globe (preferably one showing topography) and find what effects might be causing this change with time.

Figure 3.13: Group velocities for Rayleigh waves in a solid half space overlain by a water layer, with various water depths.

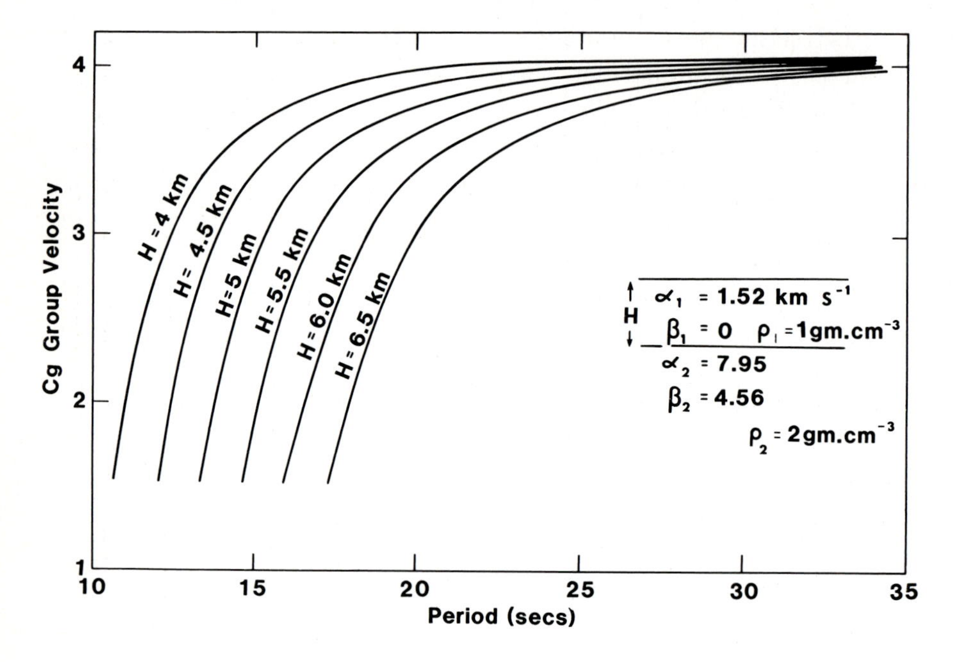

Figure 3.14: Seismogram for Practical 1. From station DUG on 22/APR/77, vertical component, long period, denoted DUG(4). This and the following 2 seismograms are reproduced one quarter actual size. They may be enlarged with an expanding xerox.

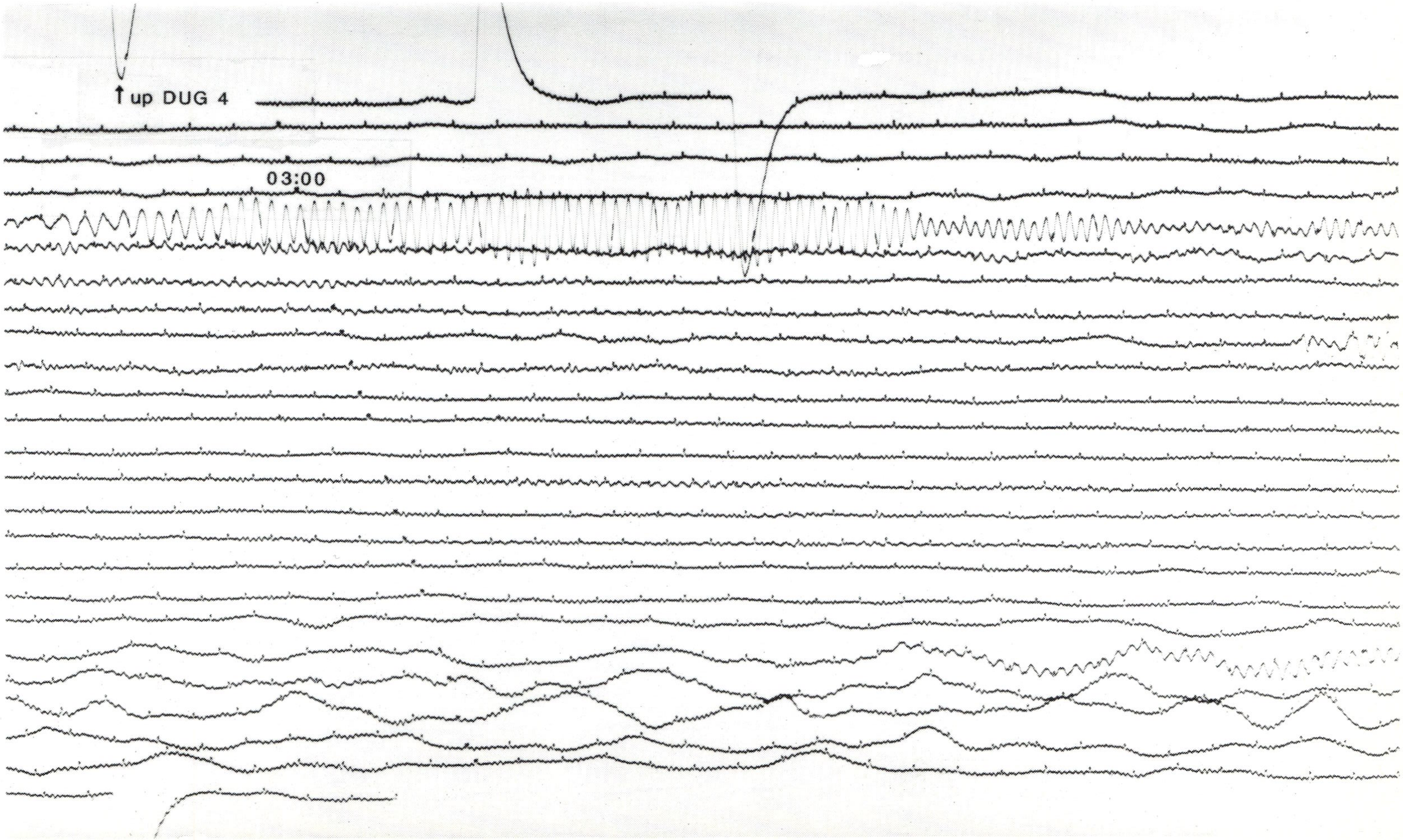

↑ up DUG 4

03:00

Figure 3.15: Seismogram for 22/APR/77, DUG(5)

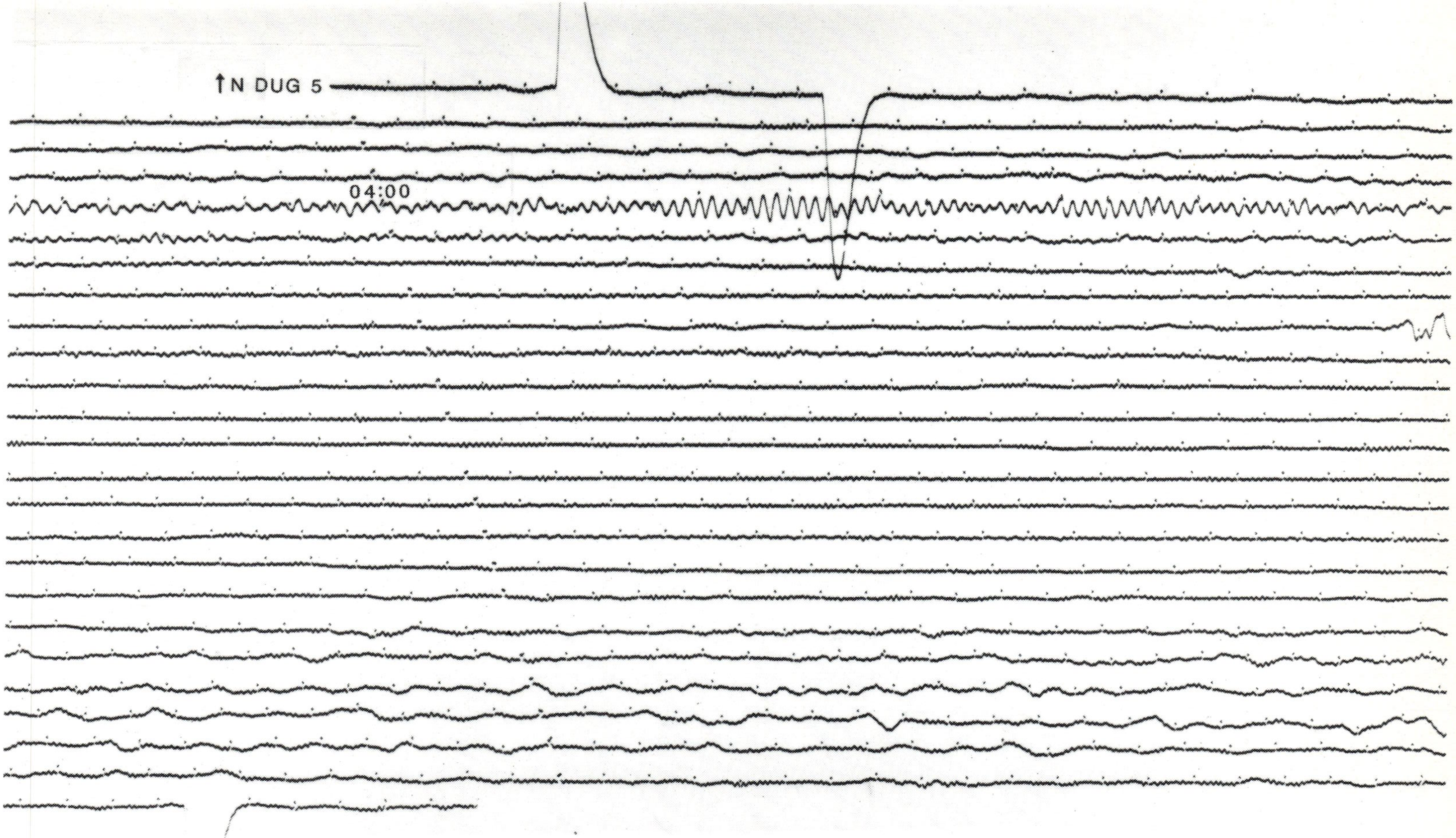

↑N DUG 5

04:00

Figure 3.16: Seismogram for 22/APR/77, DUG(6)

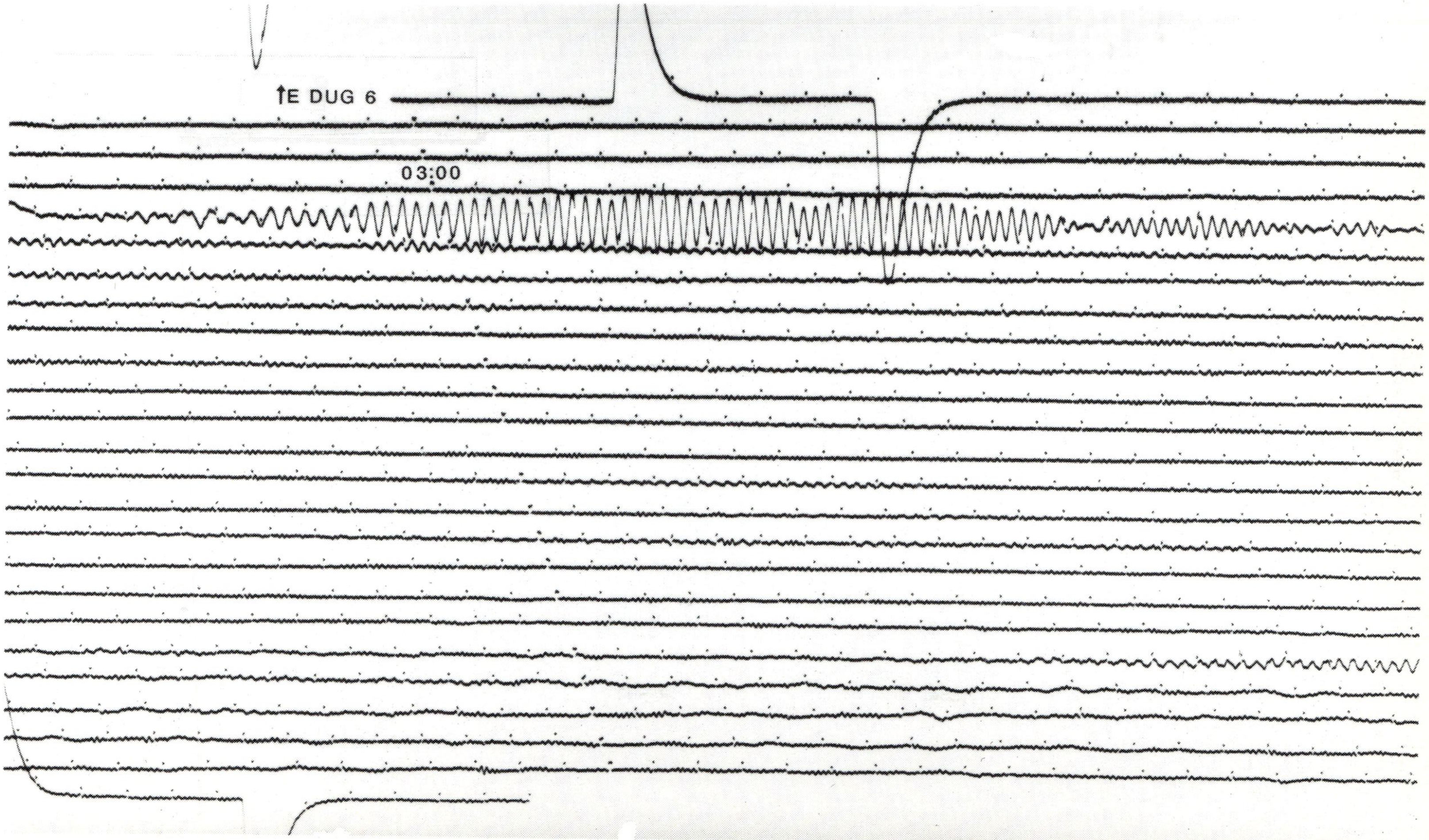

Exercises

1 Show that the elastodynamic potentials $\mathbf{\Phi}$ and $\mathbf{\Psi}$ each satisfy the equation

$$\nabla^2\left(\frac{\partial^2 y}{\partial t^2} - c^2\nabla^2 y\right) = 0$$

where y is Φ or Ψ and c the P or S wave velocity respectively. In fact the potentials can be chosen to satisfy the wave equation (see AKI & RICHARDS (Chapter 4, pp. 68ff)).

2 Choose $\mathbf{\Psi}$ to represent a plane wave travelling in the xz plane and evaluate $\mathbf{u}_S = \nabla \times \mathbf{\Psi}$. Show that the wave is transversely polarised.

3 Differentiate the expression obtained in the previous question to give the strain associated with the S wave. Verify that the strain tensor has zero determinant, and interpret your result.

4 Find the angle of incidence at a free surface for which an incident P wave is reflected totally as an SV wave.

5 Evaluate the reflection coefficients $\acute{S}\grave{P}$ and $\acute{S}\grave{S}$ by considering an incident SV wave on the free surface $z = 0$. Verify that the angle of reflection of the SV wave in this case is equal to the angle of incidence, i_s, but the P wave has a larger angle of reflection, i_p, given by an equation similar to (3.39).

6 Show that SH waves are reflected from a free surface with reflection coefficient unity.

7 Verify equations (3.63)–(3.64) for $\grave{S}\acute{S}$ and $\grave{S}\grave{S}$ for SH motion.

8 A seismometer installed on horizontal ground is used to measure the ratio of vertical to horizontal ground motion, A_v/A_h, for an incident P wave. Consider the complete system of incident and reflected waves to show that

$$2\cos^2 e = 3(1 - \sin\overline{e})$$

for a medium with $\lambda = \mu$, where e is the angle of emergence (90°– angle of incidence) of the P wave and $\overline{e}$ is the apparent angle of emergence, given by

$$\tan\overline{e} = A_v/A_h$$

This relationship shows the polarisation of the P wave is distorted by the free surface. See BULLEN & BOLT (p. 189).

9 Show that there is only one root between zero and unity of the Rayleigh wave dispersion relation (3.84) for any physical value of the parameter η (between 0 and $\frac{1}{2}$): the proof that at least one root lies in the interval (0,1) was given in the text; you must consider the turning points of the cubic.

10 Plot the dispersion relation (3.84) for a Poisson solid which has $\eta = \frac{1}{3}$. Investigate the imaginary roots of the dispersion relation, which would show the presence of leaky modes.

11 Why do the curves in Figure 3.13 asymptote to the phase velocity for Rayleigh waves in a solid elastic half space at long period?

12 Show that the energy per unit area of wavefront of a plane *SH* wave of unit amplitude and period τ is given by $2\pi\rho\lambda/\tau^2$. Supposing the geometrical optics approximation to be valid in other respects, would we be justified in taking this formula to hold for curved wavefronts in the earth, and accounting for the energy per unit wavelength by the changing area of the complete wavefront?

Summary

In a uniform medium the Navier equation separates into two wave equations, one governing compressional motions or *P* WAVES, and one governing rotational motions or *S* WAVES. *P* waves are longitudinal and travel at a speed dependent on a combination of the bulk and shear moduli. *S* waves are transverse and travel at a slower speed dependent only upon the shear modulus. A uniform medium is a poor approximation to the earth; we developed methods for coping with inhomogeneous media by examining boundaries to the medium, first with a free surface and then with an interface with a second uniform medium. At a FREE SURFACE there are no tractions: three components of the stress tensor are zero. This enables us to calculate the reflection of a *P* wave from the surface and we discover one of the peculiarities of elastic waves: because the medium supports two distinct types of wave there is conversion between the wave types. We find that *P* waves are reflected at normal incidence without conversion or loss, but that there is substantial or even complete conversion to *S* waves at other angles of incidence. At an interface with another medium the continuity of displacement and stress allows us to calculate both reflected and transmitted *P* and *S* waves. *P* waves are coupled to *S* waves polarised in the vertical plane (*SV* WAVES); the horizontally-polarised waves (*SH* WAVES) propagate independently and have their own reflection and transmission laws. Thus a separation exists between the *P-SV* and *SH* systems in a vertically inhomogeneous medium, even though the *P* and *S* waves are not separable.

The stress-free conditions that prevail at an interface with a vacuum can also be satisfied by RAYLEIGH WAVES. These waves travel along the surface of the medium at a phase speed that is slower than either the *P* or *S* wave velocities; the amplitude decays exponentially with distance from the surface and the waves are confined to the surface of the medium; they are an example of SURFACE WAVES. The particle motion is a combination of both *P* and *S* motion; it is elliptically polarised and retrograde, as for a ball rolling on the underside of the free surface. The polarisation changes with depth because the *P* component of the motion is attenuated faster than the *S* component away from the surface.

The Rayleigh wave speed is independent of frequency in a uniform medium and the waves are therefore non-dispersive. Rayleigh waves that have travelled across the ocean floor are often observed to be dispersed. This dispersion is attributed to the water layer. We studied this system with a model of a fluid layer over a solid half space and found the Rayleigh waves to be highly dispersed. This model supports higher order modes in addition to the simple Rayleigh wave. Some modes correspond to Rayleigh waves with nodal planes in the water layer; others propagate with attenuation into the solid half space and are called LEAKY MODES; and yet another type of mode travels along the solid-fluid interface with displacement which decays exponentially with distance from the interface in both solid and fluid. It is called the STONELEY WAVE. LOVE WAVES are supported by two solid layers and are very prominent on seismograms. They are equivalent to torsional free oscillations which are analysed in Chapter 5.

Considering the question of propagation in a generally homogeneous medium we found that P to S conversion is negligible provided the frequency of the wave is sufficiently high and the variation of the elastic moduli in space is small over a single wavelength of the wave. We can apply all the results from GEOMETRICAL OPTICS to P and S waves separately provided conversion of wave types can be neglected. The ray concept is valid and Fermat's principle applies. GEOMETRICAL ATTENUATION can be calculated from the change in area of a ray tube as it is propagated.

Further reading

The discussion of body waves and surface waves given in this chapter is very much in line with that to be found in standard texts on seismology. BULLEN & BOLT is thoroughly recommended for a more complete treatment than is possible in this book; see Chapter 6 and Sections 5.1–5.5. The level of mathematical difficulty is about the same as here. AKI & RICHARDS (Chapter 5) give the reflection and transmission coefficients for body waves; they treat surface waves (Chapter 7) and elastodynamic potentials (Chapter 4). The treatment is more mathematical than this book; it is a definitive work. Examples of the propagation of surface waves are given in EWING, JARDETSKY & PRESS. Propagation of body waves in layered media is given a unified treatment by KENNETT. The relevant Chapter 4 of BORN & WOLF, on the discussion of the geometrical optics approximation, is highly recommended reading. AKI & RICHARDS discuss the approximation as it is used in seismology.

4

Earth structure and earthquake location

The most common observation made in seismology is that of the arrival time of the initial onset of the disturbance. The accuracy with which this can be measured depends mainly on the shape of the initial pulse and its size relative to the background noise. For example, the onset of waves with dominant period 1 s would be difficult to read to better than 0.1 s even under ideal circumstances. Clock accuracy for instruments in remote areas used to be a problem, but this difficulty has now largely been removed by more accurate and robust clocks and by regular radio time broadcasts. The easiest arrival to read is the first onset, the P wave. Later arrivals appear on the seismogram in the midst of waves already passing and are more difficult to see in the background "noise" so formed. In particular S waves, which are particularly important in providing information about the shear modulus within the earth, are more difficult to time than P waves because they have longer periods and arrive late. S waves are often not read to better than 1 s.

If we know the variation of seismic velocity with depth within the earth then we can calculate the travel time of rays between an earthquake and a receiver using the geometrical optics approximation. We can then, in principle, locate any earthquake in both time and space by recording the arrival times of waves at a number of stations worldwide. Originally seismic velocity within the earth and earthquake locations had to be found simultaneously, but now there are good tables of travel times and we can treat the two problems separately.

4.1 Travel time tables

Several sets of *travel time tables* have been devised for global seismology. Each table is based on a very large number of observations of earthquakes in many localities. It is also common to derive a travel time table or curve for a specific locality or for a particular purpose. Table 4.1 gives a part of the Jeffreys-Bullen (J-B) tables for P waves. Times are

Δ	Depth h = Surface		0·00		0·01		0·02		0·03		0·04		0·05	
°	m s		m s		m s		m s		m s		m s		m s	
30	6 12·5	88	6 07·7	89	6 01·6	88	5 55·7	88	5 49·9	87	5 44·4	87	5 39·2	86
31	6 21·3	88	6 16·6	88	6 10·4	87	6 04·5	87	5 58·6	87	5 53·1	86	5 47·8	86
32	6 30·1	87	6 25·4	87	6 19·1	87	6 13·2	86	6 07·3	86	6 01·7	86	5 56·4	85
33	6 38·8	87	6 34·1	86	6 27·8	86	6 21·8	85	6 15·9	85	6 10·3	85	6 04·9	85
34	6 47·5	86	6 42·7	86	6 36·4	85	6 30·3	85	6 24·4	85	6 18·8	84	6 13·4	84
35	6 56·1	85	6 51·3	85	6 44·9	85	6 38·8	84	6 32·9	84	6 27·2	84	6 21·8	84
36	7 04·6	84	6 59·8	84	6 53·4	84	6 47·2	84	6 41·3	84	6 35·6	84	6 30·2	83
37	7 13·0	84	7 08·2	84	7 01·8	83	6 55·6	84	6 49·7	83	6 44·0	83	6 38·5	83
38	7 21·4	84	7 16·6	83	7 10·1	83	7 04·0	83	6 58·0	83	6 52·3	82	6 46·8	82
39	7 29·8	83	7 24·9	83	7 18·4	83	7 12·3	82	7 06·3	82	7 00·5	82	6 55·0	82
40	7 38·1	82	7 33·2	83	7 26·7	82	7 20·5	82	7 14·5	82	7 08·7	82	7 03·2	81
41	7 46·3	82	7 41·5	82	7 34·9	82	7 28·7	82	7 22·7	81	7 16·9	81	7 11·3	81
42	7 54·5	82	7 49·7	82	7 43·1	82	7 36·9	81	7 30·8	81	7 25·0	80	7 19·4	80
43	8 02·7	81	7 57·9	81	7 51·3	81	7 45·0	80	7 38·9	80	7 33·0	80	7 27·4	80
44	8 10·8	81	8 06·0	80	7 59·4	80	7 53·0	80	7 46·9	79	7 41·0	79	7 35·4	79
45	8 18·9	79	8 14·0	80	8 07·4	79	8 01·0	79	7 54·8	79	7 48·9	78	7 43·3	78
46	8 26·8	79	8 22·0	78	8 15·3	79	8 08·9	78	8 02·7	77	7 56·7	78	7 51·1	77
47	8 34·7	79	8 29·8	79	8 23·2	78	8 16·7	78	8 10·4	78	8 04·5	77	7 58·8	77
48	8 42·6	77	8 37·7	77	8 31·0	77	8 24·5	77	8 18·2	76	8 12·2	76	8 06·5	76
49	8 50·3	77	8 45·4	77	8 38·7	77	8 32·2	76	8 25·8	76	8 19·8	75	8 14·1	75
50	8 58·0	76	8 53·1	76	8 46·4	76	8 39·8	76	8 33·4	75	8 27·3	74	8 21·6	74
51	9 05·6	76	9 00·7	75	8 54·0	75	8 47·4	75	8 40·9	74	8 34·7	74	8 29·0	74
52	9 13·2	75	9 08·2	75	9 01·5	75	8 54·9	74	8 48·3	74	8 42·1	73	8 36·4	73
53	9 20·7	73	9 15·7	74	9 09·0	73	9 02·3	73	8 55·7	73	8 49·4	72	8 43·7	72
54	9 28·0	74	9 23·1	73	9 16·3	73	9 09·6	73	9 03·0	72	8 56·6	72	8 50·9	71
55	9 35·4	72	9 30·4	72	9 23·6	72	9 16·9	72	9 10·2	71	9 03·8	71	8 58·0	71
56	9 42·6	72	9 37·6	72	9 30·8	71	9 24·1	70	9 17·3	71	9 10·9	70	9 05·1	69
57	9 49·8	70	9 44·8	70	9 37·9	70	9 31·1	70	9 24·4	70	9 17·9	70	9 12·0	69
58	9 56·8	70	9 51·8	70	9 44·9	69	9 38·1	70	9 31·4	69	9 24·9	69	9 18·9	69
59	10 03·8	69	9 58·8	69	9 51·8	69	9 45·1	68	9 38·3	68	9 31·8	68	9 25·8	67
60	10 10·7	68	10 05·7	68	9 58·7	68	9 51·9	68	9 45·1	67	9 38·6	67	9 32·5	66
61	10 17·5	68	10 12·5	67	10 05·5	67	9 58·7	67	9 51·8	67	9 45·3	66	9 39·1	66
62	10 24·3	66	10 19·2	67	10 12·2	67	10 05·4	66	9 58·5	66	9 51·9	65	9 45·7	65
63	10 30·9	66	10 25·9	65	10 18·9	65	10 12·0	65	10 05·1	65	9 58·4	65	9 52·2	64
64	10 37·5	65	10 32·4	65	10 25·4	65	10 18·5	65	10 11·6	64	10 04·9	64	9 58·6	64
65	10 44·0	64	10 38·9	64	10 31·9	64	10 25·0	64	10 18·0	64	10 11·3	63	10 05·0	63
66	10 50·4	64	10 45·3	63	10 38·3	63	10 31·4	63	10 24·4	63	10 17·6	63	10 11·3	63
67	10 56·8	63	10 51·6	63	10 44·6	62	10 37·7	62	10 30·7	62	10 23·9	62	10 17·6	61
68	11 03·1	62	10 57·9	62	10 50·8	62	10 43·9	61	10 36·9	61	10 30·1	61	10 23·7	61
69	11 09·3	61	11 04·1	61	10 57·0	61	10 50·0	61	10 43·0	61	10 36·2	61	10 29·8	61
70	11 15·4		11 10·2		11 03·1		10 56·1		10 49·1		10 42·3		10 35·9	

Table 4.1: Part of the Jeffreys-Bullen travel time tables. Reprinted with permission from Jeffreys & Bullen (1970).

in minutes and seconds for distances at one degree intervals (in global seismology the distance is measured in degrees, as the angle subtended by the two points at the centre of the earth) for a range of source depths. First differences are given in tenths of seconds. The earth's crust varies so much from place to place that a general table for it would be unreliable, and so the tables contain times only for sources in the mantle. Depth = 0.0 corresponds to the base of the crust, at a radius of 6338 km. Other depths are in fractions of this radius. "Surface" in the table applies to the earth's surface that includes a two-layer model of the crust with thicknesses 18 and 15 km. For example, "65 km deep" means 65 km below the surface, so the corresponding table depth is $(65 - 33)/6338 = 0.005$. The corresponding travel time at 60°distance is found to be 10 min 2.2 s, by interpolating between depth 0.00 and 0.01. We shall see shortly that travel time tables can be used for much more than simply looking up the travel time.

These tables were derived first in 1935 and have needed little modification since. The times predicted by the tables are often slow, sometimes by as much as a few seconds, compared with the measured times for many regions. They are biassed towards continental structures because they were based on measurements made at mainly continental sites; oceanic regions are generally faster. Improved tables have appeared, but the ocean-continent differences are often comparable with the differences between tables — the concept of a spherically symmetric earth breaks down at about this level of accuracy. We use J-B tables here because they are still the standard adopted by the International Seismological Centre (ISC) for their routine earthquake locations.

4.2 The ray parameter and seismic rays

In geometrical optics a ray is refracted according to Snell's law. There are two implications for the refracted ray: it lies in the plane normal to the interface that contains the incident ray, and the angles of refraction and of incidence are related by the formula (3.56)

$$\frac{\sin i_1}{v_1} = \frac{\sin j}{v_2} \tag{4.1}$$

where v_1 and v_2 are the wave speeds of the two media. This result applies to all types of seismic waves to which the geometrical approximation is appropriate, and not just body waves. We can trace the path of a ray through a stack of layers of different homogeneous materials by simply applying Snell's law at each interface (Figure 4.1). Increasing the number of layers and reducing their thickness gives an approximation to a continuous medium. The approximation can be made exact by proceeding to the limit of infinitesimally thin layers.

For parallel interfaces the angle of refraction equals the angle of incidence on the next medium. $j = i_2$ and (4.1) becomes

$$\frac{\sin i_1}{v_1} = \frac{\sin i_2}{v_2} \tag{4.2}$$

An important consequence of this equation is that the quantity $\sin i/v$ is a constant throughout the length of the ray regardless of the local wave speed of the layer. It is therefore a convenient parameter for describing the ray, and is called the *ray parameter.* Horizontal interfaces can only be used to represent a medium varying in one direction but Snell's law can be applied to more complex media because the only properties of the medium that determine the ray's refraction are those in its immediate vicinity: we can therefore apply Snell's law across the tangent plane to the surface of constant wave speed.

Consider the refraction of rays in a spherically symmetric medium, as shown in Figure 4.2. Spherical shells 1 and 2 have wave speeds v_1 and v_2 respectively. A ray impinging on the interface at A will obey Snell's law.

Figure 4.1: Refraction of a ray path at a plane interface. The ray is bent away from the normal in the medium with higher wave speed. Here $v_3 > v_2 > v_1$, and the ray is bent progressively towards the horizontal as it descends. $v_4 < v_3$, making zone 4 a "low velocity zone"; the ray is bent towards the normal in this medium.

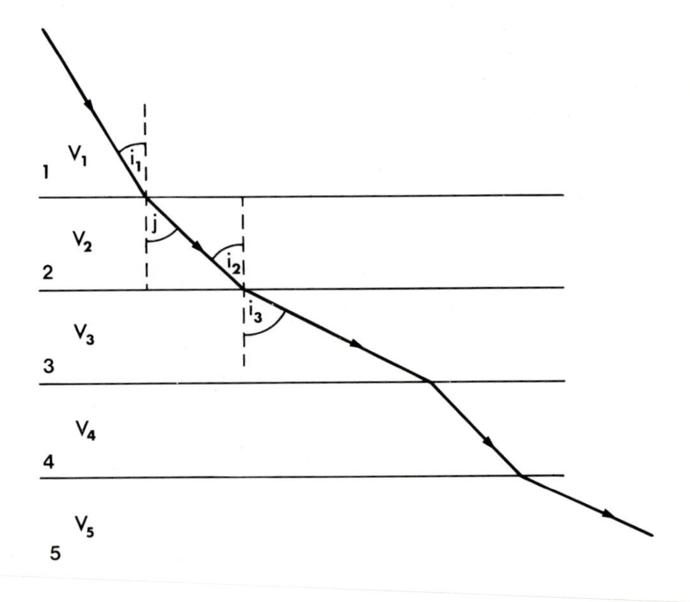

$$\frac{\sin i_1}{v_1} = \frac{\sin j}{v_2} \tag{4.3}$$

The ray will also obey Snell's law at B where it passes out through the bottom of shell 2. Unlike the plane layer case (Figure 4.1), the new angle of incidence i_2 is not equal to j. The two angles are related to each other by the geometry. From triangles AQC and BQC we see that

$$\sin j = \frac{QC}{AC}; \quad \sin i_2 = \frac{QC}{BC} \tag{4.4}$$

and therefore

$$\sin j = \sin i_2 \frac{BC}{AC} \tag{4.5}$$

Let $r_1 = AC$ and $r_2 = BC$, the radii of the two shells. Then Snell's law can be written as

$$\frac{r_1 \sin i_1}{v_1} = \frac{r_2 \sin i_2}{v_2} \tag{4.6}$$

The conserved quantity along the ray is now $r \sin i/v$; it is therefore a suitable choice for the ray parameter, p, in a spherically symmetric medium

$$p = \frac{r \sin i}{v} \tag{4.7}$$

p identifies a ray travelling between two points on or inside the earth. The other consequence of Snell's law, that the refracted ray lies in the

Figure 4.2: Snell's law in a spherically symmetric medium. Spherical shells 1 and 2 have wave speeds v_1 and v_2 respectively. Snell's law implies that $r \sin i/v$ is conserved along the ray path

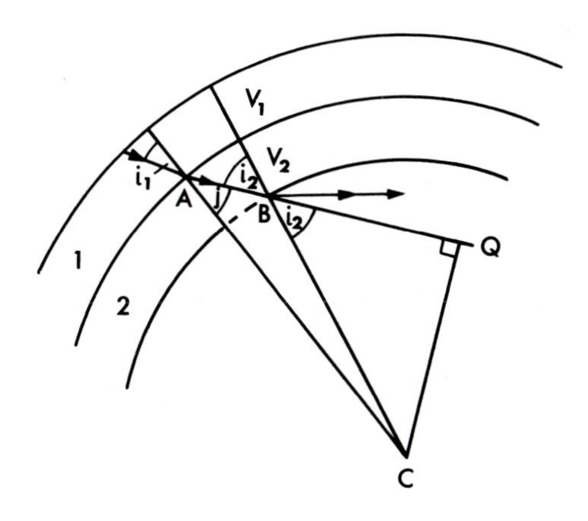

plane containing the incident ray and the normal to the plane tangent to the interface, implies in spherically symmetric media that it lies in a diametric plane (one that contains the centre of the sphere).

An important equation allows us to calculate the ray parameter from a set of travel time tables. Consider two parallel rays PP' and QQ' in Figure 4.3. PN is normal to QQ' at N. $d\Delta$ is an infinitesimal increase in the angular distance Δ between P and P'. The change in travel time associated with $d\Delta$ is

$$dT = \frac{2QN}{v} = \frac{2}{v} PQ \sin i = \frac{PC \sin i}{v} d\Delta \tag{4.8}$$

Let $PC = r$, the radius of the sphere through P, Q and P'. Then

$$p = \frac{r \sin i}{v} = \frac{dT}{d\Delta} \tag{4.9}$$

Thus knowing T for a range of distance angles, Δ, we can take differences to estimate the derivative and hence obtain the ray parameter as the derivative, using (4.9). Equation (4.7) shows that p has dimensions of time; it can only be identified with p in (4.9) if Δ is measured in radians rather than degrees (this was assumed in replacing PQ by $\frac{1}{2}PCd\Delta$ in equation (4.8)). Knowing the ray parameter and velocity as a function of radius allows calculation of the incidence angle i at the earth's surface and the maximum depth of the ray (where $i = 90°$) from the definition (4.7). The deepest point is referred to as the *turning point* of the ray.

The first differences in Table 4.1 can be used to estimate $dT/d\Delta$ for a particular distance. For a focus at the top of the mantle and 43° distance the tabulated first difference is 8.1 s deg^{-1}. Converting to radians we find $p = 464$ s. Knowing this and the wave speed enables us to find the angle of incidence everywhere. In particular the take-off angle at the source

Figure 4.3: Derivation of the equation $p = dT/d\Delta$

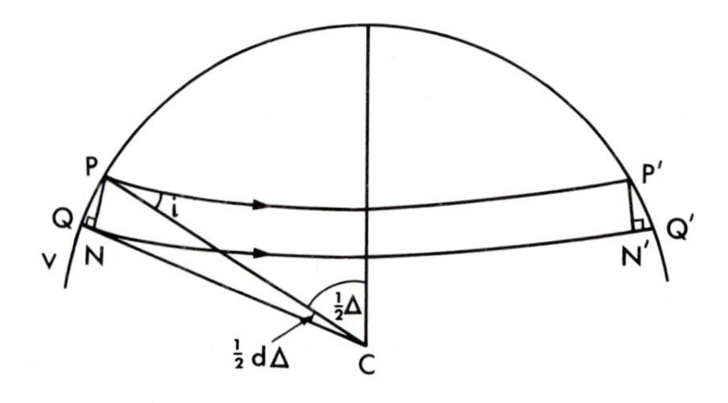

is, from (4.7), $i = \sin^{-1}(vp/r)$ where v is the velocity at the top of the mantle. With $r = 6338$ km and $v = 7.75$ km s^{-1}, from Table 4.2, we have $i = 34.6°$. The ray path at the bottom is horizontal, $i = 90°$ there, and hence $p = r_0/v(r_0)$, where r_0 is the maximum depth of the ray. Using the velocity function $v(r)$ in Table 4.2 we find this ray to bottom at a radius of approximately 5338 km, or depth 1000 km, where the velocity is 11.42 km s^{-1}.

Imagine that we can see through the earth as if it were made of glass, but instead of using light use seismic rays. Refraction distorts the image of the earth we see, just as in the optical case. If the eye is placed at the location of a receiver we "see" the source of the waves in a false position

Shell			Outer core			Inner core	
r/R	α (km./sec.)	β (km./sec.)	r/R	r/R_1	α (km./sec.)	r/R_2	α (km./sec.)
1·00	7·75	4·353	0·548	1·00	8·10	1·0	11·16
0·99	7·94	4·444	0·537	0·98	8·18	0·9	11·19
0·98	8·13	4·539	0·526	0·96	8·26	0·8	11·21
0·97	8·33	4·638	0·515	0·94	8·35	0·7	11·23
0·96	8·54	4·741	0·504	0·92	8·44	0·6	11·25
0·95	8·75	4·850	0·493	0·90	8·53	0·5	11·27
0·94	8·97	4·962	0·482	0·88	8·63	0·4	11·28
0·93	9·50	5·227	0·471	0·86	8·74	0·3	11·29
0·92	9·91	5·463	0·460	0·84	8·83	0·2	11·30
0·91	10·26	5·670	0·449	0·82	8·93	0·0	11·31
0·90	10·55	5·850	0·438	0·80	9·03		
0·88	10·99	6·125	0·427	0·78	9·11	($R_2 = 0·36R_1$	
0·86	11·29	6·295	0·416	0·76	9·20	$= 0·197R$	
0·84	11·50	6·395	0·406	0·74	9·28	= 1250 km.)	
0·82	11·67	6·483	0·395	0·72	9·37		
0·80	11·85	6·564	0·384	0·70	9·44		
0·78	12·03	6·637	0·373	0·68	9·52		
0·76	12·20	6·706	0·362	0·66	9·58		
0·74	12·37	6·770	0·351	0·64	9·65		
0·72	12·54	6·833	0·340	0·62	9·72		
0·70	12·71	6·893	0·329	0·60	9·78		
0·68	12·87	6·953	0·318	0·58	9·84		
0·66	13·02	7·012	0·307	0·56	9·90		
0·64	13·16	7·074	0·296	0·54	9·97		
0·62	13·32	7·137	0·285	0·52	10·03		
0·60	13·46	7·199	0·274	0·50	10·10		
0·58	13·60	7·258	0·263	0·48	10·17		
0·56	13·64	7·314	0·252	0·46	10·23		
0·55	13·64	7·304	0·241	0·44	10·30		
			0·230	0·42	10·37		
			0·219	0·40	10·44		
			0·208	0·38	9·92		
			0·197	0·36	9·40		

($R_1 = 0·548R = 3473$ km.)

Table 4.2: Velocity as a function of depth in the Earth. R_1 and R_2 are the radii of outer and inner cores respectively. From JEFFREYS (1976)

because of the refraction; and because rays are reversible an eye placed at the source (focus) of the rays will "see" all the receiving stations in distorted positions — in fact further away than they really are. The distorted positions are called *apparent positions.* They are calculated using spherical geometry, the azimuth of the receiver from the focus, and the angle of incidence found from travel time tables. We shall find the concept of apparent position important in dealing with fault plane solutions in Chapter 6.

The amplitude of a wave from a point source in a homogeneous medium falls off with the distance, provided no energy is lost in transmission. This is because the energy is spread out uniformly over the area of the wavefront, which is spherical and therefore increases as the square of the distance. Refraction causes focussing of the rays and distorts the wavefront so that it is no longer spherical. Again we can make use of the travel time tables to calculate the effect on the amplitude; it depends on the second derivative of time with respect to distance.

In Figure 4.4 H represents the position of a seismic source that emits energy $I(i, \phi)$ per unit solid angle, where angle ϕ measures rotation about axis HC. HS and HS' are two rays that define a pencil of rays intersecting the surface at SS'. Regard i_2 and ϕ as the two angles defining a spherical coordinate system with centre H and pole C. By rotating the whole plane in Figure 4.4 about axis HC by an angle $d\phi$ we sweep out solid angle $\sin i\ di\ d\phi$, and line SP sweeps out an area of wavefront $r_1 SP \sin\Delta d\phi = r_1 SS' \sin\Delta \cos i_1 d\phi = r_1^2 \sin\Delta \cos i_1 d\Delta d\phi$. The energy per unit area of wavefront is therefore

Figure 4.4: Geometrical spreading. The energy per unit area of wavefront depends on the refraction of the rays.

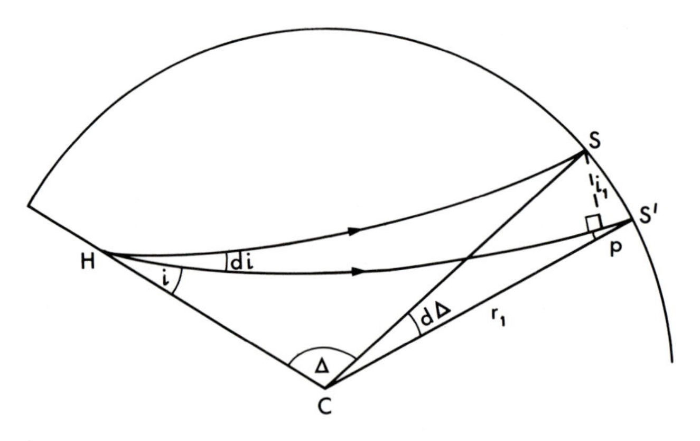

$$e(\Delta) = \frac{I \sin i}{r_1^2 \sin \Delta \cos i_1} \left| \frac{di}{d\Delta} \right| \tag{4.10}$$

Combining equation (4.7) and (4.9) for the ray parameter

$$\sin i = \frac{v}{r} \frac{dT}{d\Delta} \tag{4.11}$$

where v, r apply at H, and differentiating with respect to Δ gives

$$\cos i \frac{di}{d\Delta} = \frac{v}{r} \left| \frac{d^2T}{d\Delta^2} \right| \tag{4.12}$$

Hence (4.10) becomes

$$e(\Delta) = \frac{I \tan i}{r_1^2 \sin \Delta \cos i_1} \frac{v}{r} \left| \frac{d^2T}{d\Delta^2} \right| \tag{4.13}$$

So, armed with a travel time curve $T(\Delta)$ and a velocity model $v(r)$ we can estimate the effect of refraction on the amplitude of a wave at any distance by differencing the travel times to obtain first $dT/d\Delta$, then $d^2T/d\Delta^2$, finding i and i_1 from (4.11) and the energy per unit area from (4.13).

For some applications we need to know exactly where the ray has gone. Calculating the ray path, or *ray tracing*, in complicated media is a difficult numerical problem. For spherically symmetric media it requires only an integral. The ray path must lie in a diametric plane, and therefore we need only consider the problem in two dimensions. Once the path is found in a plane, the plane itself can be rotated into the correct orientation with respect to the earth. Within the plane the ray path is specified as a function $\theta(r)$, where θ is the angular distance from some reference point and r is the radius.

Figure 4.5 shows that $\theta(r)$ satisfies the simple differential equation

$$\frac{d\theta}{dr} = \frac{1}{r\left(r^2/p^2v^2 - 1\right)^{\frac{1}{2}}} \tag{4.14}$$

If v is a known function of r we can integrate (4.14) to give $\theta(r)$ for any choice of ray parameter p. The ray is symmetric about its lowest point (r_0, θ_0), and therefore it is only necessary to find one half of the ray path:

$$\theta - \theta_0 = \int_{r_0}^{r} r^{-1} \left(r^2/p^2v^2 - 1\right)^{-\frac{1}{2}} dr \tag{4.15}$$

Note that the integrand has a square root singularity at the bottom of the ray path, where $r/pv = 1$. This singularity is removed on integration.

These equations have been used to calculate ray paths for spherically symmetric models of the seismic velocity within the earth. For the most part seismic velocities increase with depth and rays are refracted back up towards the earth's surface. In spherical geometry v/r must increase with depth for this to happen, rather than just v as for a plane medium, because of the form of the ray parameter (4.7). If v/r decreases with depth the ray is refracted downwards until it reaches a region in which v/r increases once more and the ray is refracted back up. This usually results in a shadow zone forming on the earth's surface, a region which is not illuminated by any rays from a particular source. This absence of arrivals makes it difficult to infer the velocity structure within such a *low velocity zone*, as we discuss in Section 4.5.

4.3 Time-distance curves

The travel time for waves between two points on the surface of a plane medium increases monotonically with distance between the points provided the seismic velocity increases steadily with depth. The time-distance curve can adopt more complicated forms depending on the variation of seismic velocity with depth. The first task of any seismic experiment designed to determine the velocity at depth is the estimation of the time-distance curve. We now study some examples of typical media.

Figure 4.5: Ray tracing. PQ is an element of a ray path of length ds. From triangle NPQ we have $\tan i = rd\theta/dr = (pv/r)(1 - p^2v^2/r^2)^{-\frac{1}{2}}$. If v is known as a function of r, this equation describes the ray path in the form of $\theta(r)$.

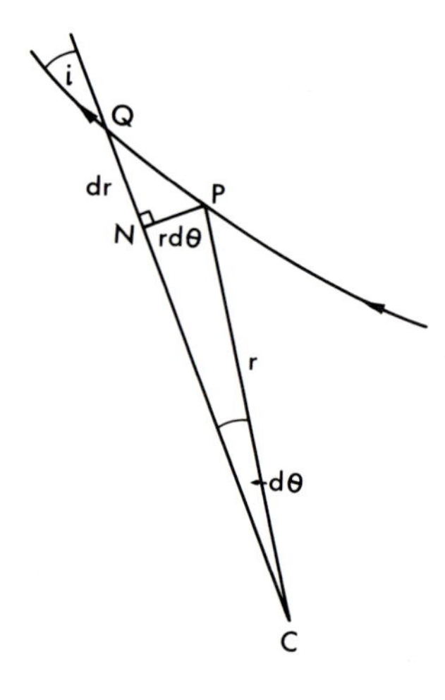

Consider media consisting of plane layers. In each layer the seismic velocity is either uniform or increases linearly with depth. Ray paths in uniform layers are straight lines; those in variable layers are arcs of circles, concave upwards. The curvature of the ray paths is proportional to the velocity gradient (exercise 6). A medium consisting of one layer with linearly increasing velocity will have a monotonically increasing time function $T(X)$. If such a layer has a uniform layer beneath it then ray paths passing through the top layer will continue downwards as straight lines and never be recorded at the surface. The $T(X)$ curve terminates at the distance corresponding to rays with turning points at the interface between the two layers. The lower layer is essentially a low velocity zone. Spherical media behave in a similar way provided $v(r)/r$ behaves like $v(z)$ in the plane medium.

When a uniform layer lies above a medium with increasing velocity then multipathing can occur: there are two or more quite different ray paths between two points. Distinguishing between different arrivals is often the most difficult part of interpretation. The rays for this two-layer model are shown in Figure 4.6(a). Ignore the wave that travels along the surface of medium I — the *direct wave*. Rays like OD, which have large take-off angles, do not penetrate medium II to any great depth and travel large horizontal distances. Rays like OE, with small take-off angles, penetrate to great depth in medium II and also travel large horizontal distances.

Figure 4.6: (a) Ray paths in a half space with two media: I has uniform velocity and II has velocity increasing linearly with depth. Ray paths are straight lines in medium I and arcs of circles in medium II. (b) The time-distance curve.

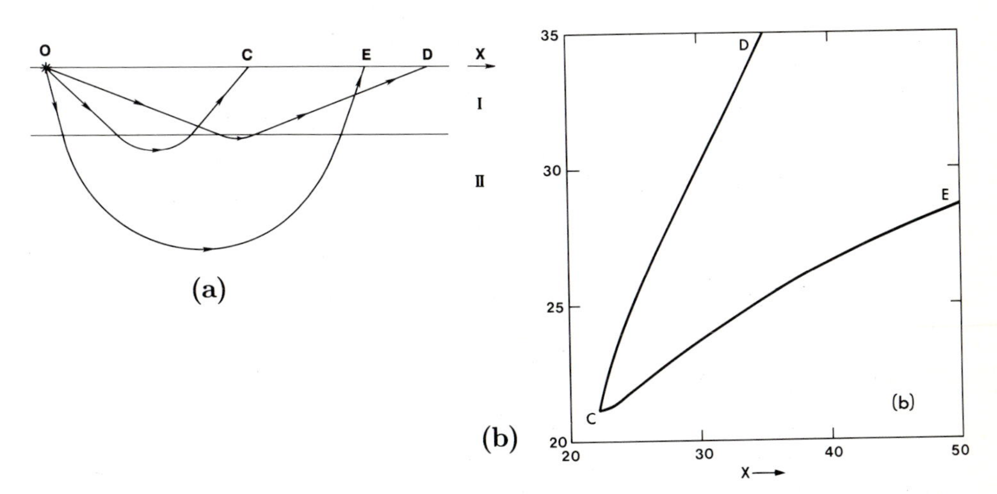

Rays with intermediate take-off angles travel shorter horizontal distances; *OC* represents the shortest distance for any ray penetrating layer II. The time-distance function is shown in Figure 4.6(b); the letters correspond to the rays in Figure 4.6(a). The travel time is not a single valued function of distance. The two branches *CD* and *CE* correspond to different sets of ray paths.

Now consider the three layers shown in Figure 4.7(a). The velocity within each layer increases linearly with depth, but in the middle layer it increases at a more rapid rate than in the other two. This produces a triplication as shown in Figure 4.7(b). The branch AD corresponds to rays travelling only in the top layer. D is at the maximum possible distance for these rays. The branch CE corresponds to rays that have been refracted out of the very bottom layer. The branch DC has rays that have travelled in the middle layer, which is not thick enough to allow for a critical point as in Figure 4.7(b). C therefore corresponds to the ray that bottoms at the top of layer III. Three pulses will be seen over the distance range from C to D.

These complications arise in the earth. At certain distances the travel time curve can be difficult to estimate accurately. For example, the core produces a shadow zone beginning at 104° (for the Jeffreys-Bullen velocity model; diffraction effects and shadows for high frequency waves occur at smaller distances). *PKP* and *PKIKP* have a triplication similar to that in Figure 4.7(b). P and S rays observed at distances of around 20° have turning points near the 650 km discontinuity; they often have loops in their travel time curves.

Time-distance functions which are multiply or partially defined are difficult to deal with theoretically and are difficult to estimate from measurements. Scatter in the measurements near a triplication like the one in Figure 4.7(b), for example, may obscure details of the triplication altogether, and the data may appear as a cloud of points in that vicinity. It is even more difficult to place error bounds on our estimate of the $T(\Delta)$ curve in these cases. Some research has made use of the function defined by

$$\tau = T(p) - p\Delta \tag{4.16}$$

τ can be regarded as a function either of Δ or of p. Choosing p as the independent variable has the advantage that both τ and T take values for all possible values of p: there are no gaps in $\tau(p)$ when shadow zones are present. A shadow zone introduces a discontinuity into the $\tau(p)$ function, since a small increase in p at the start of the shadow zone causes a finite

Figure 4.7: (a) Ray paths in a three layer medium in which the middle layer has a wavespeed that increases at a higher rate than in the other two. In (b) the travel time curve has three branches corresponding to rays that bottom in layer I (AD), layer III (CE), and layer II (DC). (c) is the corresponding τ-p curve, which is monotonic. The cusps have become the inflection points indicated.

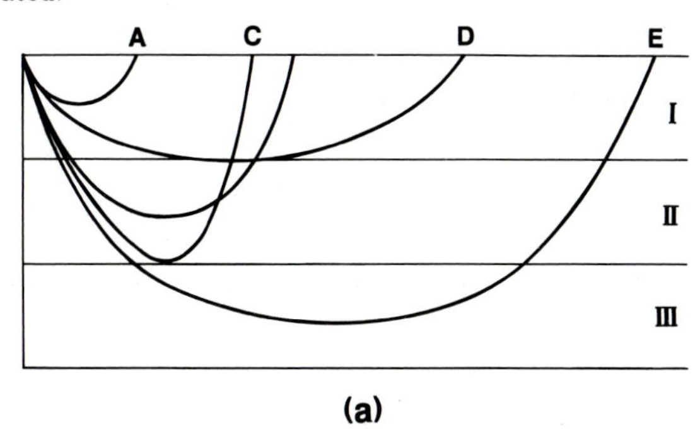

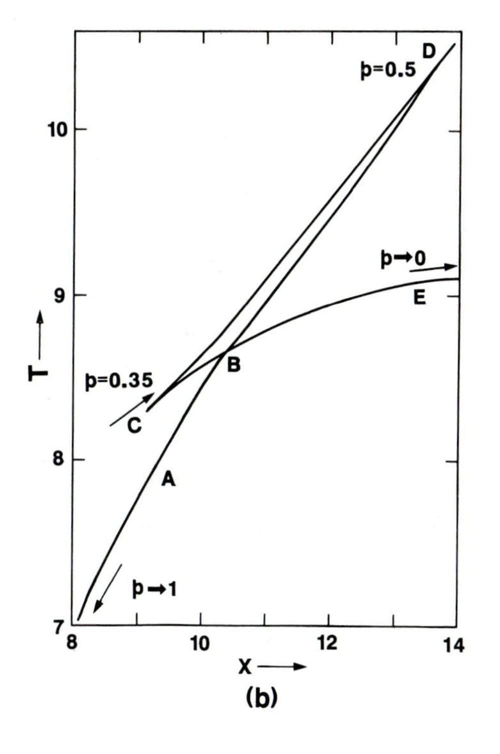

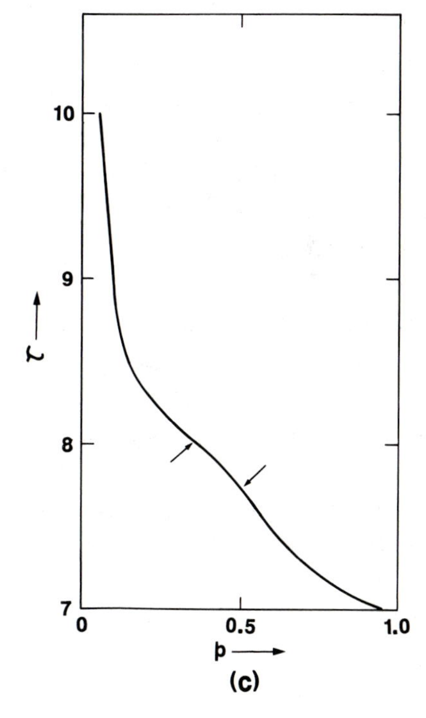

jump in Δ which, from (4.16), causes a drop in τ. Differentiating (4.16) with respect to p and using the chain rule gives

$$\frac{d\tau}{dp} = \frac{dT}{dp} - p\frac{d\Delta}{dp} - \Delta = \frac{dT}{dp} - p\frac{dT}{dp}\frac{d\Delta}{dT} - \Delta \tag{4.17}$$

and since from (4.9)

$$\frac{d\Delta}{dT} = p^{-1} \tag{4.18}$$

we have

$$\frac{d\tau}{dp} = -\Delta \tag{4.19}$$

$\tau(p)$ is therefore a monotonic function. The distance, Δ, is always positive, so that $\tau(p)$ must everywhere decrease. A triplication is "unwrapped" and appears on a $\tau(p)$ plot as a kink. The $\tau(p)$ function for the triplication in Figure 4.7(b) is shown in Figure 4.7(c). The simple nature of the $\tau(p)$ curve suggests that it would be a useful function to take the place of the traditional travel time tables, the advantages being that several branches of the $T(\Delta)$ tables can be represented in a single $\tau(p)$ table and, as is shown in Section 4.7, computation of $\tau(p)$ from a velocity model is simple.

4.4 Seismic arrays

Accurate timekeeping has been a major experimental difficulty with seismic recording. The seismometer's own clock is used to record timing marks on the traces, usually at one minute intervals, and the clock is checked regularly against a time standard to allow for drift. In this way it is possible to keep time to better than a tenth of a second. An *array* of seismometers has a single, central clock for all the instruments; the relative timing between individual members of the array is not subject to drift and is therefore very much more accurate than the absolute time. We make the distinction between a *network* of seismometers and an array: the array has central timing. Arrays allow the direct measurement of the slowness, or ray parameter, p. The principle of the method is shown in Figure 4.8. If we know the azimuth of the arriving wave we can choose a value of p and calculate the delay times at each instrument from the array geometry. Each trace is delayed and summed to form a beam. If the power in the beam is plotted as a function of slowness, p, there will be a peak at the value of p corresponding to the incoming wavefront. The procedure can also be applied to find the azimuth of the incoming wave, Z. Delay times are calculated for a range of values of p and Z and the individual traces delayed and summed. It is usual practice to compute the total power in the beam for a time window of a few seconds and plot the power as a function of p and Z. Again a peak in the power denotes the value of p and Z for the incoming wave.

This procedure is called "steering" the array: it makes it sensitive to a particular arrival direction. Other arrival directions also produce peaks, however. Consider the array in Figure 4.8, and an incoming wave with slowness p' and angular frequency ω. During time ℓp this wavefront will sweep a distance $\ell p/p'$ across the ground. If $p' < p$ then this distance will exceed ℓ, but a second wavefront will have travelled a distance $\ell p/p' - 2\pi/\omega p'$. If this distance is equal to ℓ the delay will produce traces that are in phase. The beam therefore has a large response to the slowness $p' = p - 2\pi/\omega\ell$. The response is also large for other values of p'.

Consider now an array with arbitrary two-dimensional geometry. The N sensors are placed at positions $\{\mathbf{h}_n; n = 1, 2, \ldots, N\}$. A slowness vector

Figure 4.8: A plane wavefront is incident on a linear array of equally spaced seismometers. The angle of incidence, i, is equal to the angle between wavefront and horizontal. In 1 second the wavefront sweeps along the ground a distance $AB = v/\sin i$. The inverse of this speed is called the *slowness* of the wavefront and is equal to the ray parameter, p. If each trace is delayed by ℓp they will be in phase.

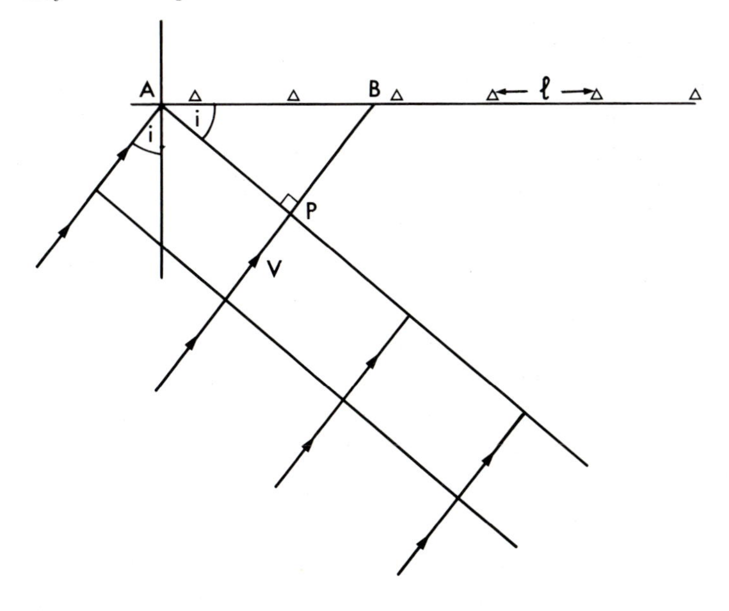

$\mathbf{p}$ is defined, where $\mathbf{p} = \mathbf{k}/\omega$; the vector is normal to the wavefronts and its magnitude is the inverse of the phase velocity. Consider an incoming wave of the form

$$\psi(t) = A(\omega) \exp i\omega\,(\mathbf{p}\cdot\mathbf{x} - t) \tag{4.20}$$

The array is steered by delaying the recording at the nth sensor by a time $\mathbf{p}_0 \cdot \mathbf{h}_n$ and summing to give the composite recording, or *beam*

$$B(t, \mathbf{p} - \mathbf{p}_0) = A(\omega) \exp -i\omega t \sum_{n=1}^{N} \exp i\omega\,(\mathbf{p} - \mathbf{p}_0)\cdot\mathbf{h}_n \tag{4.21}$$

The responses will add in phase for slowness $\mathbf{p}_0$; the array has been steered in the direction of $\mathbf{p}_0$. The steered array has a frequency response in the same sense as a single seismometer has a frequency response (cf. equation (3.3)). Note that we have assumed each individual sensor gives a perfect recording of the ground motion (i.e. its impulse response is a delta function); in practice we would have to take account of each instrument's response. From (4.21), the frequency response is

$$R(\omega, \mathbf{p} - \mathbf{p}_0) = \sum_{n=1}^{N} \exp i\omega\,(\mathbf{p} - \mathbf{p}_0)\cdot\mathbf{h}_n \tag{4.22}$$

The impulse response is the inverse Fourier transform of the frequency response (Section 3.1). In this case it is

$$\sum_{n=1}^{N} \delta(\mathbf{x} - \mathbf{h}_n) \tag{4.23}$$

Consider the simple case of N sensors equally spaced a distance ℓ apart along the x axis. Then $\mathbf{h}_n = (n-1)\ell\hat{\mathbf{x}}$, where $\hat{\mathbf{x}}$ is the unit vector in the x direction. (4.23) gives

$$R(\omega, \mathbf{p} - \mathbf{p}_0) = \sum_{n=1}^{N} \exp 2iN\phi \tag{4.24}$$

where $2\phi = \omega(\mathbf{p} - \mathbf{p}_0)\cdot\hat{\mathbf{x}}\ell$. This geometric progression has factor $\exp 2i\phi$ and sums to

$$R(\phi) = \frac{\exp 2iN\phi - 1}{\exp 2i\phi - 1} \tag{4.25}$$

Taking out the factor $\exp i\,(N-1)\,\phi$ by changing the origin for x gives

$$R(\omega, \mathbf{p} - \mathbf{p}_0) = \frac{\sin N\phi}{\sin \phi} \tag{4.26}$$

This function is plotted in Figure 4.9. Note the large side lobes, which correspond to slowness vectors $(p_0 - 2n\pi/\omega\ell)$, as referred to above.

Several arrays have been in operation since the mid 1960s. Most of them were installed initially to detect nuclear explosions. The Norwegian Seismic Array (*NORSAR*) consists of a network of *subarrays*. Each subarray consists of five instruments measuring the vertical component of ground motion and a central three-component instrument. The traces from individual instruments can be aligned and summed to form a composite trace for the subarray, a process which enhances the signal-to-noise ratio and partly eliminates effects due to local anomalies in the vicinity of the subarray. Each subarray is some 10 km across; the whole array was originally composed of 22 subarrays and was about 100 km across (it was subsequently reduced in size and later dismantled altogether).

Direct measurement of p makes it possible to separate phases that arrive simultaneously, or so close in time that they cannot be distinguished

Figure 4.9: Frequency response of a linear array of 20 equally spaced seismometers expressed as a function of ϕ, where $2\phi = (\mathbf{p} - \mathbf{p}_0)_x\ell$.

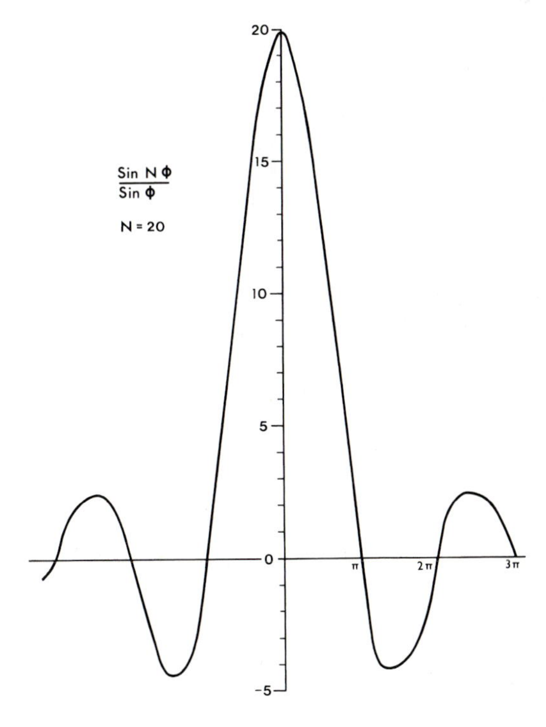

on a single seismogram. The *NORSAR* array has been used to identify both the inner core reflection *PKiKP* and the phase *PKIKP* that travels through the inner core as a body wave (Doornbos & Husebye 1972). There are troublesome precursors to these phases which can also be separated by steering the array. Figure 4.10 shows the power in the beamed wavetrain as a function of time first arrival, at a slowness of about 2.6 s (degree)$^{-1}$, is identified as *PKIKP*. The later arrival, near 4.5 s (degree)$^{-1}$, is a *PKP* phase. Examples of precursors to *PKIKP* are shown in Figure 4.11. These are the result of steering the beams using the maxima of the precursors for the early parts of the traces, and at *PKIKP* for the later part. The traces have been lined up with a *PKIKP* slowness of 1.8 s (degree)$^{-1}$. Results from seven earthquakes at a distance range of 136–142 ° are shown.

The precursors had been attributed to refraction in the *F* region around the inner core. Array studies have now shown the precursors to be associated with a quite different slowness. The precursors are now believed

Figure 4.10: Energy of the beam from the *NORSAR* array as a function of slowness and time. The source was at depth 214 km and distance 148.2°. From Doornbos & Husebye (1972).

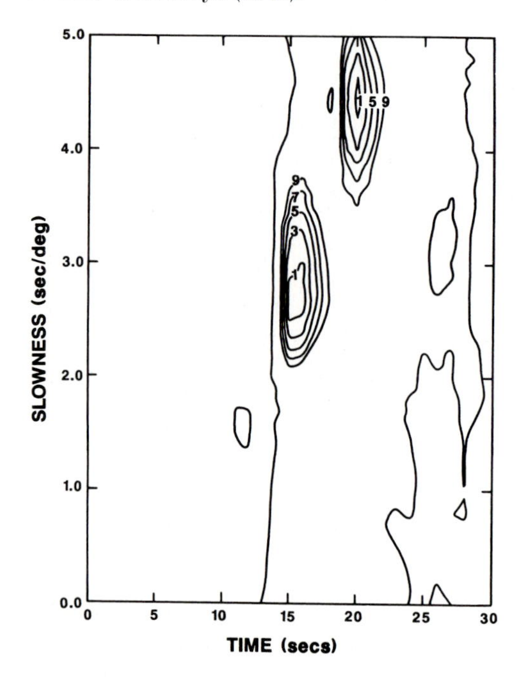

to correspond to energy scattered by inhomogeneities at the bottom of the mantle, and there is little evidence for the existence of an F region around the inner core.

Array studies have also led to a better understanding of the structure of the upper mantle, and departures from spherical symmetry within the earth. Some waves have been found to arrive at azimuths that are markedly different from the expected values, indicating that the ray had not travelled in a diametric plane. This can only be due to departures from spherical symmetry of the elastic parameters in the earth.

4.5 Inversion of travel times for earth structure

In this book we mention three types of observation that have contributed to our understanding of the earth's interior: dispersion of surface waves (Chapter 3), travel times (this chapter), and free oscillation

Figure 4.11: Composite traces formed by beaming the *NORSAR* array for a range of distances where inner core phases are expected to arrive. *PKiKP* can be seen as distinct from *PKIKP* at larger distances. From Doornbos & Husebye (1972).

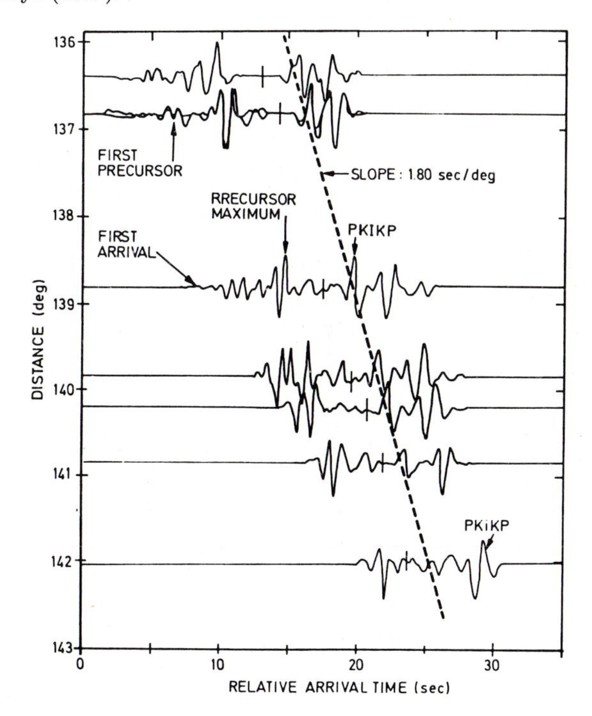

frequencies (Chapter 5). Travel time inversion is also an important part of explosion seismology, which does not suffer the complication of uncertain source locations. Suppose we have measured a travel time function $T(\Delta)$. We ignore departures from spherical symmetry in the earth and seek the seismic velocity as a function of radius, $v(r)$. This is an example of an *inverse problem.* If we know the function $v(r)$ we can calculate $T(\Delta)$ from the formula

$$T(\Delta) = \int_A^B \frac{dl}{v} \tag{4.27}$$

where A and B are two points a distance Δ apart, and dl is measured along the ray path from A to B. Finding $T(\Delta)$ from (4.27) is the forward problem: finding v from $T(\Delta)$ is the inverse problem. There is no guarantee that an inverse problem will, in general, have a unique, or indeed any, solution. Sometimes the travel time problem has no unique solution. Setting $dl = dr/\cos i$ in (4.27), and noting that $\cos i = r^{-1}\left(r^2 - v^2p^2\right)^{\frac{1}{2}}$, it becomes

$$T(p) = 2\int_{r_0}^{r_E} \frac{rdr}{v\left(r^2 - v^2p^2\right)^{\frac{1}{2}}} \tag{4.28}$$

where r_0 is the radius at the bottom of the ray path, r_E is the earth's surface, and now T is regarded as a function of ray parameter p rather than Δ. p can be found from the function $T(\Delta)$ using $p = dT/d\Delta$, so there is no problem in principle with converting $T(\Delta)$ to $T(p)$.

Before going into specific details of the travel time problem it is helpful to consider how we might go about estimating v from (4.28) by a straightforward, brute-force, approach. The integral on the right hand side of (4.28) can be approximated by the trapezium rule (there is a minor difficulty at $r = r_0$ when the denominator of the integrand vanishes, but the singularity is integrable) to give an algebraic relationship between $T(p)$ and the velocity v at the finite number of levels, $\{r = r_k; k = 1, 2, ..., P\}$. The P parameters $\{v(r_k); k = 1, 2, \ldots, P\}$ are then adjusted so that the predicted $T(p)$ fits the observations, using some standard procedure such as least squares (see Section 4.9 on earthquake location). This brute-force approach can be extended, in principle, to three-dimensional media using (4.27) rather than (4.28), but it has pitfalls. What if no solution exists for $v(r)$? Will our scheme produce a misleading result, or will it diverge, providing no clue of the time structure? The problem is discussed by Johnson & Gilbert (1972).

4.6 The method of Herglotz & Wiechert

A more subtle approach begins with the equation for distance as a function of ray parameter. For conciseness we consider only surface events; the generalisation to deep events being obvious. For the ray path we have, from (4.14),

$$\frac{d\theta}{dr} = \frac{p}{r\left(\eta^2 - p^2\right)^{\frac{1}{2}}} \tag{4.29}$$

where $\eta = r/v$ is a function of r that replaces the velocity. Integrating along the whole ray path gives

$$\Delta(p) = 2\int_{r_0}^{r_E} \frac{p\,dr}{r\left(\eta^2 - p^2\right)^{\frac{1}{2}}} = 2\int_{p}^{\eta_E} \frac{p}{r\left(\eta^2 - p^2\right)^{\frac{1}{2}}}\frac{dr}{d\eta}d\eta \tag{4.30}$$

where we have changed the integration variable from r to η. The limits become $\eta(r_0) = p$, because $pr/v = 1$ at the bottom of the ray path, and $\eta(r_E) = \eta_E$.

Now multiply both sides of (4.30) by $\left(p^2 - \eta_1^2\right)^{-\frac{1}{2}}$ and integrate over p:

$$\int_{\eta_1}^{\eta_E} \frac{\Delta}{\left(p^2 - \eta_1^2\right)^{\frac{1}{2}}}dp = \int_{\eta_1}^{\eta_E}\int_{p}^{\eta_E} \frac{2p}{r\left(p^2 - \eta_1^2\right)^{\frac{1}{2}}\left(\eta^2 - p^2\right)^{\frac{1}{2}}}\frac{dr}{d\eta}d\eta\,dp \tag{4.31}$$

and change the order of integration of the double integral. The new limits on p and η are found from the area of integration in the p,η plane (Figure 4.12). This leads to

$$\int_{\eta_1}^{\eta_E}\left(\int_{\eta_1}^{\eta} \frac{2p}{\left(p^2 - \eta_1^2\right)^{\frac{1}{2}}\left(\eta^2 - p^2\right)^{\frac{1}{2}}}dp\right)\frac{d\log r}{d\eta}d\eta \tag{4.32}$$

The integral in brackets is performed by standard methods: the reader is invited to derive the result for himself. The answer, surprisingly, is π. The remaining integral in (4.32) is now easy; it gives $\pi\log\left(r_E/r_1\right)$ where r_1 is the radius for which $(r/v) = \eta_1$.

Return now to the left hand side of equation (4.31). We change variables from p to Δ and integrate by parts

$$\begin{aligned}\int_{\eta_1}^{\eta_E} \frac{\Delta(p)}{\left(p^2 - \eta_1^2\right)^{\frac{1}{2}}}dp &= \int_{\Delta_1}^{0} \frac{\Delta}{\left(p^2 - \eta_1^2\right)^{\frac{1}{2}}}\frac{dp}{d\Delta}d\Delta \\ &= \left[\Delta\cosh^{-1}\left(\frac{p}{\eta_1}\right)\right]_{\Delta_1}^{0} - \int_{\Delta_1}^{0}\cosh^{-1}\left(\frac{p}{\eta_1}\right)d\Delta\end{aligned} \tag{4.33}$$

The first term vanishes when we put in the limits. Δ_1 is the distance for the ray path with ray parameter η_1, and $\cosh^{-1}(1) = 0$, leaving

$$\int_0^{\Delta_1} \cosh^{-1}\left(\frac{p}{\eta_1}\right) d\Delta = \pi \log\left(\frac{r_E}{r_1}\right) \tag{4.34}$$

This equation can be used to construct the velocity function $v(r)$ directly. To see this, first recall that our measurements are of times T as a function of distance, Δ. This function can be differentiated to give $p(\Delta)$ using (4.9). Since $\eta_1 = p(\Delta_1)$ we can now evaluate the integral on the right hand side of (4.34) for any choice of Δ_1, provided it is chosen so that $\eta_1 \leq p$ (the function $\cosh^{-1} y$ is complex for $y \leq 1$). The right hand side then gives r_1 for distance Δ_1: we also know η_1 corresponding to this radius, and since $\eta_1 = r_1/v(r_1)$ we have the velocity at one known depth. We can build up the velocity function for a range of depths by repeating the process for other values of Δ_1.

The only restriction imposed on the radius r_1 is that $\eta_1 \leq p$. p must therefore attain its minimum value in the distance range $(0, \Delta_1)$ at distance Δ_1 for the method to be applicable. This is the case if $\eta(r)$ is a monotonically increasing function, since $p = \eta(r_0)$ where r_0 is the bottom of the ray path. It is not the case if "low velocity zones" (zones where η decreases with r) are present: in these η may increase with depth to a point where $\eta > p$. Low velocity zones can also produce shadow zones

Figure 4.12: Changing limits in the double integral in (4.31). The shaded triangle denotes the area of integration. The area is swept out by the limits $\eta = p$ to η_E; $p = \eta_1$ to η_E. Changing the order of integration, the appropriate limits are $p = \eta$ to η_E, $\eta = \eta_1$ to η_E.

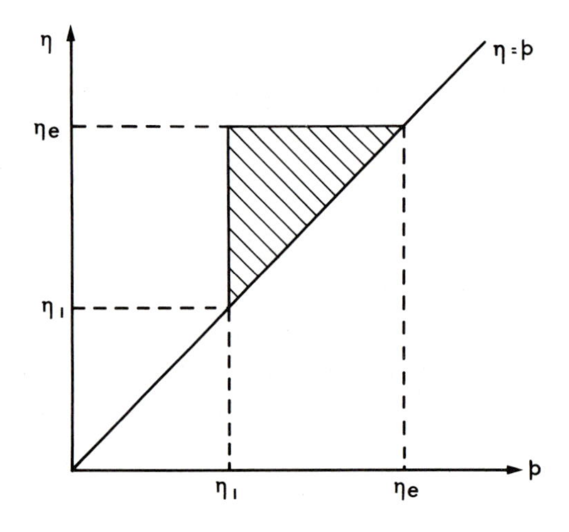

that appear as gaps in the $T(\Delta)$ function: this would appear to be a fundamental obstacle to inversion. The velocity function in Figure 4.7 gives another illustration of the problem. X replaces Δ in this plane geometry; p is a two-valued function of X in the range $X > PC$. If we restrict ourselves to the CE branch of the travel time curve then $p(X)$ is a monotonic decreasing function and the planar equivalent of (4.34) can be applied. On the CD branch $p(X)$ is an increasing function and (4.34) cannot be applied. Success of the method depends on the nature of the $T(\Delta)$ curves, which can be very complex.

4.7 $\tau - p$ inversion

The relationship between travel times and seismic velocity is non-linear; a change in velocity does not lead to a proportional change in the observed travel times. Non-linearity usually demands an iterative solution, often based on linearisation. It also makes estimation of error bounds on the velocity difficult. The τ-function is, under certain circumstances, a linear function of the depth considered as a function of seismic velocity. Since knowledge of $r(v)$ is in most cases as good as knowing $v(r)$, the discovery allows us to use the full body of theory for linear equations.

Recalling equation (4.28) for $T(p)$ and the similar equation for $\Delta(p)$

$$\Delta(p) = 2\int_{r_0}^{r_E} \frac{pv/r}{(r^2 - p^2v^2)^{\frac{1}{2}}} dr \tag{4.35}$$

we have the equation for $\tau(p)$

$$\tau(p) = T(p) - p\Delta(p) = 2\int_{r_0}^{r_E} \frac{(r^2 - p^2v^2)^{\frac{1}{2}}}{rv} dr \tag{4.36}$$

We measure the delay time τ for various values of p, and try to estimate $v(r)$ from the equation (4.36). In planar geometry there is a similar result

$$\tau(p) = T(p) - pX(p) = 2\int_0^{z_0(p)} \left[s^2(x) - p^2\right]^{\frac{1}{2}} dz \tag{4.37}$$

where $s(z) = 1/v(z)$ and z is depth. Derivation of (4.37) is left as an exercise. Here we consider surface focus events. z_0 denotes the deepest point of the ray. This will be a function of ray parameter and velocity, since $p = s(z_0)$. Change the variable of integration from z to s. If $s(z)$ is a monotonic decreasing function so that the limits change from $z = (0, z_0(p))$ to $s = (s_{\max}, p)$, (4.37) becomes

$$\tau(p) = 2\int_{s_{\max}}^{p} \left(s^2 - p^2\right)^{\frac{1}{2}} \frac{dz}{ds} ds \tag{4.38}$$

so that now the limits are independent of the functon $s(z)$, except for the single value at the surface. Integrating by parts and putting in limits gives

$$\tau(p) = 2\int_{p}^{s_{\max}} \frac{sz(s)}{(s^2 - p^2)^{\frac{1}{2}}} ds \tag{4.39}$$

This equation is a linear relationship between the functions $z(s)$ and $\tau(p)$: for example, doubling z will double τ. The linear relationship only holds if the limits remain independent of $s(z)$. This is the case when $s(z)$ is a monotonic decreasing function (or equivalently $z(s)$ is monotonic increasing) because the limits can be changed from $(0, z_0)$ to $(s_{\max}, p)$ as above. If low velocity zones are present the transformation of limits is not straightforward. Suppose the layer has increasing velocity down to depth z, where the inverse velocity is s_1. Below this depth there is a low velocity zone where the slowness increases from s_1 to s_2 at depth z_2. Beneath z_2 the velocity increases again. (4.37) transforms to

$$\tau(p) = 2\int_{s_{\max}}^{s_1} \left(s^2 - p^2\right)^{\frac{1}{2}} \frac{dz}{ds} ds \quad - \int_{s_2}^{s_1} \left(s^2 - p^2\right)^{\frac{1}{2}} \frac{dz}{ds} ds$$
$$+ \int_{s_2}^{p} \left(s^2 - p^2\right)^{\frac{1}{2}} \frac{dz}{ds} ds \tag{4.40}$$

and now the limits of integration do depend on $s(z)$ — the relationship is therefore inherently non-linear. This process has not solved the fundamental problems associated with low velocity zones.

A group of Russian seismologists have given methods for mapping error bounds in the $\tau(p)$ curve into an error corridor for the velocity function (Bessonova *et al.* 1974). We mention here the application of these methods to a marine seismic refraction profile from the East Pacific Rise at 14°44′S, 112°05′W, called *DW-34*. The data are shown in Figure 4.13. The conventional method of interpretation is to assume a uniform layered medium for which the time-distance curve takes the form of sections of straight lines. For the tau method we need to estimate $\tau(p)$ with error bounds. Kennett & Orcutt (1976) estimated the $\tau(p)$ function by fitting smooth interpolation functions (e.g. quadratics) to $T - pX$ versus X for various choice of p. This method is recommended as stable by Bessonova *et al.* (1974). The error in τ is dominated by the error in the measured time, and error bounds were assigned accordingly (Figure 4.14). The corridor of allowed values in the $\tau(p)$ plane was mapped into a corridor of possible velocities as a function of depth, also shown in Figure 4.13. The inflection in the $\tau(p)$ curve near $p = 0.147$ s km^{-1} (6.8 km s^{-1}) suggests a change in velocity gradient at that depth, and this is reflected in the velocity-depth curve. An inversion linearised in $v(z)$ leads to the model in Figure 4.14(c). There is little evidence in the τ solutions for the detailed layering of the conventional model (c).

4.8 Preliminary location of earthquakes

We shall treat an earthquake as if it was a point source radiating elastic energy; it is a reasonable but not a perfect assumption, particularly for large earthquakes. The distance to the earthquake is obtained by reading as many phases as possible on the seismogram and comparing the arrival times with standard travel time curves, and the location then found by triangulating, using distances from several stations. In practice the P-minus-S time is important in preliminary determination of distance, but not important in the final refinement because the S wave cannot be read accurately. Depth phases such as pP are useful in determining the depth of the hypocentre. A quick determination of a location might be obtained initially from an earthquake using two or three local instruments but for an event large enough to yield global coverage many tens of stations would be available. It takes time for the reports to be gathered into a central recording agency, and definitive locations are not published until months or years after the event.

There are several national agencies that make preliminary determinations of epicentres and hypocentres. The National Earthquake Information Centre (*NEIC*) of the USA is one such organisation; it publishes a regular bulletin called the "Preliminary Determination of Epicentres", or *PDEs*. The International Seismological Centre (*ISC*) gathers data from

Figure 4.13: Travel time data for a seismic profile, and linearised inversion for a layered velocity structure (after Kennett & Orcutt, 1976).

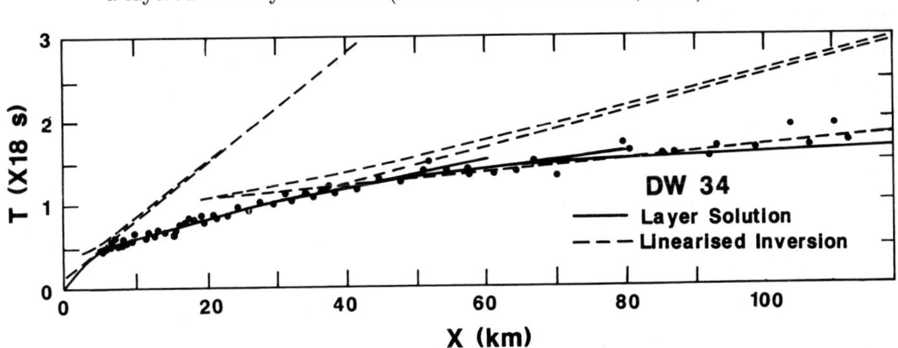

Figure 4.14: (a) Extremal bounds on the observed $\tau(p)$ function mapped into bounds on the P-wave velocity as a function of depth as shown in (b). (c) is P wave velocity calculated by conventional linearized inversion. From Kennett & Orcutt (1976).

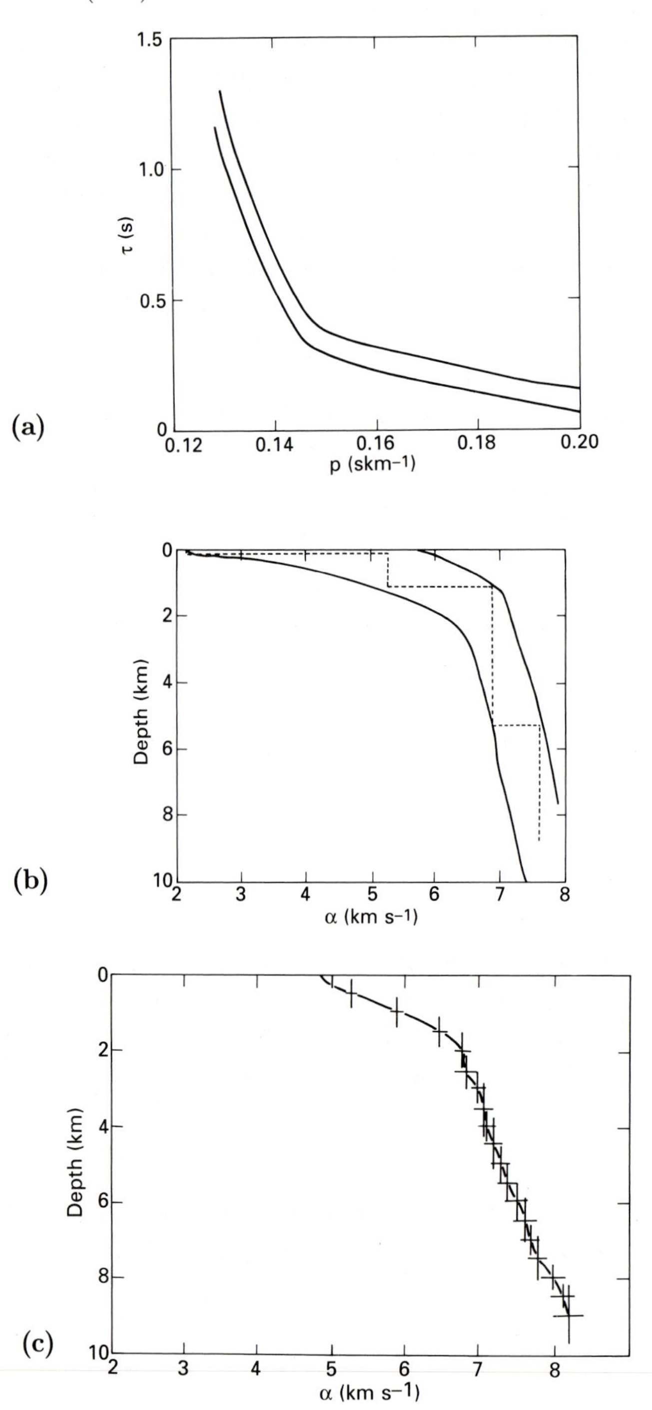

all the national networks and individual stations and produces a final location for the event by a least-squares fit to all the available data.

4.9 Refining the locations

The earthquake location is defined as the *origin time* and coordinates of the focus of the source of elastic radiation, the *hypocentre*. The *epicentre* is the point on the earth's surface directly above the hypocentre; epicentre and hypocentre therefore coincide for a surface event. The preliminary estimate of the location is adjusted so as to provide as good an agreement as possible with the predictions of a set of travel time tables.

In earthquake location we seek the hypocentral coordinates (T_0, h, θ, ϕ), where T_0 is the origin time, h the depth, and (θ, ϕ) the colatitude and longitude of the epicentre. A predicted travel time for a particular arrival is a function of the hypocentral coordinates and a small change in the location will lead to a small change dT^p in the predicted travel time, where

$$dT^p = dT_0 + \frac{\partial T}{\partial h}dh + \frac{\partial T}{\partial \theta}d\theta + \frac{\partial T}{\partial \phi}d\phi \tag{4.41}$$

The partial derivatives may be expressed in terms of the ray parameter and azimuth. The geometry is shown in Figure 4.15.

Increasing the depth by dh gives a change in T^p of

$$-\frac{\cos i}{v\,(r)}dh \tag{4.42}$$

so that

$$\frac{\partial T}{\partial h} = -\frac{\left(r^2/v^2 - p^2\right)^{\frac{1}{2}}}{r} \tag{4.43}$$

The quantity on the right hand side can be calculated by differencing travel time tables to give p (from $dT/d\Delta$) and evaluating v at the appropriate depth. A change in colatitude $d\theta$ will move H a distance $rd\theta$; the ray path is lengthened by a distance $rd\theta \cos Z \sin i$, so that

$$\frac{\partial T}{\partial \theta} = -\frac{r \sin i}{v} \cos Z \tag{4.44}$$

A change $d\phi$ in longitude moves H a distance $r \sin\theta d\phi$ in a direction perpendicular to the plane *ECN*, shortening the ray path by a distance $r \sin\theta d\phi \sin Z \sin i$, so that

$$\frac{\partial T}{\partial \phi} = \frac{r \sin i}{v} \sin Z \sin\theta \tag{4.45}$$

Writing i in terms of the ray parameter and velocity, the partial derivatives become

$$\frac{\partial T}{\partial h} = -\frac{1}{r}\left(\eta^2 - p^2\right)^{\frac{1}{2}} \tag{4.46}$$

$$\frac{\partial T}{\partial \theta} = -p\cos Z \tag{4.47}$$

$$\frac{\partial T}{\partial \phi} = p\sin Z \sin\theta \tag{4.48}$$

We make the notation more concise by introducing a hypocentre coordinate vector

$$\mathbf{m} = (T_0, h, \theta, \phi) \tag{4.49}$$

and a matrix of partial derivatives

$$A_{i\alpha} = \frac{\partial T_i^p}{\partial m_\alpha} \tag{4.50}$$

where $\{i = 1, 2, \ldots, D\}$ and $\{\alpha = 1, 2, 3, 4\}$; D is the number of separate arrival times we have measured, at times $\{T_i^0; i = 1, 2 \ldots D\}$, and predicted at times $\{T_i^p; i = 1, 2 \ldots D\}$. If the initial guess at the location is

Figure 4.15: Geometry for calculating the partial derivatives of travel time with respect to changes in hypocentral coordinates. H is the hypocentre and S the station. C is the centre of the earth. HS is the ray path that takes off with angle of incidence i. E is the epicentre and N the north pole. ES is the great circle path between epicentre and station. Z is the angle between the meridian plane ECN and the plane containing the ray path, ECS. Z is called the *azimuth* of S from E. It is measured clockwise from north by convention.

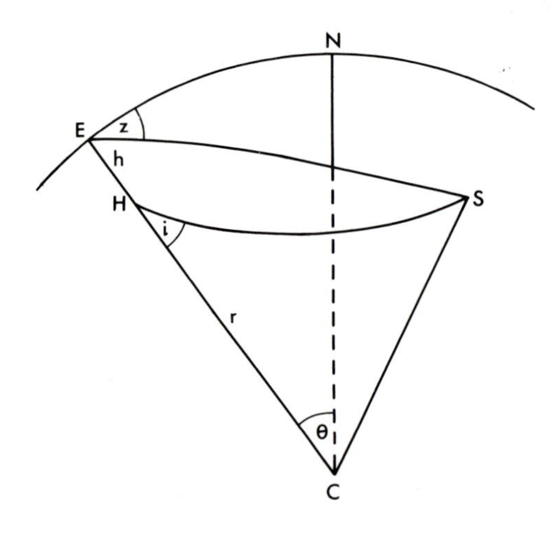

at $\mathbf{m}_0$, we seek an improved estimate $(\mathbf{m}_0 + \delta\mathbf{m})$ where $\delta\mathbf{m}$ is small. The predicted times will be given, approximately, by (4.41),

$$T_i^p(\mathbf{m}) = T_i^p(\mathbf{m}_0) + \sum_{\alpha=1}^{4} A_{i\alpha}\delta m_\alpha \tag{4.51}$$

The residual vector with respect to the location $\mathbf{m}_0$ is defined as

$$\delta\mathbf{T} = \mathbf{T}^0 - \mathbf{T}^p(\mathbf{m}_0) \tag{4.52}$$

$\delta\mathbf{m}$ is chosen to minimise the length of the residual vector for the new location:

$$\sum_{i=1}^{D}\left(\delta T_i - \sum_{\alpha=1}^{4} A_{i\alpha}\delta m_\alpha\right)^2 \tag{4.53}$$

which is accomplished by differentiating with respect to each component: $\{\delta m_\alpha; \alpha = 1, 2, 3, 4\}$ and setting the result equal to zero. This gives four simple simultaneous equations to solve for the components of $\delta\mathbf{m}$:

$$\sum_{\beta=1}^{4}\sum_{i=1}^{D} A_{i\alpha}A_{i\beta}\delta m_\beta = \sum_{i=1}^{D} A_{i\alpha}\delta T_i \tag{4.54}$$

or in matrix notation

$$\mathrm{A}^\mathrm{T}\mathrm{A}\,\delta\mathbf{m} = \mathrm{A}^\mathrm{T}\delta\mathbf{T} \tag{4.55}$$

The expression (4.51) for the predicted travel times is a good approximation only when $\delta\mathbf{m}$ is small. The new location $\mathbf{m}$ can be used as the basis of further refinement, with $\mathbf{m}$ replacing $\mathbf{m}_0$ as the starting guess. Iteration is repeated until $\delta\mathbf{m}$ is sufficiently small that the procedure can be deemed to have converged.

It is instructive to take a geometric view of this procedure. The sum of squares of residuals:

$$S = |\mathbf{T}^0 - \mathbf{T}^p(\mathbf{m})|^2 \tag{4.56}$$

represents a surface when S is plotted as a function of the components of $\mathbf{m}$. Five dimensions are needed to plot the surface, but the three-dimensional analogy of a surface plotted as height, z, above the xy plane is still helpful. We are searching for a minimum in this surface. The approximation (4.51) involves fitting a quadratic surface to S that is valid near $\mathbf{m}_0$. The new estimate is a minimum of this quadratic surface. This shows one of the problems of the procedure: there may well be several minima, and with a bad starting guess we may converge on the wrong one. A more common problem is that the arrival times may be quite insensitive to one, or a combination, of the hypocentral coordinates. Later in this chapter we discuss a common problem which arises in determining the depth: the travel times are insensitive to one particular combination of origin time and depth and therefore the same times can be produced by changing depth and compensating by changing origin time. In these cases the matrix $A^T A$ in equation (4.55) is singular and, in the absence of additional data, we have to introduce *ad hoc* information in order to obtain any result for $\mathbf{m}$ (for example by fixing the depth).

4.10 Procedures of the International Seismological Centre

The *ISC* collects data from many countries and carries out a routine programme of event location. These locations are based on more worldwide data than those from a national agency and are therefore considered more reliable, but they are not available until some considerable time after the event has occurred. At present it takes 21 months to collect the data, one month to process it, and two months to print.

The principal difficulty with such a location is the association of a particular reading by an observatory with a particular event. A preliminary association is made as described in the *ISC* Bulletin for January 1986 (broadly speaking, the observation is associated with an event if it arrives within 70 s of the expected time) and all the associations are then scrutinised and revised where necessary by a seismologist. Revision of the location is accomplished by a procedure similar to that described in the previous section, except that observations are weighted according to their presumed accuracy. Small correction is made for the earth's ellipticity and station altitude. The function minimised is not (4.56) but

$$\sum_{i=1}^{D} w_i \left[\delta T_i^0 - \delta T_i^p(\mathbf{m})\right] \tag{4.57}$$

where w_i is the weight attached to the ith travel time. The weights are specified in terms of a Gaussian function based on the size of the residual. This *method of uniform reduction*, as it is called, reduces the influence on the solution of data with large residuals. The weighted data have a distribution that is more nearly normal than the raw data because outliers have been weighted down. The procedure is iterative. Each subsequent iteration begins once again with a critical assessment of the data association before proceeding with the automated relocation. The *ISC* publish monthly bulletins containing the locations made by national agencies plus all the reported arrival times including those associated with the event but not actually used in their own location. For deep events they sometimes report a second location with a depth, where this can be determined from surface reflections, using procedures to be described in the next section. Full details of the whole procedure are given in the January and July issues of the *ISC* Bulletin.

4.11 Depth phases

How well determined are the event locations? Least squares provides a standard theory for calculating error ellipsoids around the most likely position of the hypocentre. Depth is the most poorly determined parameter if there are no nearby stations. We can see this without resorting to the full theory of error analysis, but just by inspecting equation (4.41).

The partial derivatives in (4.41) give the change in travel time for a small change in hypocentre coordinate, and are therefore a measure of the sensitivity of our data to the exact location. In particular, a small value of the partial derivative indicates the measurement is insensitive to that coordinate. But sensitivity is only one requirement for a sound location. We must also be able to separate the influence of the four coordinates on the times. Consider the difference in arrival times for two phases, 1 and 2, for the same event. From (4.41), changes in this difference are seen to be given by the formula

$$\begin{aligned} d(T^{p_1} - T^{p_2}) &= \left[\left(\frac{\partial T}{\partial h}\right)_{p_1} - \left(\frac{\partial T}{\partial h}\right)_{p_2}\right] dh \\ &+ \left[\left(\frac{\partial T}{\partial \theta}\right)_{p_1} - \left(\frac{\partial T}{\partial \theta}\right)_{p_2}\right] d\theta \\ &+ \left[\left(\frac{\partial T}{\partial \phi}\right)_{p_1} - \left(\frac{\partial T}{\partial \phi}\right)_{p_2}\right] d\phi \end{aligned} \tag{4.58}$$

Differential times are independent of the origin time for the event, and have the new partial derivatives given in (4.58). From (4.46)–(4.48) we see that the partial derivatives for the depth depend only on p, the ray parameter, and not on azimuth. If all reports are from similar distances then the ray parameter will be similar for each report, the partial derivative $[(\partial T/\partial h)_{p_1} - (\partial T/\partial h)_{p_2}]$ will be small, and the differential travel times insensitive to the depth. By eliminating the origin time we have also removed most of the dependence on depth. The travel times undergo almost equal changes for changes in either origin time or depth. This leads to a trade-off between depth and origin time.

Returning to (4.58), we see that partial derivatives for the epicentral coordinates (θ, ϕ) depend on azimuth. A good determination of the epicentre can be obtained with a good distribution of stations surrounding the event, regardless of their distance. The trade-off between depth and origin time, however, can only be broken by a wide range of ray parameter, p. Refraction makes p only a weak function of distance: by distance $\Delta = 30°$ rays for P waves already have a small angle of incidence ($i = 26°$) and this decreases to $i = 13°$ at $\Delta \approx 100°$ where the core shadow starts. This means that there is only a small range of ray parameter available for P waves at teleseismic distances. Core phases (*PKP*, *PKIKP*) provide smaller values of p and can improve the determination of depth, but they are difficult to read accurately.

Other phases are needed to provide really satisfactory depths. S waves have different ray parameters from P waves by virtue of the difference in wave speed (the ray paths for P and S waves are similar because the respective wave speeds remain in almost constant ratio with depth).

Unfortunately S waves are also difficult to read accurately because they have long periods and arrive after the compressional waves, which can obscure the onset. The best phases for depth determination are the surface reflections *sP* and *pP*. The phases are only seen at certain distances. The rays travel upwards from the event and are reflected from the earth's surface as shown in Figure 4.16. *pP* is often impulsive and quite easy to read accurately. It may be difficult to distinguish *sP* and

Figure 4.16: Surface reflections from an event at depth h. Differential times for *pP*-minus-P are almost a direct estimate of depth because travel times along AS and BS are similar, so $T^{pP} - T^P \approx BH/v$.

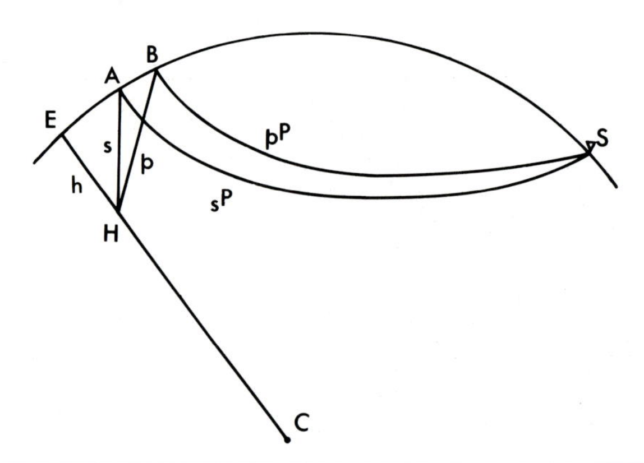

pP. Their relative amplitudes will depend on the radiation pattern for the event, which will often not be available. Most deep events occur near deep ocean trenches and there is often a strong reflection from the water surface; the time difference between this *pwP* phase and true *pP* is several seconds (Figure 1.1(d)). For relatively shallow events the surface reflections follow the *P* wave so closely that they cannot be read as a separate phase. They impart a characteristic shape to the waveform, however, and waveform modelling with two close phases can sometimes be used to find the depth very accurately.

4.12 Master event methods

For many applications it is sufficient to find locations of earthquakes relative to other earthquakes, rather than the absolute locations. For example, the thickness of deep seismic zones is uncertain. The scatter of *ISC* locations give zones 50–100 km thick, but this thickness probably still represents the size of location errors rather than any real structure. Special studies have improved the locations in certain areas, and shown that the events lie in even thinner zones, perhaps 20 km across. Only relative, not absolute, determinations are needed to establish properties like the thickness of the seismic zone.

Consider a set of events in a fairly small region of the earth. Figure 4.17 illustrates an example of deep events in a subduction zone, but these methods are equally applicable to shallow events. The essential

Figure 4.17: The master event method. Earthquakes in a small region can be located relative to each other. Relative locations have greater accuracy than absolute locations because they are insensitive to the travel time tables provided the ray paths are similar. The method involves using arrival times relative to those from a master event, *M*, the position of which is held fixed throughout.

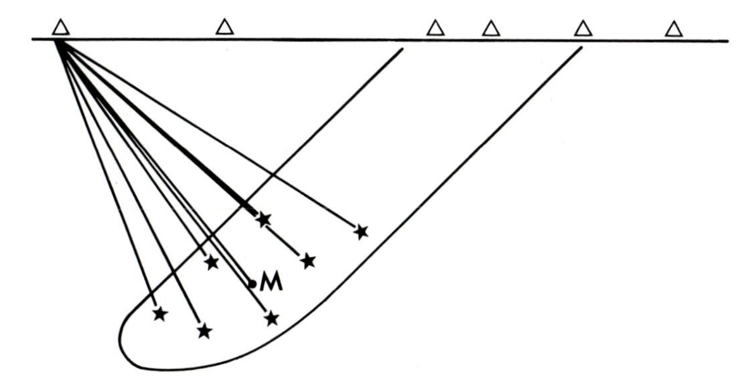

requirement is that the region containing the events be small compared with the distances to the stations, so that rays travel along similar paths. One event is denoted the *master event.* Relative arrival times for another event are calculated from the observations as

$$\{T_i^A - T_i^M; i = 1, 2, \ldots, D\} \tag{4.59}$$

where A is another event, M the master event, and subscript i denotes the phase and station. The master event needs to be one that is widely reported so that the $\{T_i^M\}$ are available for most stations. The master event is held fixed during the location procedure, so that (4.41) can be written as

$$d(T^A - T^M) = dT^A + \left(\frac{\partial T}{\partial h}\right)_A dh^A + \left(\frac{\partial T}{\partial \theta}\right)_A d\theta^A + \left(\frac{\partial T}{\partial \phi}\right)_A d\phi^A \tag{4.60}$$

The relative times are used to refine the location of A relative to M. Relative locations found in this way are insensitive to errors in the travel time table. The partial derivatives depend only on the orientation of the ray path near A and the local seismic velocity. Any error in the travel time tables will lead to a large error in the time from A to the station and hence in the absolute location of A, but this error will affect both M and A almost equally and is absent from the relative location.

If the master event itself is exceptionally well located then we can use the relative location to produce an improved absolute location. This is useful for aftershocks of a larger and therefore well-located master event, or if there is an explosion of known position in the region. Success of the method depends on the region being limited in extent and subtending a small solid angle at each receiver. Fitch (1975) applied this technique to events near the bottom of a deep seismic zone. He also produced a revised estimate of the velocity at the bottom of the zone by including it as an additional parameter to be solved for during the location procedure. Note that the method does not alleviate the problem of depth determination: relative times suffer from the same problems of a trade-off between depth and origin time, unless surface reflections are observed, as do absolute times.

Practical 2. Location of earthquakes from teleseismic records

Here we show the procedure followed by an observer when he first reads the seismogram from an event, and how one obtains a preliminary estimate of the epicentre from seismograms from just two or three sites (the accurate location of earthquakes requires the analysis of many records from around the world). First the phases are read and their arrival times noted, without using any prejudice as to the location of the event. The distance of the event is then obtained from the relative arrival times using a plot of travel time against distance. The event is then known to lie on a small circle centred on the station with radius given by the measured angular distance. If two such distances are known then two possible locations can be obtained (Figure 4.18); the choice between these two can be made using records from a third station. This practical shows how each of these steps is carried out.

The records

The first two stations are at Shillong (SHL) NE of Calcutta in India (25°N, 92°E), and at Bullowayo (BUL) in Zimbabwe (20°S, 28°E).

Figure 4.18: S_1 and S_2 are two seismic stations. The epicentre lies on a small circle centred on a station with radius the angular distance between them. If distances can be obtained from arrivals at two stations there are two possible locations of the events, F_1 and F_2, the intersection points of the two circles. A third station can be used to determine which of the two points is the true epicentre.

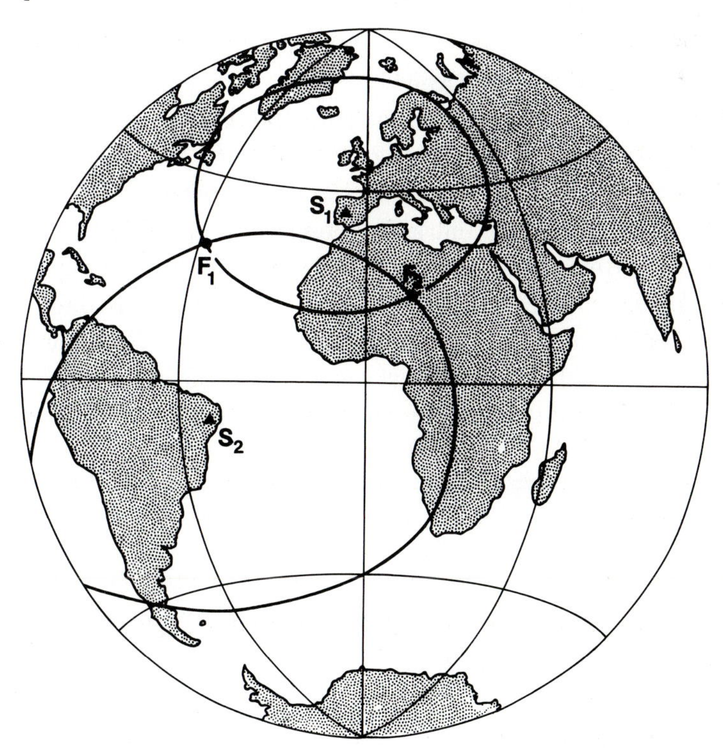

The long period records are from the same *WWSSN* instruments as were used in the first practical in Chapter 3. The short period records have timing marks every minute, as for the long period, but the drum rotates once in 15 min.

Reading the arrival times

Draw up a table of the arrival times of all the phases you can see as shown in Table 4.3. Each phase begins immediately to the right of the timing mark shown in column 2; measure the distance in mm and write the result in column 3; measure the distance between the two minute marks and use it to convert the distance to a time and enter the time in column 4; add to the time of the timing mark, subtract the time of the first arrival, and enter the result in column 5; convert this time back to a distance so that it can be compared with the travel time curves in Figure 4.19 by converting to seconds and dividing by 12 to give a length in mm; enter the result in column 6. Column 7 is reserved for use later when the phase has been identified; do not write in it yet. Read records from both sites before going on to the next part. There are several phases to be found on the long period records but it is difficult to read more than P and S arrivals on the short period.

Obtaining the distance

Mark a strip of graph paper with the distances in column 6, and align it with the time axis in Figure 4.19 with the first arrival mark level with the P wave arrival. Slide the graph paper along the distance axis until as many of the marks as possible align with the curves (see Figure 4.20). Remember that you will not have been able to read all arrivals with equal accuracy, that you may have missed some, identified some wiggles on the trace that are not true arrivals, and an inexperienced observer will probably read the arrivals *late*. The readings of P and S from the short period should be the most accurate of all because these

1	2	3	4	5	6	7
Instrument	hr:min	dist. to right of minute mark (mm)	time (s)	time after 1st arrival	convert to mm.	phase
1	5:25	31	:17	0	0	
2	5:33	80	:44	8:27	42.3	

Table 4.3: How to record the arrivals on a seismogram

Figure 4.19: Travel time curves for various phases for a surface focus, from the Jeffreys-Bullen tables. R and L denote Rayleigh and Love waves respectiveley.

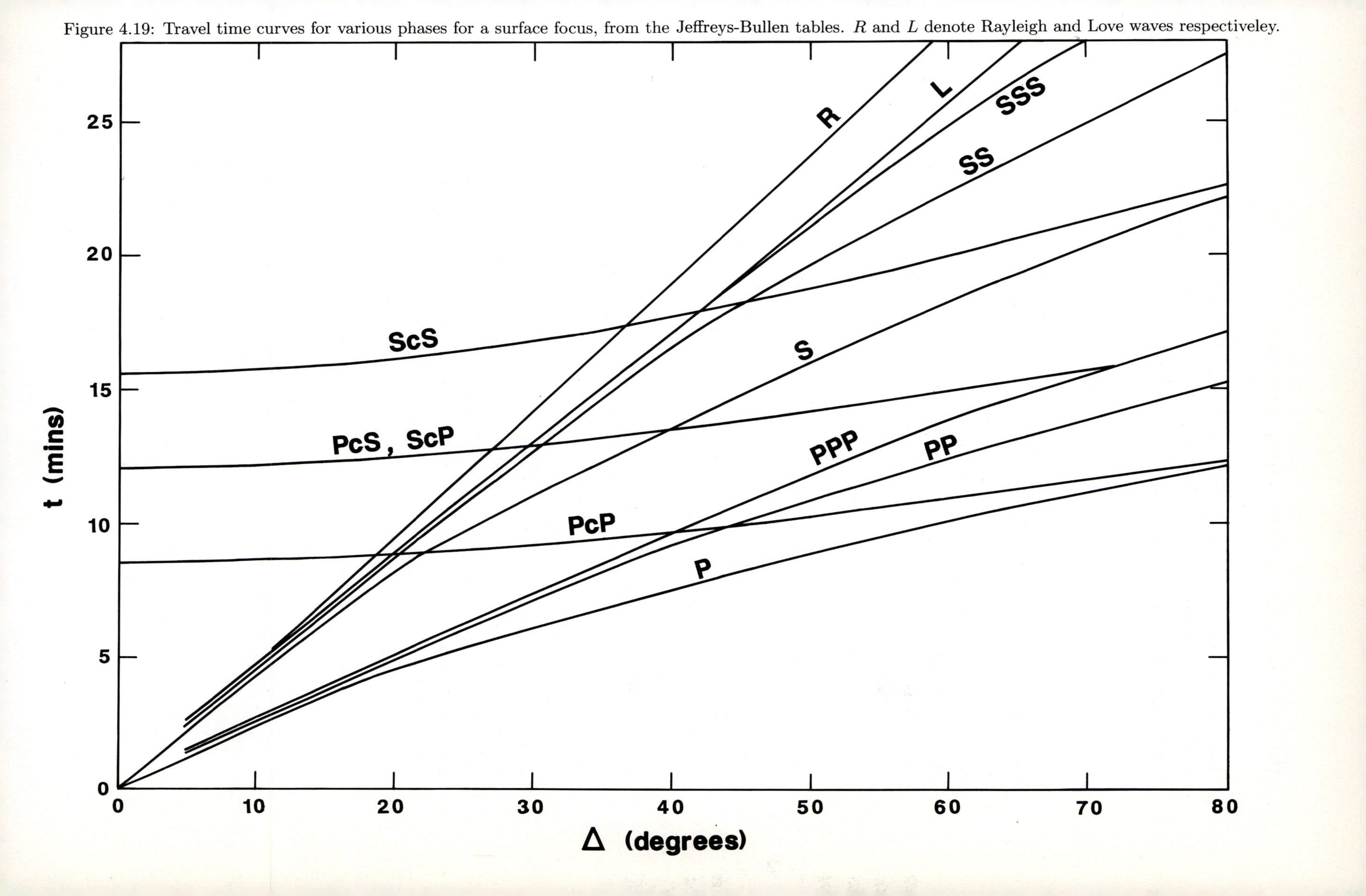

records afford better time resolution: you should therefore take these times in preference to those from other records in determining the final, accurate, measurement of distance.

Finding the two possible locations

We need to draw two small circles on a globe, one centred on each station, with radii corresponding to the measured distances. This can be achieved with a globe and two pieces of string, but a more accurate method is to use a projection of the globe. The *stereographic projection* is used because it has the property that small circles plot onto circles on the projection: we can therefore draw the small circles on the projection with a compass. The distance of a point an angular distance θ from the centre of the projection is at distance ($R \tan \theta/2$) mm. Take a piece of A3 tracing paper and put the centre of the net in Figure 4.21 underneath at 160 mm from the left edge of the tracing paper and 100 mm above the bottom edge — this will enable you to fit everything on. Mark the

Figure 4.20: Determining the distance to the earthquake having measured several arriving phases. Mark the arrival times on graph paper and lay it parallel to the time axis in Figure 4.19. Slide the graph paper along the distance axis until the marks line up.

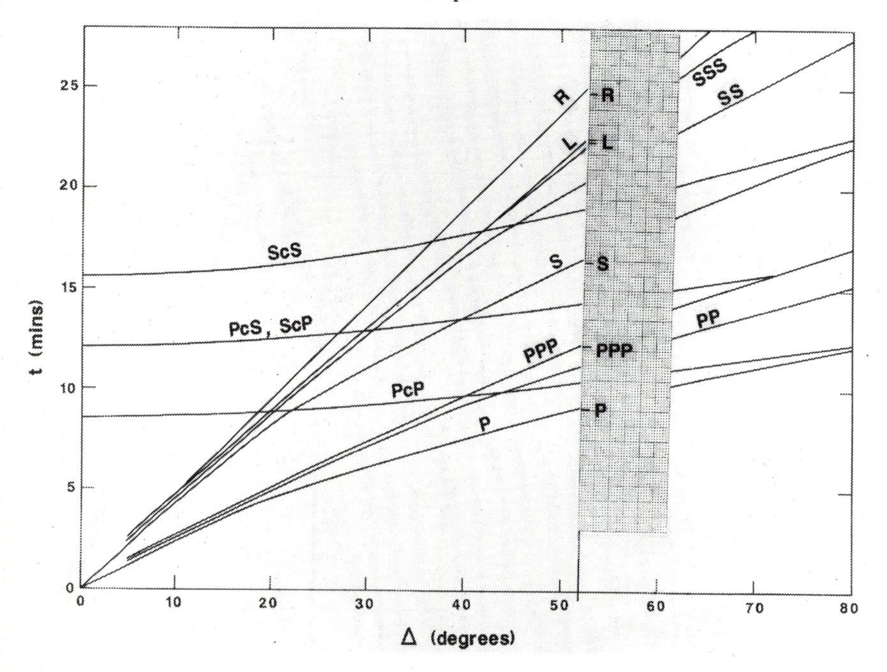

positions of SHL and BUL using the grid and taking the centre of the grid to be 0°N, 0°E.

The great circle through the origin and BUL maps into a straight line on the projection: use the measured angular distance to find the two points where the small circle centred on BUL cuts this great circle. Although the small circle maps into a circle on the projection, the centre of the circle on the sphere does not map into the centre of the circle on the projection: the circle must therefore be drawn using as centre the mid-point between the two intersections found above. Repeat the procedure for SHL. In this case one of the intersections is off the grid; its position must be calculated using the formula $d = R \tan\theta/2$ where R is the radius of the stereonet in Figure 4.21 in mm. The two points of intersection are the possible locations.

The third station

The remaining records are from College (COL) in Alaska. It plots on the projection at a distance of 112° from the origin at an azimuth of 345°. Measure the P and S times at COL and use the difference to measure the angular distance. Work out the position of the centre of the circle as before, which you will find to be very large, and approximate the circle as a straight line perpendicular to that joining COL to the origin (or use a globe and a piece of string!). Hence find which of the two possible locations is the true one. How good do you think the location is?

The surface waves

Determine the arrival direction of waves at BUL from the net. Identify the Rayleigh waves on the 4 and 5 records at BUL and use the fact that they have retrograde particle motion to verify this arrival direction. What are the large waves arriving on 6 between 5:40 and 5:55? Why are the S wave frequencies lower than those of the P waves?

Figure 4.21: The stereographic net. The lines are projections of lines of equal latitude and longitude at 2° intervals on the sphere.

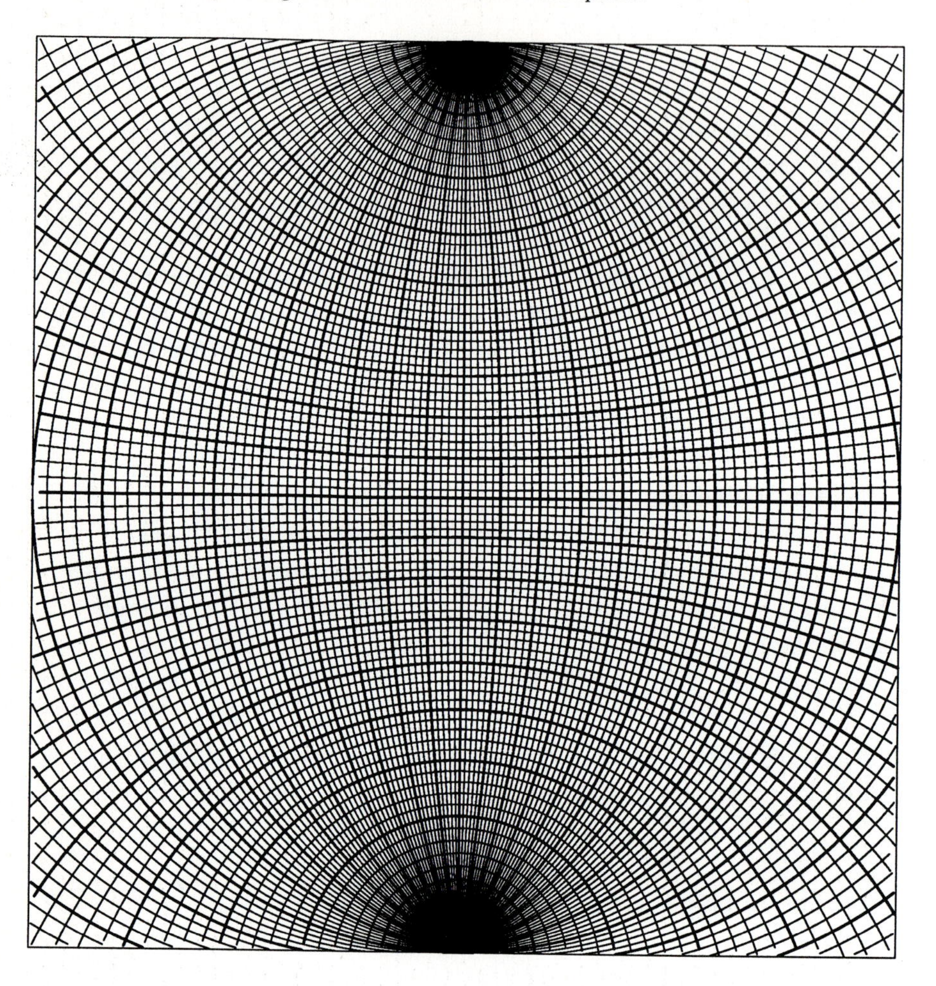

Figure 4.22: BULL(1), date 27/3/75, half actual size. To read the seismogram produce a full size copy by running it through an expanding xerox.

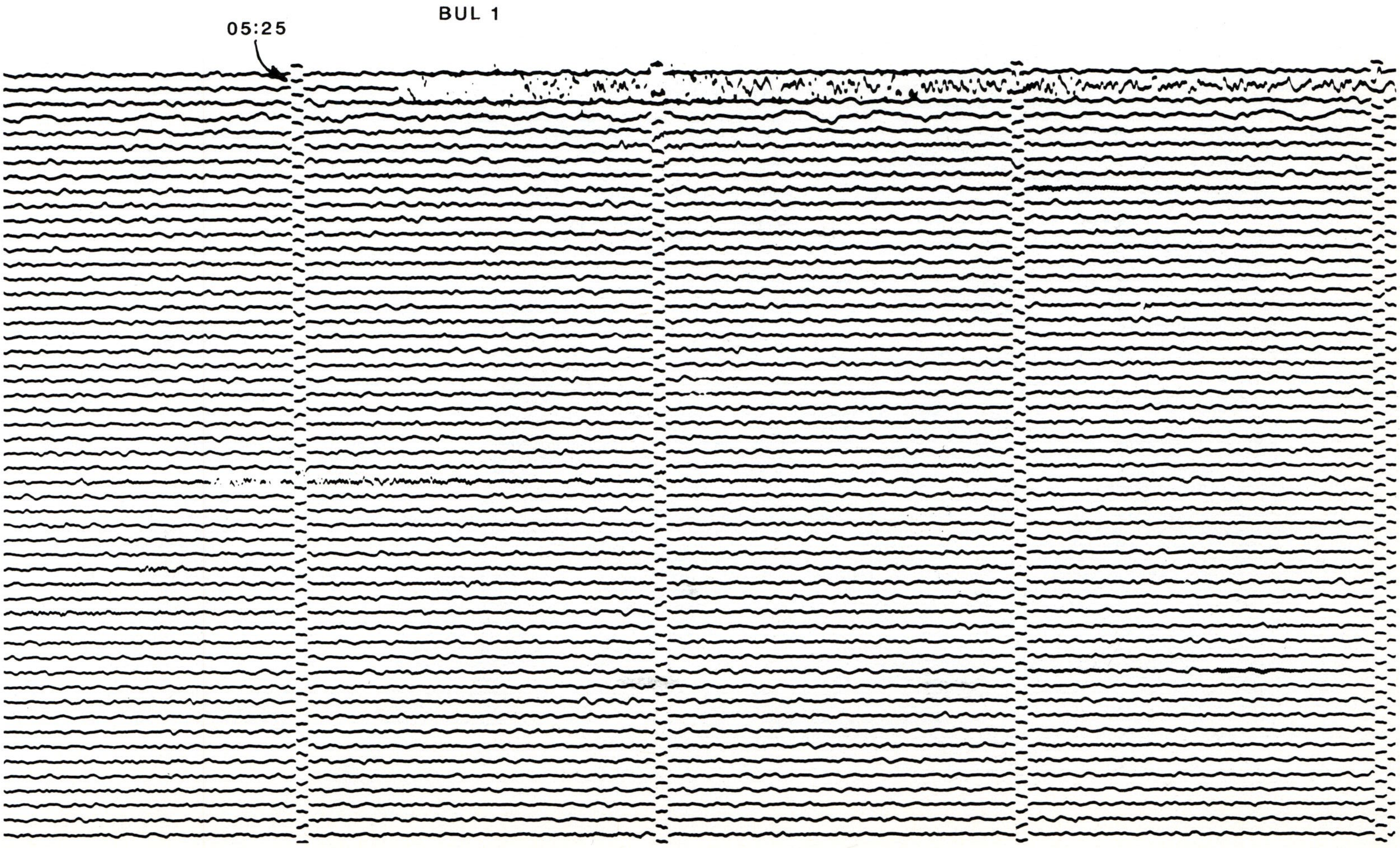

BUL 1

05:25

Figure 4.23: BUL(2), date 27/3/75, half linear dimension.

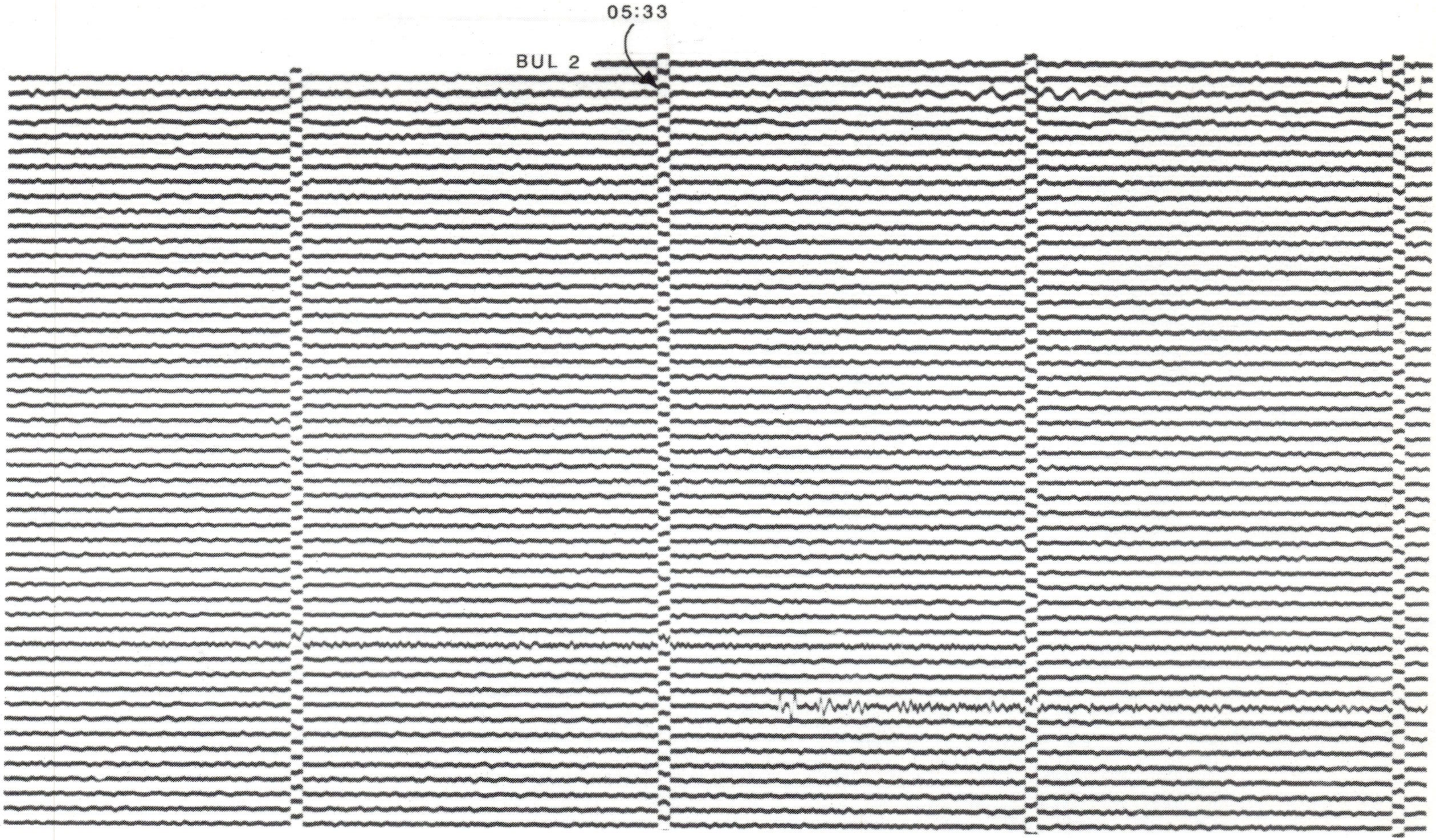

05:33

BUL 2

Figure 4.24: BUL(4), date 27/3/75, quarter linear dimension.

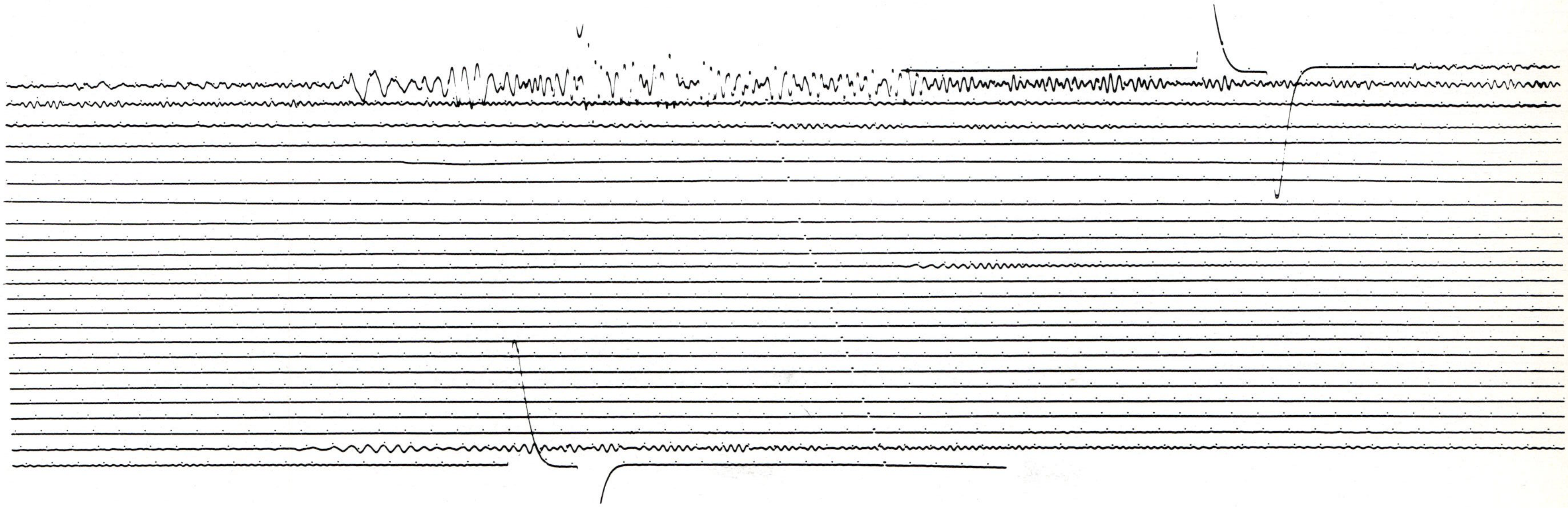

Figure 4.25: BUL(5), date 27/3/75, quarter linear dimension.

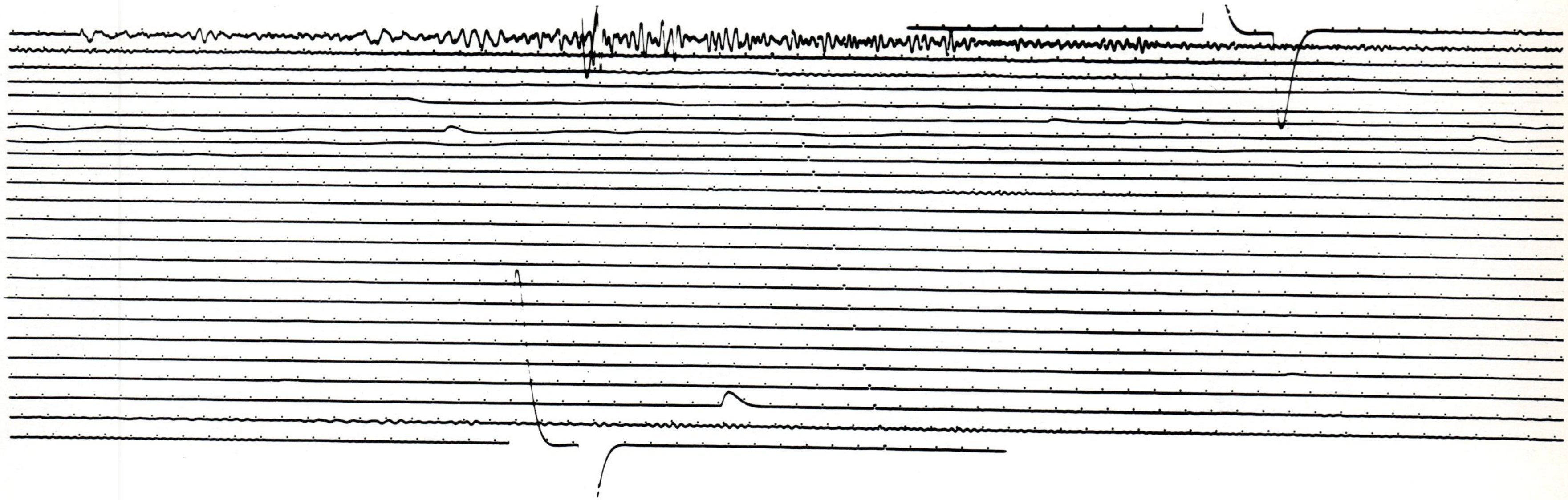

Figure 4.26: BUL(6), date 27/3/75, quarter linear dimension.

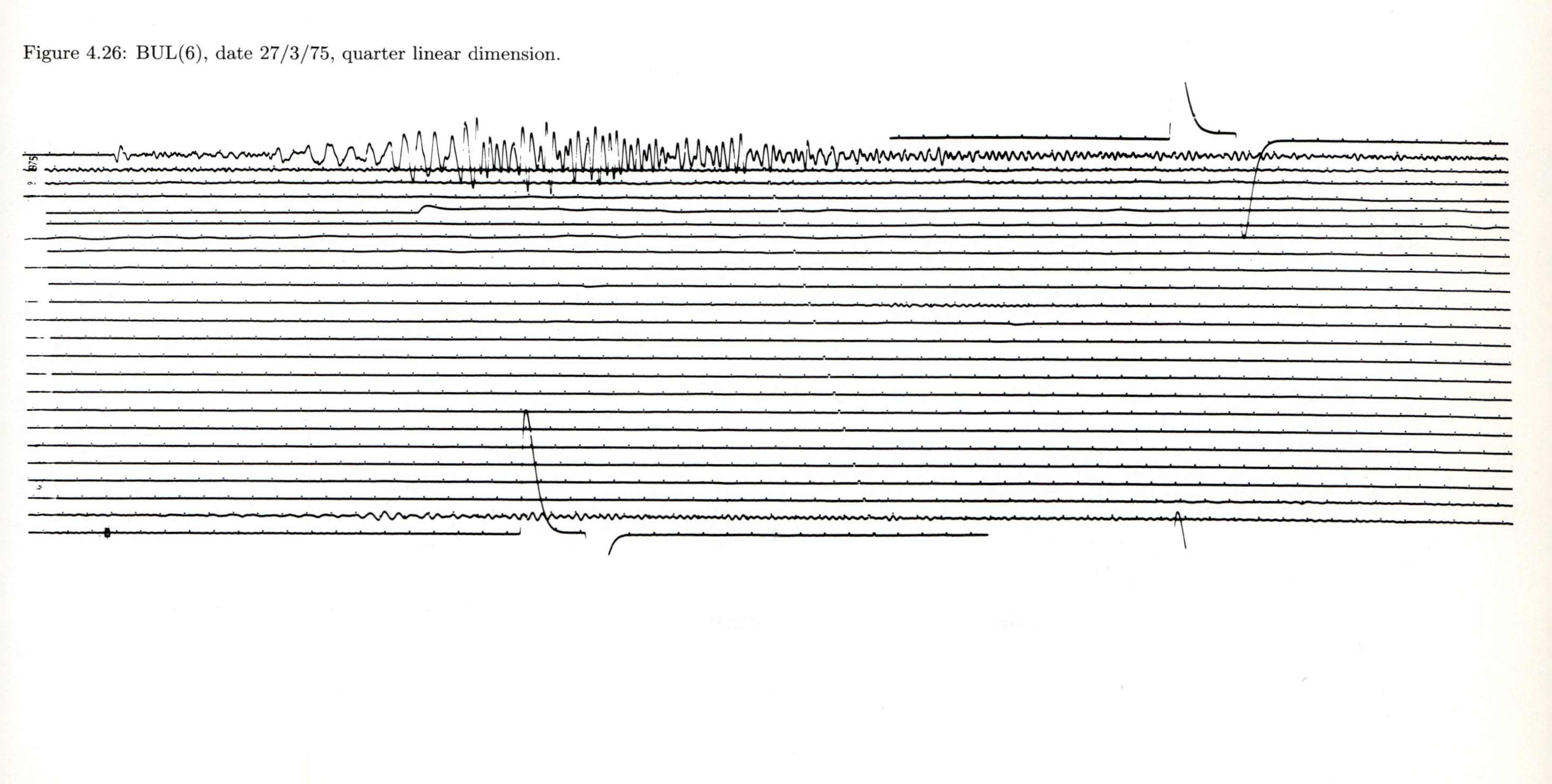

Figure 4.27: SHL(1), date 27/3/75, half linear dimension.

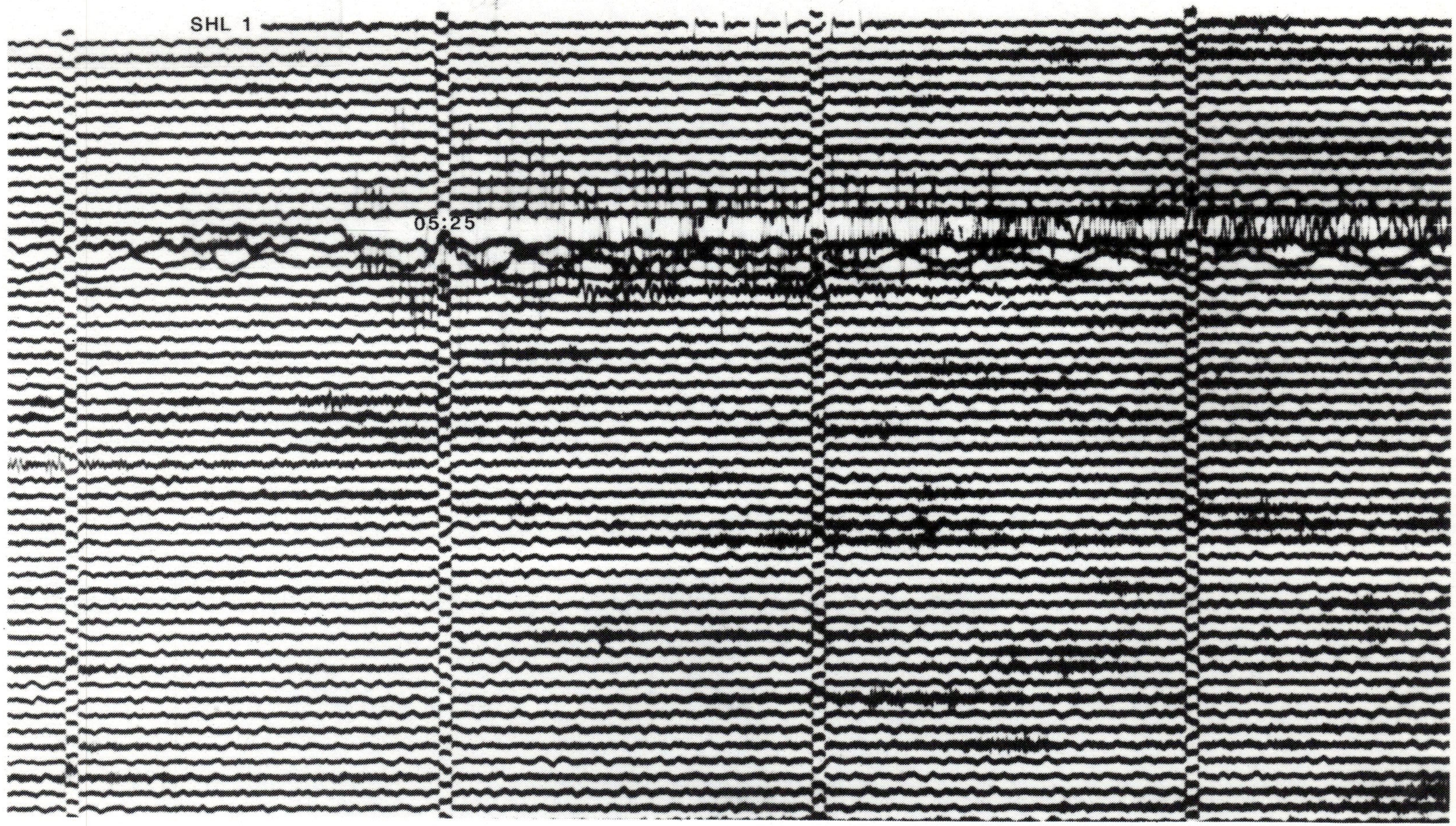

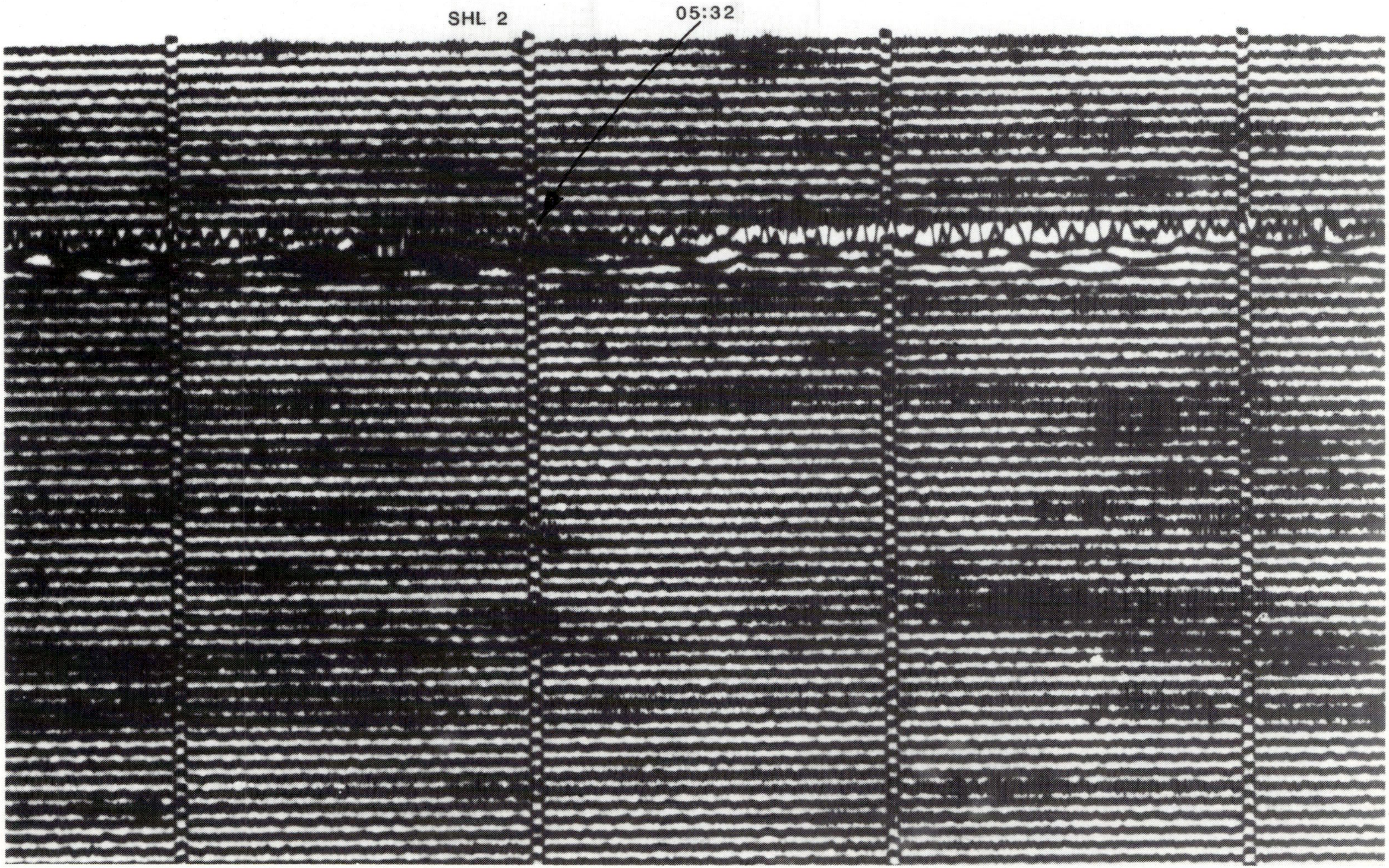

Figure 4.28: SHL(2), date 27/3/75, half linear dimension.

Figure 4.29: SHL(4), date 27/3/75, quarter linear dimension.

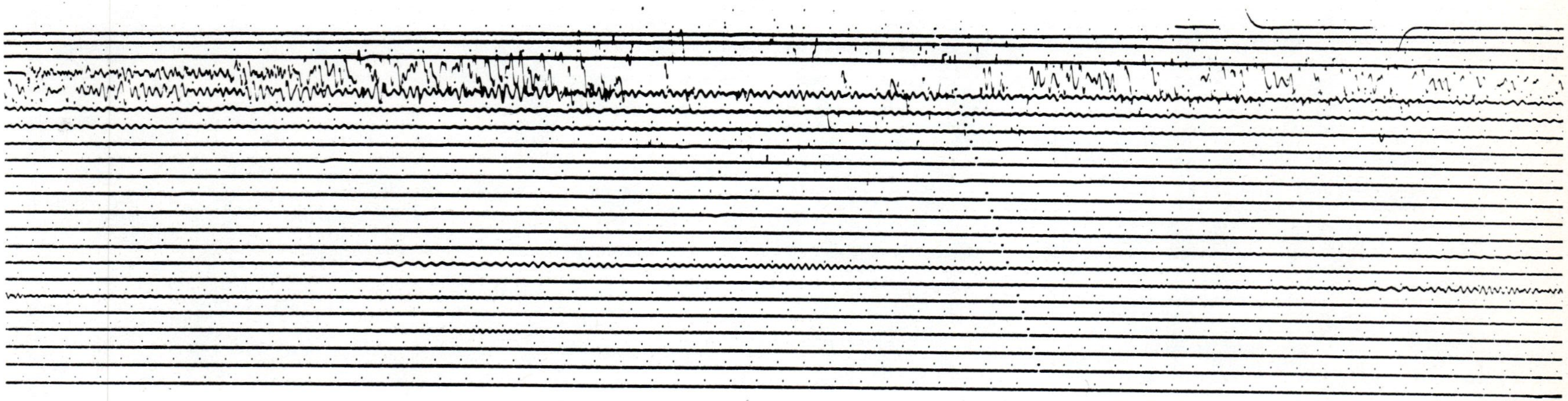

Figure 4.30: SHL(5), date 27/3/75, quarter linear dimension.

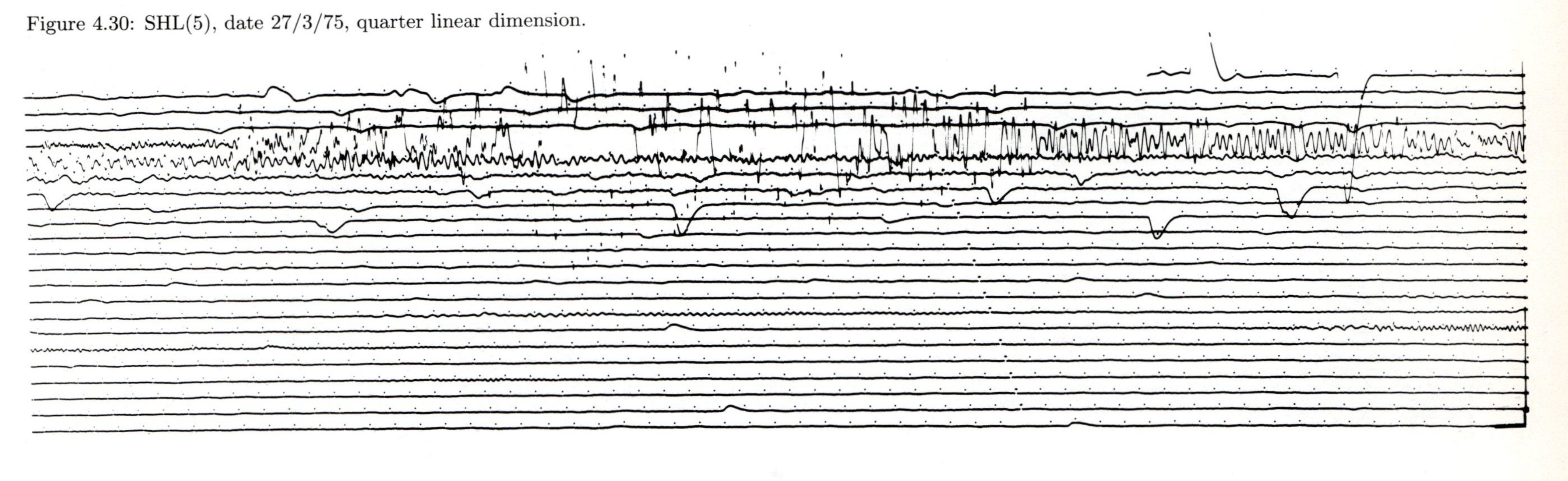

Figure 4.31: COL(1), date 27/3/75, half linear dimension.

Figure 4.32: COL(4), date 27/3/75, quarter linear dimension.

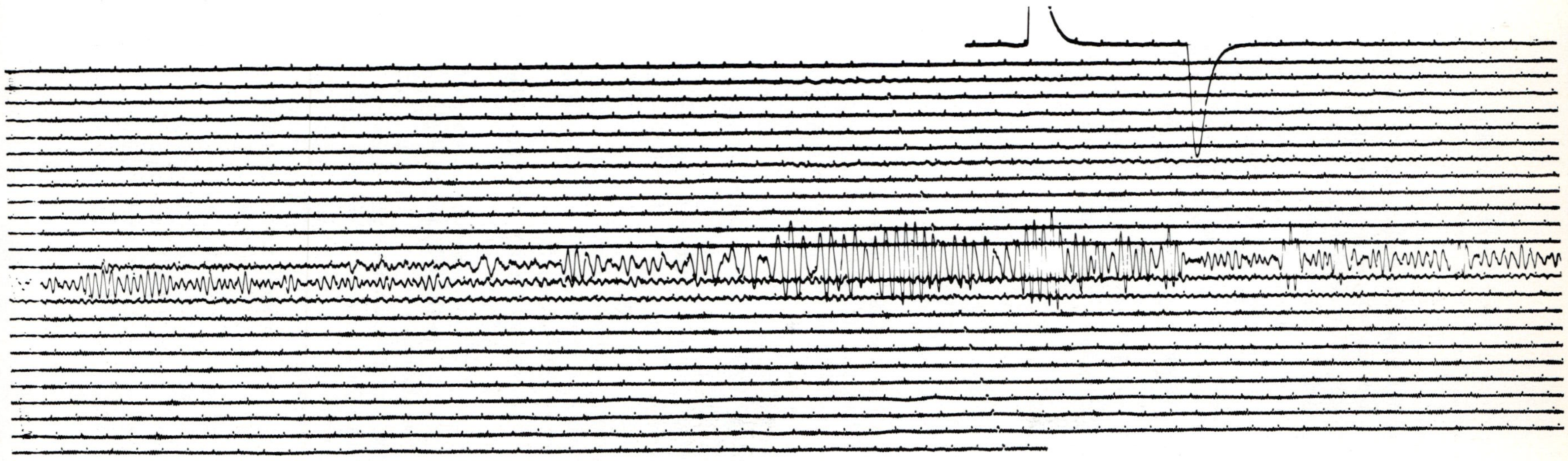

Exercises

1 An event at depth 65 km is recorded at an epicentral distance of 60°. Find the travel time, ray parameter, deepest point on the ray path, and angle of incidence at 33 km depth for the P wave.

2 Use the P wave table to find, to the nearest second, the time for PP at distance 60° from an event 160 km deep. You may assume the ray path forms a straight line from the surface to 160 km. What is the greatest source of inaccuracy in this calculation?

3 Calculate the apparent position of a seismic station at 50°S, 180°E as "seen" along the seismic ray path of a P wave from a shallow earthquake at 15°S, 180°E.

4 Which leg of the phase PS travels the deeper; the P or S wave leg?

5 For a laterally homogeneous medium with $v = a + bz$, where z is the depth, show that

$$\frac{dT}{dz} = \frac{1}{v\left(1 - p^2v^2\right)^{\frac{1}{2}}}$$

$$T = \frac{2}{b}\cosh^{-1}\left(\frac{1}{pa}\right)$$

and that the ray paths are arcs of circles.

6 Show in general that a 1D medium with velocity dependent only on z, the depth, the curvature of the rays is proportional to the velocity gradient.

7 Show that a wavefront, with ray parameter p, traverses the earth's surface with speed p^{-1}. p is sometimes called the *slowness* of the wavefront for this reason.

8 Investigate the function $T(X)$ and the configuration of the ray paths for a medium consisting of four layers with wave speeds as follows:

a Layer 1: uniform velocity v_0

b Layer 2: velocity increases linearly from v_0 to v_1

c Layer 3: velocity decreases linearly from v_1 to v_2

d Layer 4: velocity increases linearly.

9 If $v(r) = a - br^2$, show that the ray paths are circles. Find their radii and centres.

10 Body waves give good information about the seismic velocity in the earth near the bottom of their path. Why might SKS be a better phase than PKP for studying the upper regions of the outer core?

11 The core radius is to be estimated by an inversion of travel times of P and PcP phases. Show that the differential times of P-PcP are relatively insensitive to uncertainties in the seismic velocity in the mantle but very sensitive to earth radius. Illustrate how the method would be implemented for an earth model with constant wave speed in the mantle. Can you think of another reason why the differential time is more easily measured than either of the absolute times?

12 A seismic event occurs at time T_0 and depth h in a half space $z > 0$. The velocity v of P waves depends only on z. The epicentre is known to be at the origin of coordinates. Seismometers are set up at points

A $(x_A, 0, 0)$ and B $(x_B, 0, 0)$. Show that, for differential changes in the depth and origin time, the arrival time of P waves at A changes by

$$dT_A = [v^{-2}(h) - p_A^2]^{\frac{1}{2}} dh + dT_0$$

where p_A is the parameter for rays travelling from O to A, and $v(h)$ is the velocity at depth h. What values of x_A and x_B would be best for depth determination? Demonstrate the result by considering $dT_A - dT_B$.

13 An earthquake occurs at the centre of a circular array of seismometers. Show the trade-off between depth and origin time is complete.

14 A depth phase *pP* is observed at seismometer A of question (12). Approximate the ray parameter by taking the *P*-leg of the path to be much greater than the upward-travelling leg, and find the depth in terms of the differential travel time *pP-P*.

15 Deep earthquakes are recorded at a small network of stations. The events all occur within a small region with constant seismic velocity α. The hypocentral coordinates are (T_0, h, θ, ϕ). Show that, approximately,

$$\begin{aligned} &\delta T - \delta T_M \\ &= \delta T_0 - \delta T_{0M} \\ &+ \left(\frac{\partial T}{\partial h}\right)_M (\delta h - \delta h_M) \left(\frac{\partial T}{\partial \theta}\right)_M (\delta \theta - \delta \theta_M) \left(\frac{\partial T}{\partial \phi}\right)_M (\delta \phi - \delta \phi_M) - \frac{\delta \alpha}{\alpha^2} \ell \end{aligned}$$

where ℓ is a distance and subscript M refers to the master event. Draw a diagram of the ray paths to show the length ℓ.

16 The homogeneous station method involves subtracting the average arrival time from each station for a group of earthquakes in the same locality. What advantage might this procedure offer in separating path effects from mislocation?

Summary

Travel times of seismic waves are the most important measurements for determining seismic velocity in the earth and locating earthquakes. A table of observed times versus distance can be differentiated to provide the RAY PARAMETER, a constant for the ray. Simple integral expressions relate the ray parameter to the ray path. Time-distance functions can be complicated; low velocity zones produce shadows which manifest themselves as gaps in the time-distance curve. Variable velocity gradients lead to multipathing and triplications in the curve. A seismic ARRAY is a network of seismometers with central timing: a single clock allows very accurate relative timings between the members of the array, which in turn allows determination of arrival direction and slowness of the wavefront. The array can be steered to give an optimum response to a particular direction. The time-distance curve can be inverted to give the seismic velocity as a function of depth. Three basic methods were described: brute-force parameter fitting, HERGLOTZ-WIECHERT which uses times at short distance to determine shallow velocities and then proceeds with larger distances and greater depths, and the TAU METHOD which makes use of the monotonic function $T - p\Delta$. The tau method has the advantages that the function $\tau(p)$ is easier to estimate from data and it is linearly related to the function $r(v)$ (or $z(v)$ in the plane case) in the absence of low velocity zones.

Preliminary estimates of the location of an earthquake are made using P and S arrivals, and any other prominent phases, from a few stations. The distance is determined from time-distance curves for the phases identified. After distances have been obtained from several different stations it is possible to find a rough location of the EPICENTRE (Practical 2). Locations are refined using as much data as possible, weighted according to the accuracy with which each phase can typically be read, and minimising the rms difference between observed and predicted times. The *ISC* produces routine definitive locations. The focus of the earthquake is the HYPOCENTRE, which lies directly beneath the epicentre. The depth of an earthquake is often quite difficult to determine. Depth phases like pP are important in establishing an accurate depth because the difference in arrival times of pP and P is almost proportional to depth. When several events occur in the same region (an aftershock sequence of a large earthquake for example) it is possible to determine their relative locations more accurately than their absolute locations. This is because the relative locations are insensitive to errors in the travel time table. If the absolute location of one MASTER EVENT is good the relative locations of the others can be converted to absolute locations.

Further reading

The use of Snell's law of refraction in seismology is discussed fully in Chapter 7 of BULLEN & BOLT. The use of travel time tables is described in Jeffreys & Bullen. The first-hand discussions of the calculation of travel time tables in GUTENBERG & RICHTER and JEFFREYS(1976) are well worth reading. Arrays are described in BULLEN & BOLT but for more details the reader must consult the original papers quoted in the text. SIMON gives a guide to reading seismograms. Travel time inversion is discussed in most books on seismology. See Buland & Chapman (1983) for a discussion of the tau function. It is worth looking at the *PDE* Bulletin published by the *NEIC*, just to see the level of seismicity (and bomb testing!) that occurs. The *ISC* published their detailed location procedure in the January and July issues of their bulletin. Buland (1976) gives a further discussion of the computational aspects of earthquake location; Fitch (1975) gives an example of the master event method; and

Ansell & Smith (1975) give a related method called the "homogeneous station" method in which only relative arrival at each station from all earthquakes are used. Joint hypocentre-seismic velocity determination is reviewed by Gubbins (1980).

5

Free oscillations

Elastic disturbances in the earth can take the form of standing waves as well as travelling waves. Waves can propagate along a stretched elastic string (Figure 5.1(a)). If the string is clamped at both ends as in Figure 5.1(b) then it can support standing waves. The wavelength of the standing waves must be an integer fraction of the total length of the

Figure 5.1: The analogy between standing waves and normal modes

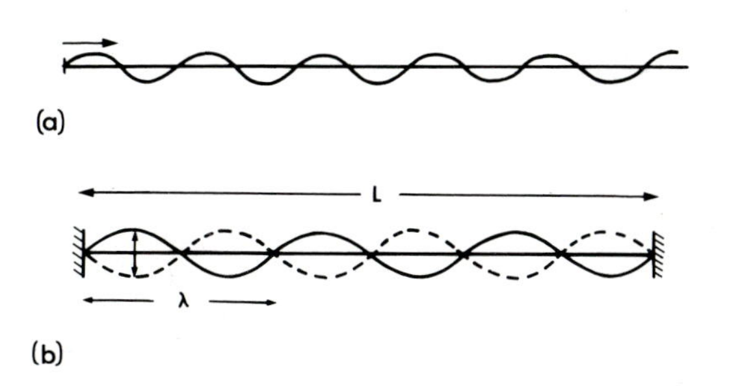

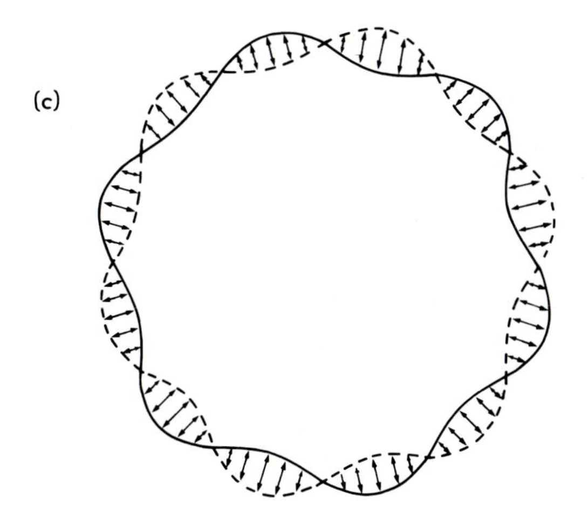

string. Thus only certain wavelengths and frequencies are permitted, determined by the boundary conditions. The spectrum is discrete, although the number of allowed frequencies is still infinite. These standing waves are called the *normal modes* or *free* or *natural oscillations* of the system.

A spherical body can also support standing waves as in Figure 5.1(c). For example, Rayleigh waves propagate around the earth's surface; its spherical shape is only important for very long wavelengths. If an integral number of wavelengths can be fitted around the circumference then a standing wave can be set up. Like the stretched string the spectrum is discrete, and the allowed frequencies are determined by the boundary conditions. All types of elastic displacement have equivalent modes, not just surface waves, although the surface wave-equivalent modes are among the easiest to detect on a seismogram.

The complete set of free oscillations forms a basis for describing any general elastic displacement undergone by the earth. Put another way, any elastic displacement may be written as a superposition of normal modes. This includes propagating Rayleigh waves, P waves, or any other sort of displacement. In practice the concept of normal modes is most useful when dealing with very long period disturbances. There are two distinct classes of free oscillations called *spheroidal* and *torsional* (or toroidal). The separation is the spherical analogue of that between P-SV and SH motion for body waves in a half space, and Rayleigh and Love motion for surface waves. Torsional modes depend only upon the shear velocity, spheroidal modes depend upon both shear and compressional velocities. The simplest spheroidal mode is expansion and contraction of the earth and is called the "breathing" mode for obvious reasons; in the earth it has a period of 20 minutes. The "football" mode represents deformation of the earth into the shape of a rugby ball; it has a period of 54 minutes, the gravest mode. The simplest torsional mode is in fact the earth's rotation, but this mode is of no interest in seismology. The next mode is one in which two hemispheres rotate in opposite directions, justifying the name "torsional". Particle motions for these modes are illustrated in Figure 5.2.

The earth's gravest modes are difficult to detect on conventional seismometers because of their very long periods and because they are excited only by very large earthquakes. Their discovery therefore had to await the development of reliable long-period instruments, and then a large earthquake. In 1952 Benioff announced the observation of a free oscillation with a period of 57 min. This announcement came over half a century after the development of the theory for free elastic oscillations in a sphere. Free oscillations set up by a very large earthquake in Chile in 1960 were observed on both strain meters and gravity meters. Torsional modes, which have no radial displacement, were not seen on the gravity meter recordings because they produce no change in gravity. The longest observed period was about one hour. These observations marked the beginning of a new and productive branch of seismology, now known as *very low frequency seismology.*

5.1 Spheroidal and torsional modes

We take the density, ρ, and elastic constants, λ and μ, to be functions only of radius. The gravitational force also depends on the radius. Under these conditions of spherical symmetry it is possible to separate the displacement into torsional and spheroidal parts, the two parts being completely uncoupled. The result does not hold if there are departures from spherical symmetry, or if there is any force (such as the Coriolis force) that lacks spherical symmetry. The analytic proof of the separation is very lengthy and will not be given here — references are given at the end of the section. A brief discussion will bring out the important elements of the proof.

The separation depends on the mathematical identity

$$\mathbf{u} = \nabla \times (T\hat{\mathbf{r}}) + \nabla \times \nabla \times (S\hat{\mathbf{r}}) + \nabla\Phi \tag{5.1}$$

where $\hat{\mathbf{r}}$ is the unit vector in the radial direction and T, S, Φ are scalar functions of position. Any vector $\mathbf{u}$ can be written in this form. S, T and Φ are the *poloidal*, *torsional* and *scaloidal* potentials for the displacement, $\mathbf{u}$, respectively. Φ is the scalar potential and $\nabla \times (S\hat{\mathbf{r}}) + T\hat{\mathbf{r}}$ the vector

Figure 5.2: Particle motions for (a) the breathing mode (b) the football mode (c) the second torsional mode.

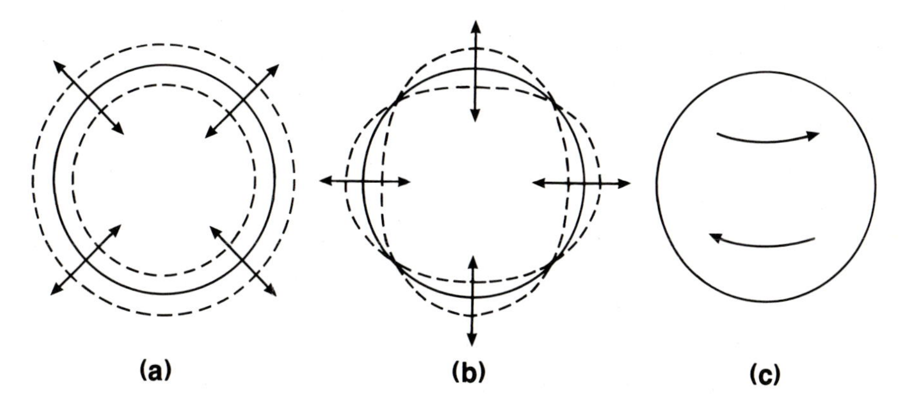

potential. From the properties of spherical coordinates we have

$$\nabla \times \mathbf{A} = \frac{1}{r}\left\{\frac{1}{\sin\theta}\left[\frac{\partial}{\partial\theta}(\sin\theta A_\phi) - \frac{\partial A_\theta}{\partial\phi}\right], \frac{1}{\sin\theta}\frac{\partial A_r}{\partial\phi} - \frac{\partial}{\partial r}(rA_\phi), \frac{\partial}{\partial r}(rA_\theta) - \frac{\partial A_r}{\partial\theta}\right\} \tag{5.2}$$

Applying this with $\mathbf{A} = T\hat{\mathbf{r}}$ gives

$$\nabla \times (T\hat{\mathbf{r}}) = \left\{0, \frac{1}{r\sin\theta}\frac{\partial T}{\partial\phi}, -\frac{1}{r}\frac{\partial T}{\partial\theta}\right\} \tag{5.3}$$

and we see that the torsional displacement has no radial component — hence its name. Applying (5.2) again gives

$$\hat{\mathbf{r}} \cdot \nabla \times \nabla \times (S\hat{\mathbf{r}}) = -\frac{\nabla_{\mathrm{h}}^2 S}{r^2} \tag{5.4}$$

where

$$\nabla_{\mathrm{h}}^2 = \frac{1}{\sin\theta}\left[\frac{\partial}{\partial\theta}\left(\sin\theta\frac{\partial S}{\partial\theta}\right) + \frac{1}{\sin\theta^2}\frac{\partial^2 S}{\partial^2\phi}\right] \tag{5.5}$$

The operator ∇_{h}^2 is the horizontal part of $r^2\nabla^2$:

$$\nabla^2 S = \frac{1}{r^2}\frac{\partial}{\partial r}\left(r^2\frac{\partial S}{\partial r}\right) + \frac{1}{r^2}\nabla_{\mathrm{h}}^2 S \tag{5.6}$$

The displacement $\nabla \times \nabla \times (S\hat{\mathbf{r}})$ does have a radial component. The torsional motion is the counterpart, in spherical geometry, of *SH* motion; the poloidal motion, $\nabla \times \nabla \times (S\hat{\mathbf{r}})$, is the counterpart of *SV* motion, and the scaloidal motion, $\nabla\Phi$, that of *P* motion.

For further reference we note the following mathematical results, which are lengthy but straightforward to prove.

$$\hat{\mathbf{r}} \cdot \nabla \times \nabla \times \nabla \times (S\hat{\mathbf{r}}) = 0 \tag{5.7}$$

Thus the curl of a poloidal vector is toroidal. This property and the extensions

$$\hat{\mathbf{r}} \cdot (\nabla\times)^{2n+1}(S\hat{\mathbf{r}}) = 0 \tag{5.8}$$

$$\hat{\mathbf{r}} \cdot (\nabla\times)^{2n}(S\hat{\mathbf{r}}) = \nabla^{2n-2}r^{-2}\nabla_{\mathrm{h}}^2 S \tag{5.9}$$

are useful properties in effecting the separation of spheroidal and torsional terms in the equation of motion. The equation of motion for a spherically symmetric earth is, from (2.93)

$$\rho \frac{\partial^2 \mathbf{u}}{\partial t^2} = \nabla (\lambda \nabla \cdot \mathbf{u}) + 2\nabla \mu \cdot \mathrm{e} + \mu \nabla^2 \mathbf{u} + \mu \nabla (\nabla \cdot \mathbf{u}) + \rho \mathbf{g} \quad (5.10)$$

The gravitational potential satisfies Poisson's equation (2.24). Free oscillations are the only disturbances in seismology that have long enough periods to be affected significantly by changes to the gravitational potential.

Calculating the frequencies and displacements associated with a free oscillation involves solving (5.10) and the equation for the gravitational potential (2.24) together with the zero stress boundary condition at the earth's surface for a displacement with exponential time dependence $\exp i\omega t$. It is an eigenvalue problem with eigenvalues $-\omega^2$ and eigenfunctions equal to the displacement. (5.10) and (2.24) represent four coupled partial differential equations for the three components of the displacement, $\mathbf{u}$, and the perturbed gravitational potential, ψ. Writing $\mathbf{u}$ in terms of potentials T, S and Φ leads to a separate equation for T. The remaining equations involve S, Φ and ψ. The scalar equations are solved by separation of variables in the usual way.

5.2 Torsional modes in a uniform sphere

To simplify the mathematics we consider a uniform sphere and ignore self-gravitation. The equation of motion (5.10) is then

$$\rho \ddot{\mathbf{u}} = (\lambda + 2\mu) \nabla (\nabla \cdot \mathbf{u}) - \mu \nabla \times \nabla \times \mathbf{u} + \rho \nabla \psi \quad (5.11)$$

We seek an oscillation with frequency ω and displacement proportional to $\exp i\omega t$. We isolate the torsional part of the displacement by taking the radial component of the curl of equation (5.11). The gravitation term vanishes since $\nabla \rho = 0$. Using the identities (5.4), (5.7), (5.9) and $\nabla \times \nabla \Phi = 0$ we obtain the scalar equation

$$\rho \omega^2 \nabla_{\mathrm{h}}^2 T + \mu \nabla^2 \nabla_{\mathrm{h}}^2 T = 0 \quad (5.12)$$

This proves that the torsional modes are independent of the spheroidal motion. We can integrate (5.12) to give

$$\frac{\omega^2}{\beta^2} T + \nabla^2 T = f(\theta, \phi) \quad (5.13)$$

where f is any solution of the equation $\nabla_{\mathrm{h}}^2 f = 0$. The only solution for f that is single valued on a sphere is a constant, and since from (5.3) adding a constant to T does not change the displacement we choose $f = 0$. (5.13) becomes Helmholtz' equation for T. It must be solved subject to zero stress on the surface of the sphere, which is done by seeking a solution that is separable in spherical coordinates. Setting

$$T(r,\theta,\phi) = R(r)\Theta(\theta)\Phi(\phi) \tag{5.14}$$

we obtain the simple harmonic motion equation for Φ, the associated Legendre equation in $\cos\theta$ for Θ, and a Bessel equation for R. The physical requirement that the solution be finite and single valued leads to

$$\Theta(\theta)\Phi(\phi) = Y_\ell^m(\theta,\phi) = P_\ell^m(\cos\theta)\exp im\phi \tag{5.15}$$

where Y_ℓ^m is a spherical harmonic and P_ℓ^m an associated Legendre function.

We shall require two standard results for normalised spherical harmonics:

$$\nabla_{\mathrm{h}}^2 Y_\ell^m = -\ell(\ell+1)Y_\ell^m \tag{5.16}$$

$$\oint Y_{\ell_1}^{m_1}\left(Y_{\ell_2}^{m_2}\right)^* dS = \delta_{\ell_1\ell_2}\delta_{m_1m_2} \tag{5.17}$$

where the surface integral is taken over the unit sphere. Here ℓ and m must be integers: ℓ because otherwise the associated Legendre function is singular at the poles, and m so that the displacement is single valued in ϕ. ℓ and m are like quantum numbers — they specify precisely the angular dependence of the displacement for a particular mode. ℓ is called the *angular order number* of the mode and m the *azimuthal order number.* ℓ is equal to the total number of nodal lines (circular tracks on the unit sphere on which the displacement is zero). For example, $Y_1^0(\theta,\phi) = \cos\theta/2\sqrt{\pi}$, which gives one nodal line on the equator, where $\cos\theta = 0$. m is equal to the number of nodal lines that run from pole to pole. The real part of $Y_1^1(\theta,\phi) = \sin\theta\cos\phi/2\sqrt{\pi}$ has a nodal line crossing the equator at $\phi = \pi/2$ and $3\pi/2$.

In general a spherical harmonic Y_ℓ^m will have ℓ nodal lines, m of which travel from pole to pole, the remaining $\ell - m$ of which travel along lines of latitude. Rotating our coordinate system cannot change the angular order number ℓ, the total number of nodal lines, but it can change m. For example, rotation by 90° will transform Y_1^0 to Y_1^1. In general rotation will transform Y_ℓ^m into a linear combination of spherical harmonics with the same ℓ but different values of m. The same result applies to the vector eigenfunctions to the displacements because the differential equation governing the vector displacement (5.10) is invariant under rotation when the elastic constants depend only upon radius.

Returning to equation (5.13) for the torsional potential T, we take

$$T(r,\theta,\phi) = t_\ell^m(r) Y_\ell^m(\theta,\phi) \tag{5.18}$$

and use (5.6) and (5.16) to find the differential equation governing the radial function

$$\frac{\omega^2}{\beta^2} t_\ell^m + \frac{1}{r^2}\frac{d}{dr}\left(r^2 \frac{dt_\ell^m}{dr}\right) - \frac{\ell(\ell+1)}{r^2} t_\ell^m = 0 \tag{5.19}$$

This differential equation contains ℓ explicitly but not m, a direct demonstration that the frequency cannot depend on m in a uniform sphere. The frequencies of all modes with the same angular order number are equal; they differ only in their displacement patterns. This is known as *degeneracy*: several eigenfunctions share the same eigenvalue. The result is not limited to a uniform sphere but applies to any spherically symmetric body because modes with the same angular order number may be converted into each other simply by rotating axes: rotation can change the displacement patterns (eigenfunctions) but not the frequencies (eigenvalues). For each value of ℓ there are $2\ell+1$ possible values of m: the degeneracy is said to be $(2\ell+1)$-fold.

Separation of variables to give the spherical harmonic dependence in θ and ϕ also generalises to the case when ρ, λ and μ depend on radius and self-gravitation is included. Spherical symmetry is the essential property that allows this representation. The earth's normal modes may therefore be characterised by the angular and azimuthal order numbers. All the results we have derived so far generalise to the real earth, the only complication arising being that the differential equation in governing $t_e^m(r)$ involves the variation of the density and elastic parameters with depth.

5.3 Overtones

The radial equation (5.19) must be solved subject to zero radial traction at the earth's surface, $r = r_E$. Consider a mode with $m = 0$, which by (5.3) is axisymmetric and has displacement only in the ϕ direction. To fit the boundary conditions we must use the elements of the stress tensor in spherical coordinates. These are to be found in AKI & RICHARDS. This particular case is easy enough to allow derivation of the relevant component of the strain tensor, $e_{r\phi}$, from first principles. Consider a point on the surface and set up cartesian axes x, y, z parallel to the local directions of the spherical coordinate vectors $\hat{\boldsymbol{\theta}}, \hat{\boldsymbol{\phi}}, \hat{\mathbf{r}}$. Hooke's law can be used locally to relate the traction to $e_{zy} = e_{r\phi}$, using (2.83), as $2\mu e_{zy}$. The cartesian expression for the strain (2.3) must be expressed in terms of spherical components of the displacement, which involves some additional curvature terms. Writing the strain as

$$e_{r\phi} = \frac{1}{2}\left[\frac{\partial\left(\mathbf{u}\cdot\hat{\mathbf{z}}\right)}{\partial y} + \frac{\partial\left(\mathbf{u}\cdot\hat{\mathbf{y}}\right)}{\partial z}\right] \tag{5.20}$$

The change in $\mathbf{u}\cdot\hat{\mathbf{z}}$ in movement a small distance $\sin\theta d\phi$ will be $-\mathbf{u}\cdot\hat{\boldsymbol{\phi}}\sin\theta d\phi$ since $u_r = 0$ everywhere on the surface. y is in the ϕ direction and z in the r direction, so that $\partial\left(\mathbf{u}\cdot\hat{\mathbf{z}}\right)/\partial y$ can be replaced with $-u_\phi/r$. The second derivative in (5.20) is simply $\partial u_\phi/\partial r$. The boundary condition becomes

$$\mu\left(\frac{\partial u_\phi}{\partial r} - \frac{u_\phi}{r}\right) = 0 \tag{5.21}$$

The stress elements σ_{rr} and $\sigma_{r\theta}$ vanish automatically as they did in the case of reflection and transmission of *SH* waves at a plane interface (Section 3.8). Substituting the expression (5.3) for u_ϕ into (5.21), integrating with respect to θ, and dropping the constant term, gives

$$\frac{dt_\ell^0}{dr} - \frac{2t_\ell^0}{r} = 0 \tag{5.22}$$

on $r = a$.

The general solution of (5.19) is a spherical Bessel function of degree ℓ. These functions have simple analytic forms. The solution that is finite at the origin is written j_ℓ while the second independent solution, which is singular at the origin, is called y_ℓ. For a uniform solid sphere we need only the regular solution. In the earth torsional modes are restricted to the solid mantle (and the solid inner core, but it is difficult to see how such modes could ever be excited or observed) because there is no shear displacement in the liquid core. The solution is obtained only for $r > r_c$ where r_c is the core radius; solutions that are singular at the origin are therefore allowed. Stress-free boundary conditions must be applied at the base of the mantle which allows determination of the singular part of the solution. The solution of (5.19) is

$$t_\ell^0 = j_\ell\left(\frac{\omega r}{\beta}\right) \tag{5.23}$$

Spherical Bessel functions are defined by the Rayleigh formula

$$j_\ell(x) = x^\ell\left(-\frac{1}{x}\frac{d}{dx}\right)^\ell \frac{\sin x}{x} \tag{5.24}$$

Now we apply the boundary conditions by substituting (5.23) into (5.22). This procedure will determine the frequency ω, for there are no arbitrary constants left in the general solution (5.23) (save an unhelpful multiplicative factor) and a non-zero solution is attained only for certain frequencies. To illustrate the procedure for finding ω consider the case $\ell = 1$. (5.22) and (5.24) give, after dropping a multiplicative constant (β/ω)

$$t_1^0(r) = \frac{\beta}{\omega r^2}\sin\frac{\omega r}{\beta} - \frac{1}{r}\cos\left(\frac{\omega r}{\beta}\right) \tag{5.25}$$

and substituting into (5.22) with $r = r_E$ yields the equation

$$\tan x = \frac{4x}{(4 - x^2)} \tag{5.26}$$

where $x = \omega r_E/\beta$. This equation admits an infinite number of solutions for x, as may be seen in Figure 5.3. The lowest frequency, apart from the trivial one $\omega = 0$, is $\omega = 5.597\beta/a$. Higher frequencies approach $n\pi\beta/r_E$. Taking a shear velocity of 5 km s^{-1} and $r_E = 6000$ km gives $\omega \approx 5$ mHz,

Figure 5.3: Solution of the equation (5.26) for frequencies of the torsional modes for a uniform solid sphere, where $x = \omega r_E/\beta$. The roots occur when the curve $y = 4x/(4 - x^2)$ intersects the graph of $y = \tan x$. At large x the roots approach $n\pi$.

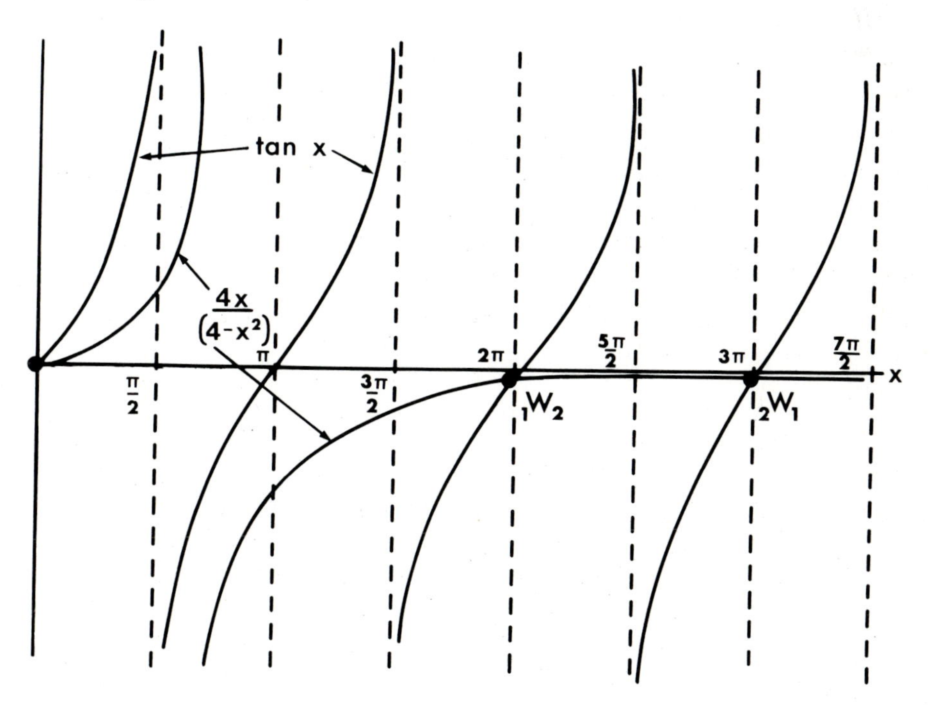

about the right order of magnitude for the earth's free oscillations despite the very crude model.

The radial variation of the torsional potential function is plotted in Figure 5.4. The radial variation for the higher frequencies has exactly the same form but the x axis has been stretched: more oscillations are packed into the interval $(0, r_E)$. The frequencies can be ordered with the *overtone number*, n, which is one less than the number of spherical nodal surfaces. Each value of n corresponds to a different free oscillation. The mode with $n = 0$ is the *fundamental mode* for that particular angular order number; modes with $n \neq 0$ are *overtones.*

A free oscillation is therefore classified according to its type (torsional or spheroidal), and three "quantum numbers": n, ℓ, and m. The modes are degenerate in m and this number is usually omitted from the description. A spheroidal mode is written as ${}_nS_\ell$ and a torsional mode as ${}_nT_\ell$, and when there is no confusion between spheroidal and torsional modes the frequency is written as ${}_n\omega_\ell$.

The roots of equation (5.26) approach the values $n\pi$ (n integer) for large values of ω. The higher overtones are therefore equally spaced in frequency with spacing

$$\Delta\omega \sim \frac{\pi\beta}{a} = \frac{2\pi}{T_S} \tag{5.27}$$

Figure 5.4: Eigenfunctions for torsional modes in a uniform sphere from equation (5.23) with $\ell = 1$. Solutions for overtones all have the same shape; they are determined by stretching the x axis to fit the appropriate number of nodes into the sphere.

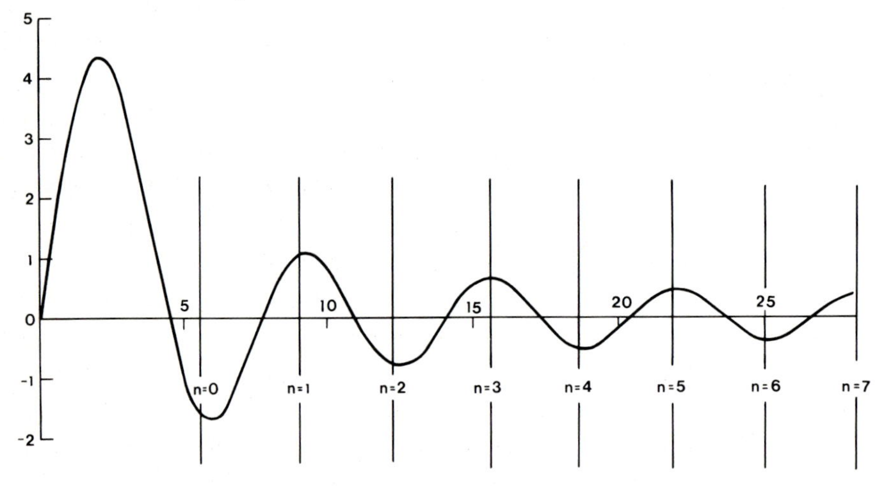

where $T_S = 2r_E/\beta$ is the time taken by an S wave to traverse the sphere. A very similar result holds for the real earth:

$$\Delta\omega \sim \frac{2\pi}{T_{ScS}} \tag{5.28}$$

where T_{ScS} is the time for an S wave to travel vertically downwards to the core boundary and back. The equations for the radial functions form a Sturm-Liouville system. This remains true for any spherically symmetric earth model and therefore the normal modes of the earth exhibit all the properties of solutions to Sturm- Liouville systems; for example the modes can be ordered by the number of nodes in radius (Figure 5.4), to give a sequence of increasing frequency. It may also be shown that high overtones of the spheroidal modes are equally spaced in frequency with separation $\Delta\omega \approx 2\pi/T_{PKIKP}$ where T_{PKIKP} is the time for a P wave to travel a diameter of the earth. The full calculation of the eigenfunctions and frequencies of the normal models involves numerical integration of the radial equations. References are given at the end of this chapter.

5.4 Fundamental modes

As stated earlier, any ground motion can be represented as a combination of the earth's free oscillations; the free oscillations are said to form a *complete set* of modes. This property of completeness means that any seismic wave, a P wave for example, can be represented as a sum of normal modes. Some modes will contribute more to the sum than others: for example a P wave with a period of 10 s would involve a great many spheroidal modes with about that period. There are tens of thousands of modes with periods of 10 s or longer, so that the representation would be a very clumsy one. However, surface waves lend themselves very readily to a normal mode representation.

It is easy to show that Love waves, which involve only *SH* motion, can be represented by torsional modes alone; no spheroidal modes appear in the expansion. Consider a surface event at the north pole that excites a Love wavetrain with frequency ω and wave number k_L. The waves travel south and the amplitude falls off with distance from the pole as the length of the wavefront increases. After passing the equator the wavefront begins to shrink and the amplitude grows, reaching a peak at the south pole (large amplitudes are observed at the point antipodal to an earthquake). The wavefront spreads to a length proportional to $\sin\theta$, and the ground motion will be proportional to $\sin^{-\frac{1}{2}}\theta\cos(kr_E\theta+\epsilon)$ where θ is colatitude and ϵ is a phase factor, provided the wavelength $2\pi/k$ is short enough for the earth's sphericity to be neglected.

Axisymmetric torsional modes have ground motion that varies with latitude as a Legendre polynomial. At large angular order the Legendre function approximates to a cosine of the form

$$P_\ell(\cos\theta) \sim \left[\frac{2}{\pi\left(\ell+\frac{1}{2}\right)\sin\theta}\right]^{\frac{1}{2}} \left\{\cos\left[\left(\ell+\frac{1}{2}\right)\theta-\frac{\pi}{4}\right]+O\left(\ell^{-1}\right)\right\} \tag{5.29}$$

This has the same form as the ground motion for Love waves, with $kr_E = (\ell+\frac{1}{2})$ and $\epsilon = -\pi/4$. The torsional mode is entirely equivalent to a standing Love wave, which can be written as two travelling Love waves. The wavelength of the Love waves is

$$\lambda_L = \frac{2\pi}{k_L} = \frac{2\pi a}{\left(\ell+\frac{1}{2}\right)} \tag{5.30}$$

and their frequency is

$$\omega_L = c_L k_L = \frac{\left(\ell+\frac{1}{2}\right)c_L}{r_E} \tag{5.31}$$

where c_L is the Love wave speed. It follows from (5.31) that fundamental torsional modes of high angular order will be equally spaced in frequency with a separation

$$\Delta\omega_L = c_L/r_E \tag{5.32}$$

The torsional modes of high angular order are standing Love waves. In the same way spheroidal modes of high angular order may be shown to be standing Rayleigh waves. The frequency spacing between spheroidal modes approaches

$$\Delta\omega_R = c_R/r_E \tag{5.33}$$

where c_R is the Rayleigh wave speed.

Fundamental modes have their energy concentrated near the earth's surface, while overtones have a greater proportion of their energy at depth. The radial functions for a uniform sphere in Figure 5.4 illustrate the point. Not only are the fundamental modes easier to detect at the earth's surface, they are also more readily excited by earthquakes, which are confined to the uppermost 700 km of the mantle. Natural oscillations of any physical system are set up most readily when forced at a place where they have large amplitude; conversely they cannot be set up by forcing at a node. Shallow earthquakes therefore excite fundamental modes far more readily than overtones. Overtones are set up by deep earthquakes, although many are only very weakly excited. Torsional modes in the inner core would need inner core earthquakes to excite them (or, more seriously, fluctuations in fluid flow in the liquid outer core), and even then we would not detect them with surface seismometers! The raw spectrum of a long period seismogram therefore shows a preponderance of fundamental modes, equally spaced in frequency. The practical at the end of this chapter gives a good illustration of this effect.

5.5 The Earth's normal modes

It is not possible to discern discrete free oscillation frequencies in a seismogram with the naked eye. Some processing is required first, the simplest form of which is to take the Fourier transform of a part of the seismogram and to look at the amplitude spectrum (modulus of the complex Fourier transform as a function of frequency), as is done in the practical at the end of this chapter. Normal modes then reveal themselves as peaks in the spectrum.

For most events the spectrum is dominated by the fundamental modes. The identification of overtones usually requires the processing of records from many seismometers for each earthquake; the records can be combined to enhance the signal-to-noise ratio for one particular mode. The simplest procedure involves using the source mechanism of the earthquake (Chapter 6) and its location to predict the phase of a normal mode, adjusting all the seismograms until they are (theoretically) in phase and summing them to give a composite record. Sometimes it is adequate simply to weight each seismogram with the predicted sign of the oscillation (i.e. to calculate the phase to within π). The selected mode will be enhanced in the composite record whereas noise and other modes will be out of phase and will tend to cancel in the sum. The process is called *stacking*: it has led to the identification of many new modes. It is also possible to process records to extract the source mechanism as well as to enhance spectra for mode identification. The process is called *stripping* and is described thoroughly in references given at the end of this chapter.

Well over a thousand modes have now been identified. Identification of a mode is based mainly on comparison of the observed frequency with that calculated for a given mode from some model of the earth's elastic parameters, but other characteristics of the mode are important also. For example, torsional modes are not seen on vertical instruments, and high overtones are not expected from shallow earthquakes. Modes for a realistic earth model are shown as large dots in Figures 5.5 and 5.6; the small dots represent calculated modes. The diagram is very crowded and it is clear that frequency alone is an inadequate criterion for mode identification.

Consider the appearance of the torsional mode diagram shown in Figure 5.5 at very high frequency and angular order number. The wavenumber of a high order mode is given approximately by $(\ell + \frac{1}{2})/r_E$. We define a slowness associated with a particular mode to be $(\ell + \frac{1}{2})/r_E\omega$. The line marked *ScS* in Figure 5.5 corresponds to the slowness of an *S* wave just grazing the core-mantle boundary. Modes above and to the left of this line have slowness smaller than the grazing *S*, and are therefore in the range of those of *ScS* phases, which are all steeper than the grazing S, whereas those modes below and to the right of the *ScS* line have slowness corresponding to mantle *S* waves that arrive at shallower angles than *ScS*. Normal modes are only an alternative description of general ground motion and we can classify them by the body wave (or surface wave) phases they resemble. For example, modes above and to the left of the *ScS* line in Figure 5.5 may be classified as *ScS* type because they sample

Figure 5.5: Torsional normal modes in the $\omega\ell$ plane. The large dots indicate observed modes used in the inversion of Gilbert & Dziewonski. The dashed line, designated *ScS*, divides the modes into two groups according to the normal mode-body wave analogy: modes to the left of this line correspond to $S_H c S_H$ reflections, those to the right correspond to mantle *SH* waves.

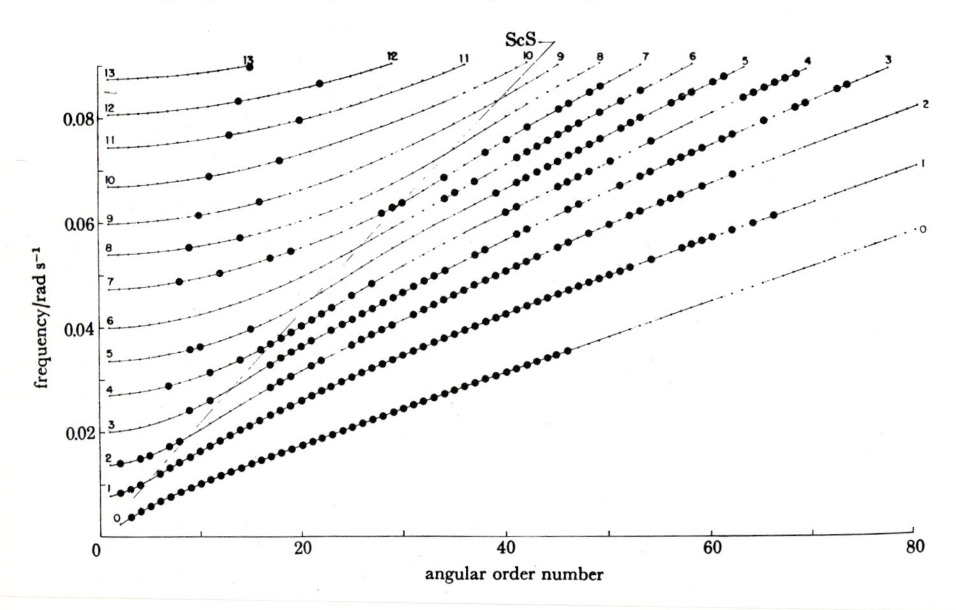

the mantle right down to the core boundary; those below and to the right of the *ScS* line as *S* type because they sample only the upper part of the mantle.

Ray equivalences for spheroidal modes are shown in Figure 5.6. Both *P* and *S* lines are drawn because spheroidal modes have both *P* and *SV* motion. Modes to the left of the *PKIKP* line sample the inner core. These ray equivalences give an indication of where the energy of a particular mode may be concentrated, and therefore of which parts of the earth they sample.

The question of which parts of the earth are sampled by a particular mode is only roughly estimated by the ray-mode equivalences given in Figures 5.5 and 5.6. By "sampling" is meant that the mode frequency depends on the properties of that part of the earth; the dependency cannot exist if the displacement for that mode is very small in that part of the earth. Thus we expect an "inner core mode" to have an eigenfunction with significant values in the inner core. The only way to determine how a mode samples the earth is to look at the form of the eigenfunction itself; such a study requires calculating and plotting all the solutions for a given earth model

If we could observe the body wave PKJKP which travels as an *S* wave in the earth's inner core we would have direct evidence of its solidity. Without such an observation we cannot discriminate between a solid and

Figure 5.6: Spheroidal normal modes in the $\omega\ell$ plane.

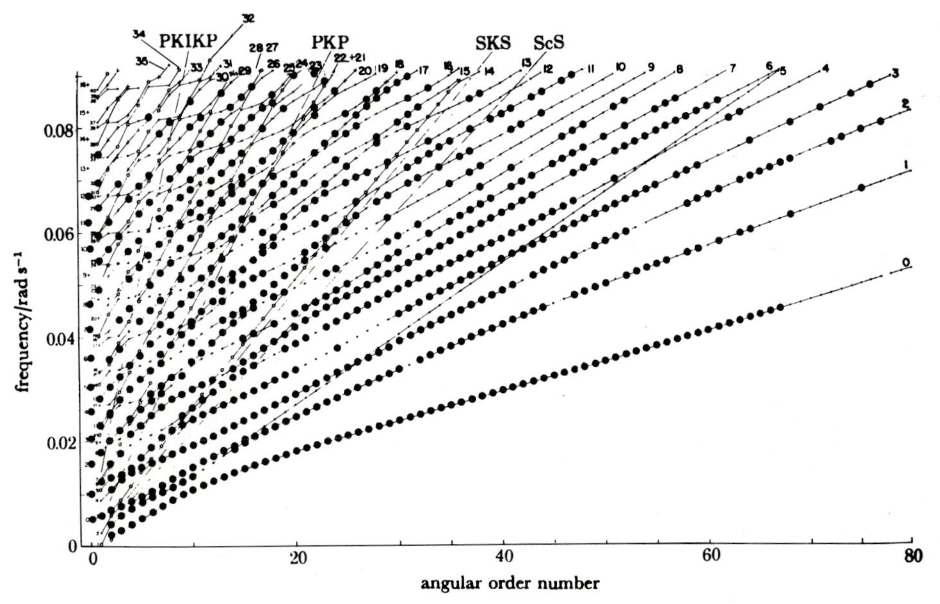

a liquid with sharply different properties from those of the outer core. When we examine the eigenfunctions for some modes we find them to have shear in the inner core. They are termed *core modes* because they sample the core; two examples are ${}_6S_2$ and ${}_7S_3$ which have 84.2% and 93.4% of their shear energy in the inner core respectively, according to the eigenfunctions based on the elastic parameters for model 1066B in Table 5.1. Observation of these modes gives direct evidence of solidity of the inner core, as well as an estimate of the shear velocity. They were observed on gravimeter records from an earthquake in the Tonga region on 22 June 1977 by Masters & Gilbert (1981). The measured frequencies give an inner core shear velocity of 3.5 km s^{-1} and the width of the spectral peaks a high value of Q of 3000 ± 2000. This low shear velocity and high Q implies an unusually high Poisson's ratio, which may be typical of solids at high pressure near their melting point.

5.6 Splitting — terrestrial spectroscopy

Free oscillations are only lightly damped — in fact the gravest modes remain detectable for up to a month after a large earthquake. Their slight decay with time is due to anelasticity. Damping also broadens the spectral peaks. However, the observed peaks are significantly broader than would be expected on the basis of attenuation alone. What causes this broadening? Recall that each frequency is associated with $(2\ell + 1)$ degenerate normal modes, each one with a different azimuthal order number. As explained above, the degeneracy is a result of the spherical symmetry of the earth: any departures from spherical symmetry will remove the degeneracy. The earth is not perfectly spherical and each individual mode will have a slightly different frequency. The spectral peak is said to be *split*; each individual member of the degenerate set is called a singlet, the whole collection of degenerate modes is a multiplet. Each singlet is a peak slightly broadened by attentuation. In some cases it is possible to resolve individual singlets, but in most cases the frequency separation is not sufficiently large to produce distinct peaks and the resulting spectrum has the appearance of a very broad, single, peak.

Splitting may be the result of lateral variations in the earth, of ellipticity of figure, or ocean-continent differences, or it may also be due to rotation. Rotation of the earth causes a Coriolis force to act on any moving particle. The force is small but noticeable, and it has axial, rather than spherical, symmetry. The Coriolis force has the form $2\mathbf{\Omega} \times \dot{\mathbf{u}}$, where $\mathbf{\Omega}$ is the rotation vector. It has no effect on axisymmetric modes with azimuthal order number zero ($m = 0$), and changes the frequencies of all modes with azimuthal order number m by the same amount. Rotational splitting is analogous to the splitting of spectral lines of the hydrogen atom by a magnetic field by the Zeeman effect. The potential for the hydrogen atom has spherical symmetry which is destroyed by an applied magnetic field with axial symmetry, thus resolving the degeneracy of a multiplet. Coriolis force splits the degenerate modes of a spherically symmetric earth in the same way.

Splitting of the low order modes ${}_0S_2$, ${}_0S_3$, and ${}_0S_4$ was detected from gravimeter records of a large earthquake that occurred on 19 August 1977 200 km south-west of the island of Sumbawa in Indonesia by Buland *et al.* (1979). Power spectra for the narrow frequency band 0.43–0.52 mHz, including the mode ${}_0S_3$, is shown in Figure 5.7. The splitting of the single mode into a group of separate peaks is very clear. All seven singlets are visible in the records, but only a few of them appear at any individual station. This is because of the different spatial form of each singlet: some will be large at a particular station and some small. These authors find the observed splitting to be consistent with Coriolis (rotational) splitting with an effect due to the earth's ellipticity. Lateral variations in the elastic constants also cause splitting. Ellipticity of the earth's shape is a large part of the departure from sphericity, and it contributes to the splitting. The study of frequency shifts is now an important means of determining lateral variations within the earth; an example of one study is given in Section 5.7.

Frequency shifts can occur that are comparable with the frequency spacing between modes. This is likely when the modes are densely packed (Figures 5.5 and 5.6). In this case the modes become coupled and oscillate together. Such mode coupling has been observed by Masters & Gilbert (1981). The spheroidal-toroidal separation also breaks down, and spheroidal modes can couple to toroidal modes, but the consequent motion is still accessible to mathematical analysis because only a few different modes participate in the motion.

The study of the earth's free oscillations is sometimes called "terrestrial spectroscopy". The split spectral peaks are the fine structure of the spectrum, and very accurate measurements of the frequencies of spectral lines can be used to determine structure within the earth just as optical spectra can be used to understand the electronic structure of an atom.

5.7 Using free oscillations to refine earth models

Improving the fit between calculated and observed frequencies has led to an improvement in models of the variation with depth within the earth of the elastic parameters. Free oscillation frequencies are dependent

Figure 5.7: Splitting of the spheroidal mode ${}_0S_3$ into singlets. Spectra are derived from 155 hours of *IDA* records beginning 25 h after the event. Reprinted with permission from Buland *et al.* (1979), copyright © 1979, Macmillan Magazines Ltd.

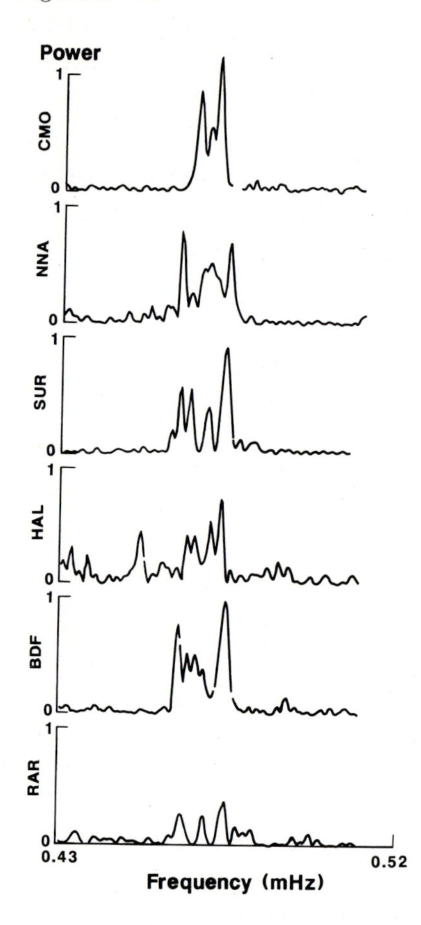

on the structure of the whole earth, unlike travel times of short period waves which sample along very localised paths through the earth. The normal mode frequencies give a picture that is relatively unbiassed by the geographical locations of sources and receivers, which is an advantage when we wish to extract a model of the earth that is in some way an "average".

A major study was undertaken and published in 1975 by Gilbert & Dziewonski. Most of the records came from the *WWSSN*. They used 1064 modes with angular order number up to 76 and angular frequencies down to 0.090 s^{-1}, using either mode identifications previously published by others or identifying them themselves. The 1064 mode frequencies were used together with the earth's mass and moment of inertia to refine the elastic parameters. One of their earth models is given in Table 5.1. The pressure has been calculated using the hydrostatic approximation as described in Chapter 2.

There is a small difference between the seismic parameters estimated from free oscillation frequencies, such as those for model 1066b in Table 5.1 and those derived from body wave data. This was originally attributed to a continent-ocean difference in the earth. The body wave data was thought to be biassed towards the continental structures where the majority of the seismometers are located. However, body waves and free oscillations have markedly different frequencies and may therefore be governed by different elastic parameters. In fact a very fundamental result ensures that dispersion must accompany attenuation always; it is best known as the *Kramers-Kronig relations* for dielectric media (LANDAU & LIFSHITZ, vol. 8). The seismological case is discussed in AKI & RICHARDS). The relations can be used to relate the attenuation of any travelling disturbance to its dispersion; in fact if the attenuation is specified as a function of frequency then the phase speed can be computed from it as a function of frequency. Thus we are not at liberty to assert that dispersion will be unimportant in the earth; it must depend on the attenuating properties of the medium. Dispersion is now recognised to be significant in the earth, but it is still not clear whether discrepancies between earth models based on free oscillations and travel times are due to it or to lateral variations, such as continent-ocean differences.

Lateral variations in seismic parameters, such as continent-ocean differences, can cause splitting of degenerate normal mode multiplets. Departures from spherical symmetry in the earth are quite small and the frequency shifts can be calculated by first-order perturbation theory. An important result is that the mean frequency of a set of singlets split by non-spherical earth structure equals the frequency of the multiplet for

level	radius km	depth km	ρ g cm^{-3}	V_p km s^{-1}	V_s km s^{-1}	ϕ km^2 s^{-2}	κ kbar	μ kbar	λ kbar	σ	pressure kbar	g cm s^{-2}
1	0.0	6371.0	13.077	11.281	3.550	110.45	14444	1648	13345	0.4450	3660.4	0.0
2	38.0	6333.0	13.074	11.280	3.550	110.45	14440	1647	13342	0.4451	3659.4	13.9
3	75.9	6295.1	13.069	11.279	3.550	110.42	14430	1646	13332	0.4451	3657.6	27.7
4	113.9	6257.1	13.064	11.277	3.549	110.37	14418	1645	13321	0.4450	3655.2	41.6
5	151.9	6219.1	13.060	11.273	3.548	110.30	14404	1643	13308	0.4451	3652.1	55.4
6	189.8	6181.2	13.053	11.269	3.546	110.22	14386	1641	13292	0.4451	3648.4	69.3
7	227.8	6143.2	13.040	11.263	3.545	110.11	14358	1638	13266	0.4450	3643.9	83.1
8	265.8	6105.2	13.024	11.257	3.543	109.99	14325	1634	13235	0.4451	3638.8	96.9
9	303.7	6067.3	13.010	11.250	3.540	109.85	14291	1630	13204	0.4451	3633.0	110.6
10	341.7	6029.3	13.001	11.242	3.538	109.69	14260	1627	13175	0.4450	3626.5	124.4
11	379.7	5991.3	12.994	11.233	3.535	109.52	14230	1623	13148	0.4451	3619.4	138.1
12	417.7	5953.3	12.987	11.224	3.532	109.33	14198	1620	13118	0.4450	3611.5	151.8
13	455.6	5915.4	12.977	11.213	3.529	109.14	14163	1616	13085	0.4450	3603.0	165.6
14	493.6	5877.4	12.969	11.203	3.526	108.93	14126	1612	13051	0.4450	3593.9	179.3
15	531.6	5839.4	12.963	11.191	3.523	108.70	14090	1609	13017	0.4450	3584.0	192.9
16	569.5	5801.5	12.958	11.180	3.520	108.47	14055	1605	12985	0.4450	3573.5	206.6
17	607.5	5763.5	12.951	11.169	3.516	108.26	14021	1601	12953	0.4450	3562.4	220.3
18	645.5	5725.5	12.942	11.158	3.510	108.07	13986	1594	12923	0.4451	3550.6	234.0
19	683.4	5687.6	12.931	11.146	3.503	107.88	13949	1586	12891	0.4452	3538.1	247.6
20	721.4	5649.6	12.923	11.132	3.496	107.63	13908	1579	12855	0.4453	3524.9	261.3
21	759.4	5611.6	12.917	11.117	3.489	107.35	13856	1572	12817	0.4454	3511.1	274.9
22	797.3	5573.7	12.911	11.102	3.484	107.07	13823	1567	12778	0.4454	3496.6	288.5
23	835.3	5535.7	12.902	11.091	3.482	106.84	13784	1564	12741	0.4453	3481.5	302.1
24	873.3	5497.7	12.891	11.081	3.481	106.64	13747	1561	12706	0.4453	3465.7	315.7
25	911.2	5459.8	12.878	11.072	3.478	106.45	13709	1558	12670	0.4452	3449.3	329.2
26	949.2	5421.8	12.863	11.062	3.474	106.27	13669	1552	12634	0.4453	3432.3	342.8
27	987.2	5383.8	12.843	11.053	3.468	106.15	13632	1544	12602	0.4454	3414.6	356.3
28	1025.2	5345.8	12.817	11.050	3.461	106.13	13602	1535	12578	0.4456	3396.3	369.7
29	1063.1	5307.9	12.784	11.051	3.456	106.19	13575	1526	12557	0.4458	3377.4	383.1
30	1101.1	5269.9	12.747	11.051	3.452	106.24	13543	1518	12531	0.4460	3357.9	396.3
31	1139.1	5231.9	12.708	11.049	3.449	106.22	13499	1511	12491	0.4460	3337.9	409.5
32	1177.0	5194.0	12.666	11.045	3.447	106.14	13443	1505	12439	0.4460	3317.3	422.6
33	1216.8	5154.2	12.620	11.038	3.447	106.00	13377	1499	12377	0.4460	3295.9	436.2
34	1216.8	5154.2	12.057	10.366	0.000	107.45	12954	0	12954	0.5000	3295.9	436.2
35	1285.9	5085.1	12.029	10.335	0.000	106.81	12848	0	12848	0.5000	3257.0	456.7
36	1356.9	5014.1	12.000	10.297	0.000	106.03	12723	0	12723	0.5000	3215.5	477.9
37	1427.8	4943.2	11.968	10.255	0.000	105.16	12585	0	12585	0.5000	3172.3	499.4
38	1498.7	4872.2	11.932	10.211	0.000	104.27	12441	0	12441	0.5000	3127.3	521.0
39	1569.7	4801.3	11.894	10.169	0.000	103.40	12298	0	12298	0.5000	3080.7	542.6
40	1640.6	4730.4	11.859	10.120	0.000	102.41	12144	0	12144	0.5000	3032.4	564.3
41	1711.6	4659.4	11.827	10.062	0.000	101.24	11974	0	11974	0.5000	2982.4	586.1
42	1782.5	4588.5	11.791	10.002	0.000	100.03	11794	0	11794	0.5000	2930.7	607.8
43	1853.4	4517.6	11.749	9.945	0.000	98.90	11620	0	11620	0.5000	2877.4	629.5
44	1924.4	4446.6	11.703	9.892	0.000	97.86	11452	0	11452	0.5000	2822.6	651.1
45	1995.3	4375.7	11.654	9.839	0.000	96.80	11281	0	11281	0.5000	2766.2	672.7
46	2066.2	4304.8	11.601	9.778	0.000	95.61	11091	0	11091	0.5000	2708.4	694.0
47	2137.2	4233.8	11.537	9.715	0.000	94.38	10888	0	10888	0.5000	2649.2	715.3
48	2208.1	4162.9	11.466	9.652	0.000	93.17	10682	0	10682	0.5000	2588.7	736.3
49	2279.1	4091.9	11.398	9.589	0.000	91.96	10481	0	10481	0.5000	2526.8	757.0
50	2350.0	4021.0	11.323	9.535	0.000	90.92	10295	0	10295	0.5000	2463.8	777.5
51	2420.9	3950.1	11.244	9.482	0.000	89.90	10108	0	10108	0.5000	2399.6	797.8
52	2491.9	3879.1	11.166	9.417	0.000	88.67	9901	0	9901	0.5000	2334.3	817.7
53	2562.8	3808.2	11.085	9.342	0.000	87.27	9674	0	9674	0.5000	2267.9	837.4
54	2633.7	3737.2	11.001	9.268	0.000	85.90	9450	0	9450	0.5000	2200.6	856.8
55	2704.7	3666.3	10.917	9.196	0.000	84.57	9232	0	9232	0.5000	2132.3	875.9
56	2775.6	3595.4	10.833	9.116	0.000	83.10	9002	0	9002	0.5000	2063.0	894.8
57	2846.6	3524.4	10.759	9.029	0.000	81.53	8771	0	8771	0.5000	1992.9	913.3

Table 5.1: Elastic parameters for the earth model 1066B, based on free oscillation frequencies. From Gilbert & Dziewonski, (1975).

level	radius km	depth km	ρ g cm^{-3}	V_p km s^{-1}	V_s km s^{-1}	ϕ km^2 s^{-2}	κ kbar	μ kbar	λ kbar	σ	pressure kbar	g cm s^{-2}
58	2917.5	3453.5	10.686	8.933	0.000	79.80	8527	0	8527	0.5000	1921.8	931.7
59	2988.4	3382.6	10.607	8.827	0.000	77.92	8264	0	8264	0.5000	1849.9	949.8
60	3059.4	3311.6	10.532	8.725	0.000	76.13	8017	0	8017	0.5000	1777.3	967.6
61	3130.3	3240.7	10.452	8.613	0.000	74.19	7754	0	7754	0.5000	1703.9	985.2
62	3201.3	3169.8	10.365	8.489	0.000	72.06	7468	0	7468	0.5000	1629.9	1002.6
63	3272.2	3098.8	10.280	8.368	0.000	70.02	7197	0	7197	0.5000	1555.2	1019.6
64	3343.1	3027.9	10.189	8.243	0.000	67.95	6924	0	6924	0.5000	1480.1	1036.4
65	3414.1	2956.9	10.088	8.110	0.000	65.77	6634	0	6634	0.5000	1404.6	1052.8
66	3485.5	2885.5	9.977	7.967	0.000	63.48	6333	0	6333	0.5000	1345.2	1069.0
67	3485.5	2885.5	5.563	13.670	7.242	116.93	6504	2917	4559	0.3049	1345.2	1069.0
68	3519.6	2851.4	5.542	13.670	7.243	116.92	6479	2907	4541	0.3048	1325.2	1064.1
69	3554.2	2816.8	5.530	13.663	7.235	116.87	6463	2895	4533	0.3051	1305.0	1059.3
70	3588.8	2782.2	5.517	13.653	7.220	116.91	6450	2875	4533	0.3060	1284.9	1054.9
71	3623.4	2747.6	5.503	13.639	7.203	116.84	6429	2854	4526	0.3066	1265.0	1050.7
72	3658.0	2713.0	5.488	13.616	7.190	116.47	6392	2837	4500	0.3067	1245.2	1046.7
73	3692.7	2678.3	5.474	13.586	7.180	115.85	6341	2821	4460	0.3063	1225.5	1042.9
74	3727.3	2643.7	5.459	13.551	7.168	115.11	6284	2805	4414	0.3057	1205.9	1039.3
75	3761.9	2609.1	5.444	13.511	7.157	114.27	6220	2788	4361	0.3050	1186.4	1036.0
76	3796.5	2574.5	5.428	13.468	7.146	113.30	6149	2771	4301	0.3041	1167.1	1032.8
77	3831.1	2539.9	5.411	13.422	7.135	112.27	6075	2754	4239	0.3031	1147.9	1029.8
78	3865.7	2505.3	5.394	13.375	7.121	111.27	6002	2735	4178	0.3022	1128.7	1027.0
79	3900.3	2470.7	5.376	13.327	7.104	110.32	5930	2713	4121	0.3015	1109.7	1024.3
80	3934.9	2436.1	5.358	13.279	7.089	109.32	5857	2692	4062	0.3007	1090.8	1021.8
81	3969.5	2401.5	5.339	13.230	7.077	108.25	5780	2674	3997	0.2996	1072.1	1019.4
82	4004.1	2366.9	5.321	13.182	7.067	107.19	5702	2656	3931	0.2984	1053.4	1017.2
83	4038.7	2332.3	5.301	13.136	7.056	106.17	5628	2639	3868	0.2972	1034.8	1015.1
84	4073.4	2297.6	5.282	13.088	7.042	105.18	5555	2618	3809	0.2963	1016.3	1013.2
85	4108.0	2263.0	5.261	13.038	7.024	104.20	5481	2595	3751	0.2955	998.0	1011.3
86	4142.6	2228.4	5.241	12.990	7.006	103.30	5413	2572	3698	0.2949	979.7	1009.6
87	4177.2	2193.8	5.224	12.951	6.991	102.56	5358	2553	3656	0.2944	961.5	1008.0
88	4211.8	2159.2	5.209	12.918	6.977	101.96	5310	2535	3620	0.2941	943.4	1006.5
89	4246.4	2124.6	5.191	12.884	6.961	101.39	5262	2514	3586	0.2939	925.4	1005.1
90	4281.0	2090.0	5.169	12.848	6.941	100.83	5212	2490	3552	0.2939	907.5	1003.9
91	4315.6	2055.4	5.148	12.814	6.921	100.33	5164	2465	3520	0.2941	889.7	1002.7
92	4350.2	2020.8	5.127	12.782	6.902	99.86	5119	2442	3491	0.2942	872.0	1001.6
93	4384.8	1986.2	5.105	12.751	6.886	99.36	5072	2420	3458	0.2941	854.3	1000.5
94	4419.4	1951.6	5.083	12.720	6.874	98.81	5022	2401	3421	0.2938	836.8	999.6
95	4454.1	1916.9	5.063	12.689	6.861	98.25	4974	2383	3385	0.2934	819.3	998.7
96	4488.7	1882.3	5.046	12.658	6.846	97.73	4931	2364	3355	0.2933	801.9	997.9
97	4523.3	1847.7	5.029	12.625	6.829	97.21	4888	2345	3324	0.2932	784.6	997.2
98	4557.9	1813.1	5.008	12.591	6.812	96.66	4841	2324	3291	0.2931	767.4	996.6
99	4592.5	1778.5	4.985	12.556	6.797	96.04	4787	2303	3251	0.2927	750.2	996.0
100	4627.1	1743.9	4.963	12.517	6.784	95.31	4730	2284	3207	0.2920	733.2	995.5
101	4661.7	1709.3	4.942	12.477	6.772	94.53	4671	2266	3160	0.2912	716.2	995.0
102	4696.3	1674.7	4.920	12.437	6.760	93.76	4612	2248	3113	0.2903	699.3	994.6
103	4730.9	1640.1	4.899	12.398	6.746	93.03	4557	2229	3071	0.2887	682.5	994.2
104	4765.5	1605.5	4.882	12.357	6.727	92.35	4508	2209	3035	0.2894	665.7	993.9
105	4800.2	1570.8	4.868	12.313	6.706	91.65	4461	2188	3002	0.2892	649.0	993.7
106	4834.8	1536.2	4.854	12.266	6.687	90.83	4408	2170	2961	0.2885	632.4	993.5
107	4869.4	1501.6	4.837	12.215	6.672	89.84	4345	2153	2909	0.2873	615.8	993.4
108	4904.0	1467.0	4.820	12.159	6.654	88.79	4279	2134	2856	0.2862	599.2	993.3
109	4938.6	1432.4	4.805	12.099	6.630	87.77	4216	2111	2808	0.2854	582.7	993.3
110	4973.2	1397.8	4.791	12.039	6.605	86.77	4156	2090	2762	0.2846	566.3	993.4
111	5007.8	1363.2	4.777	11.984	6.586	85.78	4098	2072	2716	0.2836	549.9	993.5
112	5042.4	1328.6	4.763	11.929	6.568	84.78	4037	2054	2667	0.2825	533.5	993.6
113	5077.0	1294.0	4.744	11.868	6.545	83.73	3972	2032	2617	0.2815	517.2	993.8
114	5111.6	1259.4	4.724	11.804	6.520	82.64	3903	2008	2564	0.2804	501.0	994.0

level	radius km	depth km	ρ g cm^{-3}	V_p km s^{-1}	V_s km s^{-1}	ϕ km^2 s^{-2}	κ kbar	μ kbar	λ kbar	σ	pressure kbar	g cm s^{-2}
115	5146.2	1224.7	4.705	11.741	6.500	81.52	3835	1987	2510	0.2791	484.8	994.3
116	5180.9	1190.1	4.689	11.681	6.484	80.40	3770	1971	2456	0.2774	468.7	994.6
117	5215.5	1155.5	4.673	11.625	6.468	79.36	3708	1954	2405	0.2759	452.7	994.9
118	5250.1	1120.9	4.655	11.570	6.449	78.42	3650	1936	2359	0.2746	436.7	995.3
119	5284.7	1086.3	4.634	11.518	6.426	77.60	3595	1913	2319	0.2740	420.7	995.7
120	5319.3	1051.7	4.609	11.466	6.398	76.89	3543	1886	2285	0.2739	404.9	996.1
121	5353.9	1017.1	4.586	11.414	6.366	76.25	3497	1858	2258	0.2734	389.1	996.5
122	5388.5	982.5	4.567	11.365	6.334	75.66	3455	1832	2233	0.2747	373.4	996.9
123	5423.1	947.9	4.548	11.317	6.304	75.10	3415	1807	2210	0.2751	357.7	997.4
124	5457.7	913.3	4.526	11.274	6.279	74.54	3373	1784	2183	0.2751	342.1	997.8
125	5492.3	878.7	4.503	11.235	6.259	73.98	3330	1764	2154	0.2749	326.6	998.3
126	5526.9	844.1	4.482	11.197	6.244	73.39	3289	1747	2124	0.2743	311.1	998.8
127	5561.6	809.4	4.462	11.159	6.227	72.82	3249	1730	2095	0.2739	295.7	999.3
128	5596.2	774.8	4.442	11.118	6.205	72.27	3209	1710	2069	0.2737	280.4	999.8
129	5630.8	740.2	4.420	11.078	6.183	71.75	3171	1689	2045	0.2738	265.1	1000.3
130	5665.4	705.6	4.397	11.043	6.155	71.43	3140	1665	2030	0.2747	249.9	1000.9
131	5700.0	671.0	4.372	11.013	6.117	71.39	3121	1635	2031	0.2770	235.2	1001.4
132	5700.0	671.0	4.100	9.945	5.475	58.94	2416	1229	1596	0.2825	235.2	1001.4
133	5731.4	639.6	4.049	9.900	5.415	58.91	2385	1187	1593	0.2865	222.6	1001.1
134	5762.7	608.2	4.003	9.859	5.363	58.86	2356	1151	1588	0.2899	210.1	1000.8
135	5794.1	576.9	3.958	9.805	5.306	58.61	2319	1114	1576	0.2929	197.8	1000.4
136	5825.5	545.5	3.906	9.762	5.259	58.41	2281	1080	1561	0.2955	185.6	999.9
137	5856.9	514.1	3.865	9.703	5.203	58.05	2243	1046	1545	0.2981	173.5	999.4
138	5888.3	482.7	3.821	9.639	5.161	57.40	2193	1017	1515	0.2992	161.6	998.8
139	5919.6	451.4	3.779	9.581	5.112	56.95	2151	987	1493	0.3010	149.9	998.2
140	5951.0	420.0	3.746	9.515	5.052	56.49	2116	956	1478	0.3036	138.3	997.5
141	5951.0	420.0	3.605	8.893	4.796	48.42	1745	829	1192	0.2949	138.3	997.5
142	5975.9	395.1	3.593	8.832	4.765	47.72	1714	815	1170	0.2947	129.5	996.7
143	6000.9	370.1	3.562	8.761	4.748	46.70	1663	802	1128	0.2922	120.6	995.9
144	6025.8	345.2	3.535	8.688	4.710	45.90	1622	784	1099	0.2918	111.9	995.1
145	6050.8	320.2	3.512	8.616	4.657	45.32	1591	761	1083	0.2937	103.2	994.2
146	6075.7	295.3	3.488	8.544	4.599	44.79	1562	737	1070	0.2961	94.6	993.4
147	6100.6	270.4	3.461	8.470	4.532	44.34	1534	711	1060	0.2993	86.1	992.5
148	6125.6	245.4	3.434	8.392	4.453	43.98	1510	681	1056	0.3040	77.6	991.6
149	6150.5	220.5	3.409	8.315	4.382	43.54	1484	654	1048	0.3079	69.3	990.7
150	6175.4	195.6	3.387	8.245	4.343	42.83	1450	638	1024	0.3081	60.9	989.8
151	6200.4	170.6	3.364	8.176	4.344	41.68	1402	634	979	0.3035	52.7	988.9
152	6225.3	145.7	3.351	8.098	4.355	40.28	1349	635	925	0.2965	44.4	987.9
153	6250.3	120.7	3.338	8.021	4.416	38.34	1279	650	845	0.2826	36.2	987.0
154	6275.2	95.8	3.323	7.947	4.514	35.99	1195	677	743	0.2616	28.1	986.1
155	6300.1	70.9	3.321	7.881	4.624	33.61	1116	710	642	0.2374	19.9	985.3
156	6325.1	45.9	3.309	7.808	4.725	31.21	1032	738	540	0.2113	11.8	984.4
157	6350.0	21.0	3.293	7.742	4.759	29.74	979	745	482	0.1964	4.3	983.6
158	6350.0	21.0	2.802	6.200	3.400	23.03	645	323	429	0.2852	4.3	983.6
159	6360.5	10.5	2.802	6.200	3.400	23.03	645	323	429	0.2852	1.4	982.8
160	6371.0	0.0	2.803	6.200	3.400	23.03	645	323	429	0.2852	0.0	982.0

the spherical average of the earth structure. This result is called the *angular momentum sum rule* (from quantum mechanics); it tells us that the observed peaks, broadened by splitting, sample an "average" earth. The angular momentum sum rule encourages us to value earth models based on free oscillation frequencies as more representative of the average earth structure than those based on body waves. But matters are not quite so straightforward: the angular momentum sum rule is only useful if all singlets are present with equal amplitude in the observed degenerate peak. The constitution of each multiplet is heavily dependent on the source-receiver geometry and the source mechanism; the mix of individual modes will be inhomogeneous and the peak may therefore appear shifted. This is particularly noticeable in Figure 5.7, where all singlets of the mode ${}_0S_3$ are observed, but only selected singlets are seen at individual stations. If the distribution of source-receiver pairs is persistently biassed then some members of the multiplet set will not be observed at all, and the average frequency of the multiplet peak will be biassed.

The study of Masters, Jordan, Silver & Gilbert (1982) illustrates this dependence on source-receiver geometry well. Fundamental modes are surface-wave equivalents; they are seen on spectra of seismograms which in the time domain consist of successive passes of surface waves (see the practical at the end of this chapter). The surface waves travel around the world on that great circle path which includes both source and receiver and consequently sample earth structure more strongly there than on any other part of the earth. The spectral lines will reflect this bias and have frequencies appropriate to earth structure along one particular source-receiver path.

The great circle path is conveniently represented by a single point: its pole, the point where the normal to the diametric plane containing the great circle intersects the earth's surface. Each pole corresponds to a source-receiver pair; we may measure mode frequencies for each pole and compare the results with frequencies predicted from an average earth model such as 1066B. The discrepancies give an indication of departures from spherical symmetry.

The frequencies are obtained from a single seismic station; the frequency differences are plotted in Figure 5.8 at the pole of the great circle path joining source and receiver. A plus indicates a positive difference, a diamond a negative difference, the size of the symbol is proportional to the size of the difference. The results show a remarkable pattern. Positive frequency shifts cluster in two patches near the equator and are separated by 180° in longitude. This effect can only be caused by systematic lateral variations in earth structure. The anomaly is believed to lie in the transition zone between upper and lower mantle. Surprisingly the variations do not reflect difference in elastic properties between oceans and continents.

Practical 3. Long period seismograms and normal mode identification

Detection of free oscillations requires a network of seismometers that are sensitive to long period ground motions. Gravimeters respond equally to all frequencies and are suitable for studying spheroidal modes. Project *IDA* (*I*nternational *D*eployment of *A*ccelerometers) has installed a global network of gravimeters. The distribution as of 1987 is shown in Figure 5.9. The *IDA* records include earth tides and the seismic disturbance rides on top of a sinusoidal tidal oscillation as shown in Figure 5.10. The tides can be predicted theoretically and subtracted from the record to leave a level background.

Digital recording is an important feature of all the new instruments. It facilitates processing of the data and gives a very much larger *dynamic range* than the paper or optical recording. With the recording system of the *WWSSN* instruments the gain must be turned up to give reasonable sensitivity, but not so far that the record goes off-scale for modest ground motions. Digital recording suffers the same limitation but in a much less severe form; the dynamic range is given by the maximum number that

Figure 5.8: Departures of measured fundamental mode frequencies ($\ell = 8$ and 9) from those predicted by a spherically symmetric earth model. From Masters *et al.* (1982), Copyright ©Macmillan Magazines Ltd. Reprinted with permission.

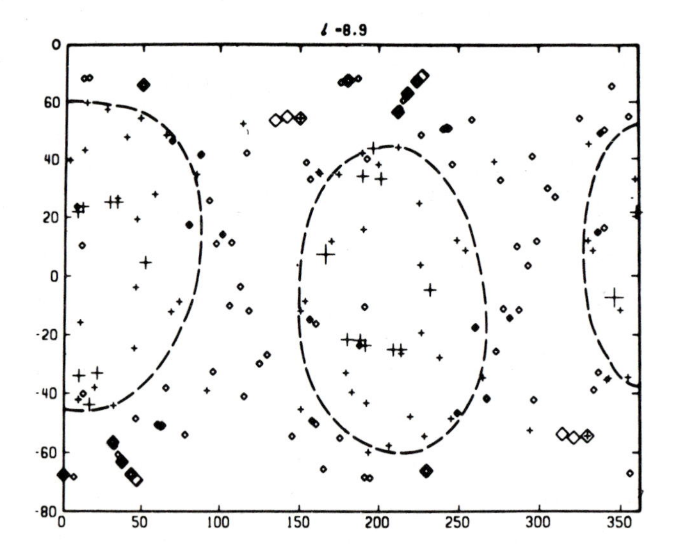

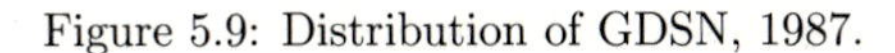

Figure 5.9: Distribution of GDSN, 1987.

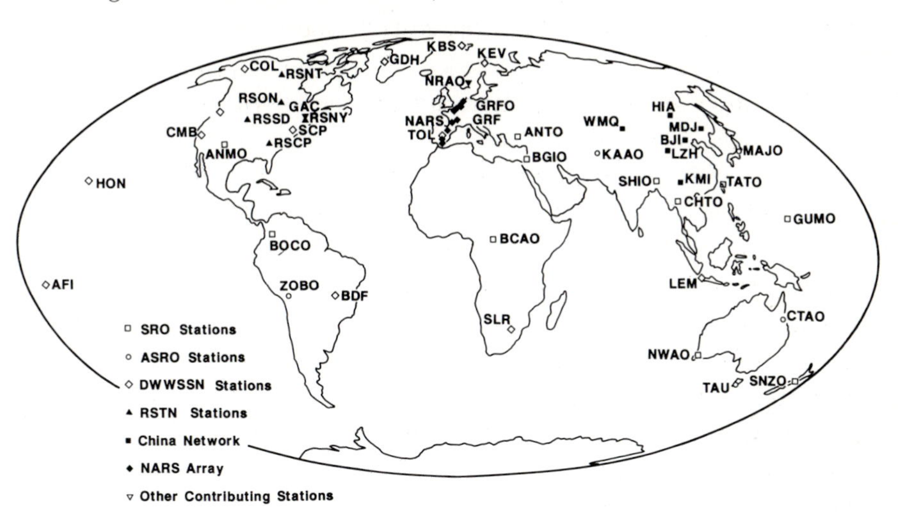

Figure 5.10: *IDA* gravimeter records of an earthquake in the Solomon Islands, 20 July 1975. Note the way in which the signal rides on the earth tide, which appears as a long period sine wave. From Agnew *et al.* (1976)

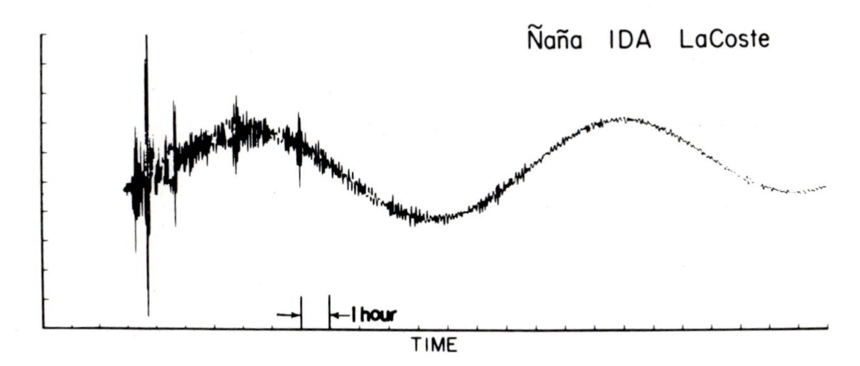

can be recorded. There is also a mechanical restriction on the dynamic range: some accelerations are so large that the mass in the gravimeter may hit the stops (or even get stuck). This is not a problem for normal mode studies because the oscillations continue for many hours and only the first part of the record is lost. The *IDA* gravimeters sample every 10 s; therefore they cannot be used to study short period waves.

Another type of digital instrument is the *SRO* (*S*eismological *R*esearch *O*bservatory). These are seismometers rather than gravimeters but the mechanical part of the device is broad-band. The instrument is located in a bore hole in order to reduce wind noise. The signal from the mechanical device is filtered electronically to produce two separate records - – a short period response that is then sampled 20 times per second, and a long period response sampled once per second. Filtering of some sort is necessary to remove the noise which dominates the ground motion at about a 6 s period. Unfortunately for normal mode studies the long period response has been designed to remove the very longest periods. The large dynamic range offered by the digital recording allows some of this very long period signal to be recovered. The short and long period responses are recorded separately on magnetic tape. There are three long period components and a vertical short period component.

The two horizontal components (north and east) are usefully recombined into the *radial* (away from the event location along the great circle path) and *transverse* (perpendicular to the great circle path, positive to the left when facing away from the event), as shown in Figure 5.11. Vertical, radial, and transverse components (VRT) are the same as the (r, θ, ϕ) components in spherical polar coordinates if the event is located at the pole of the coordinate system. These are called *epicentral coordinates.* *SH* motion is expected only on the transverse component, *P-SV* motion on the radial and vertical components. This practical is designed to provide some familiarity with these long period instruments.

1 The first two plots (Figures 5.12 and 5.13) are seismograms recorded in Charters Towers, Australia, at the *ASRO* named *CTAO*, for shallow earthquake near Oaxaca in Mexico. The two plots are for the same record but plotted on different scales. Note the difference in vertical scales on the plot. The distance is $\Delta = 120.5°$. Identify what phases you can; the *JB* travel time tables give the following times (in minutes) for this distance:

P_{diff}	15.4
PKP	19.0
PP	20.4
PKP/SKP	22.5
PPP	23.0
SKS	25.9
SKKS	27.7
S_{diff}	28.4
PKKP	29.5
PS/SP	30.4

The record is discussed in detail by Ward (1980). The *P* phase would be expected to be seen only on the vertical and radial components. What causes energy to arrive at this time on the transverse component?

2 The next three records, shown in Figures 5.14–5.16, are from *IDA* gravimeters. CMO is at College, Alaska (65°N, 148°W); GAR is at Garm, USSR (39°N, 70°E); ESK is at Eskdalemuir, Scotland (56°N, 4°W). Most noticeable are the surface wavetrains. Observe

Figure 5.11: Epicentral coordinates. *S* is the station, *E* the epicentre, *C* the centre of the earth, and *N* the north pole. *ES* is the great circle path from source to receiver. *R* lies in the direction of the radial component, *T* the transverse component.

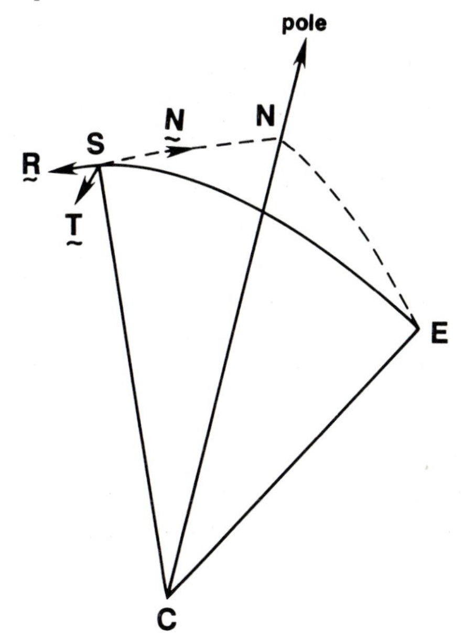

the bursts of energy corresponding to repeated passage of Rayleigh waves. The Rayleigh wavetrains are denoted $R_1, R_2, R_3, \ldots$ R_1 is the first wavetrain to arrive, having travelled a distance Δ. R_2 has travelled the long way round, a distance $360° - \Delta$. R_3 is the first wavetrain on its second passage past the receiver, having travelled a distance $360° + \Delta$. R_4 is the second passage of R_2, and so on. The surface waves from this large event cause the instrument to go off scale and it is not possible to identify the early passes. Identify the arrival times of the surface wavetrains and hence calculate the Rayleigh wave velocity and approximate distance to the earthquake. Verify that the event is in the vicinity of Mexico.

3 The next plot, Figure 5.17, is for 4 hours of data for the same earthquake, recorded at CTAO. 20 hours of data from this instrument were Fourier transformed to produce the next three plots in Figure 5.18. The Fourier transform of N discrete values $\{a_k; k = 1, 2, \ldots, N\}$ is given by

$$F_\ell = \sum_{k=0}^{N-1} a_k \exp\left(2\pi i k \ell / N\right)$$

The amplitude spectrum is $\{|F_\ell|; \ell = 0, 1, \ldots, N-1\}$. The part of the seismogram that is off scale has been removed, and the remainder multiplied by a broad Gaussian curve. Multiplication in the time domain is equivalent to convolution in the frequency domain and the convolution has the effect of smoothing irregularities in the transform. The process is called windowing. The last two plots (Figure 5.19) shows the power spectra of 20 hours of data from CMO, an *IDA* gravimeter, for different frequency ranges.

4 Measure the frequencies of the peaks and identify the modes by comparing them with the calculated frequencies in Table 5.2. All frequencies are in mHz. Estimate the asymptotic frequency spacing between high order fundamental spheroidal modes and deduce the Rayleigh wave velocity using (5.33). Repeat for the fundamental torsional modes and estimate the Love wave velocity.

5 This was a shallow event and most of the modes are fundamentals. The amplitude of each mode at any receiver will depend on the event location and source mechanism. The "scalloped" appearance of the peaks is a source effect. At *CMO* every third mode is large: the distance corresponds to an exact number of wave-

lengths for these modes and so Rayleigh waves will be in phase on repeated passes. The fall in amplitude at very low frequencies on the *CTAO* records reflects the instrument response of the *SRO* instrument.

Exercises

1 Derive equations (5.3), (5.4) and (5.7) from the formula for the curl in spherical coordinates, and hence obtain equation (5.12).

2 Show that the triplet of modes ${}_0T_1$ represents rotation of the earth about each of three orthogonal axes, respectively.

3 Sketch the angular variation of the displacement of a mode ${}_0S_2$ with $m = 0$, and show that it justifies its name of the "football" mode. Sketch the other four members of the multiplet.

4 Derive the frequencies of radial oscillations of a non-gravitating, homogeneous sphere with wave speed α and radius a. Show that for radial oscillations the frequencies are ${}_n\omega_0 = (n+1)\pi\alpha/a$.

5 Show the frequencies of radial modes in a self-gravitating, homogeneous fluid sphere are

$$ {}_n\omega_\ell = \left[\frac{2\ell\,(\ell-1)\,g}{(2\ell+1)\,a}\right]^{\frac{1}{2}} $$

6 Obtain the group velocity for the modes in exercise (5) when ℓ is large. Compare these oscillations with those of a solid non-gravitating sphere (see AKI & RICHARDS).

Figure 5.12: Long period *SRO* record from Charters Towers in Australia (*CTAO*). The three components are, from the top, vertical (V), radial (R), and transverse (T). Time is measured in minutes. The time origin on the graph is arbitrary.

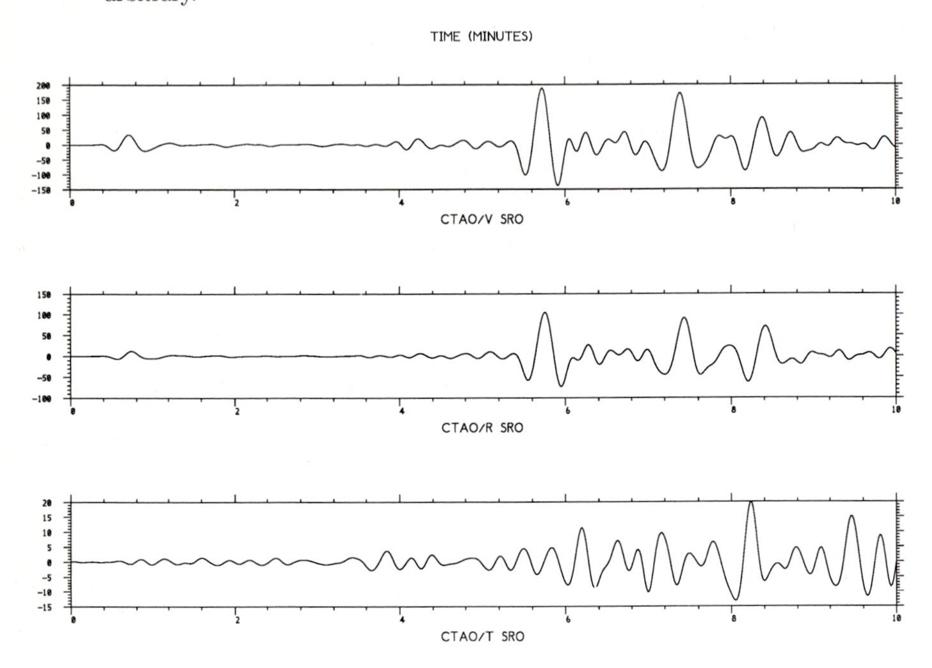

Figure 5.13: The same seismograms as in Figure 5.12 but with both time and amplitude scales compressed. The first ten minutes of these traces are the same as in Figure 5.12. This shows the advantage of digital recording and the large dynamic range of the *SRO* recording system: small events are revealed simply by plotting on a different scale.

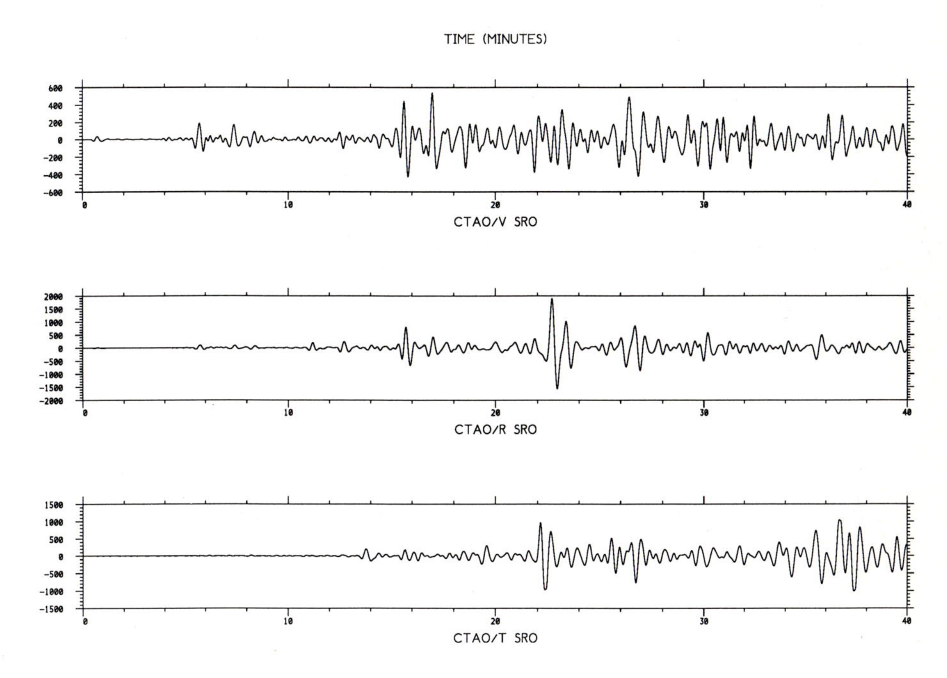

Figure 5.14: 20 hours of an *IDA* gravimeter record from College in Alaska for an earthquake in Mexico. The earth tides have been removed but no other processing has been done. Note the repeated surface wavetrain arrivals: these wavetrains have passed around the world several times.

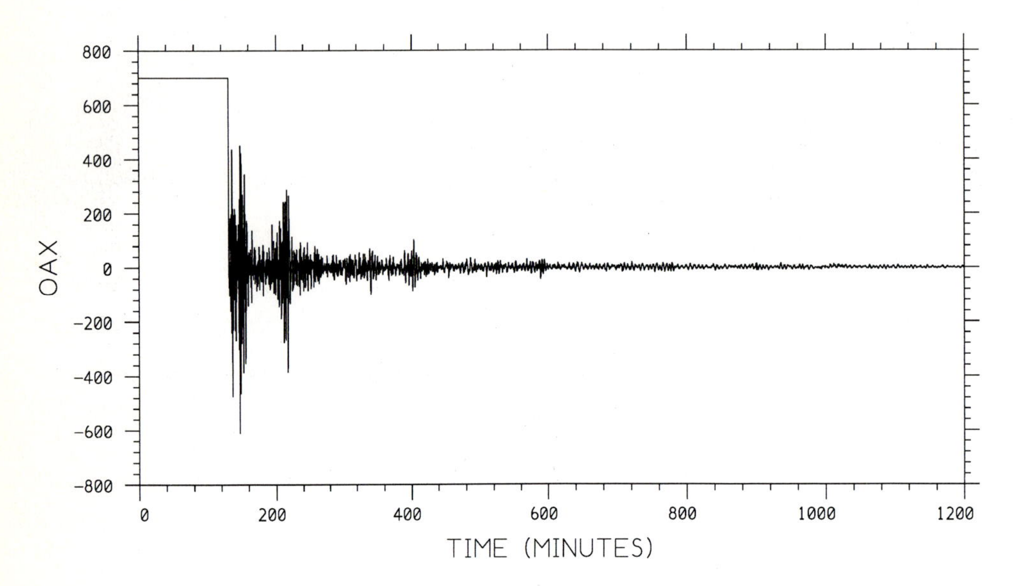

Figure 5.15: As Figure 5.14 for *IDA* instrument at Garm in USSR.

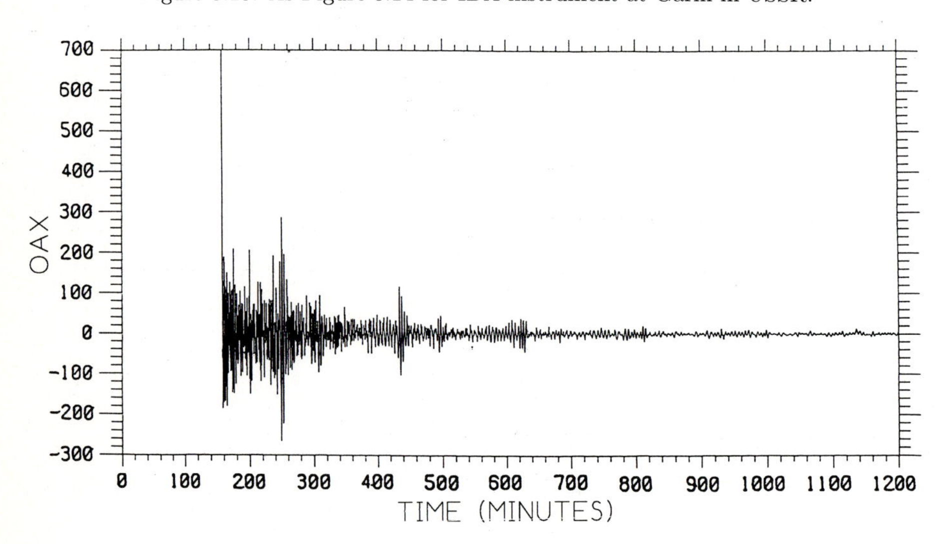

Figure 5.16: As Figure 5.14 for *IDA* instrument at Eskdalemuir in Scotland.

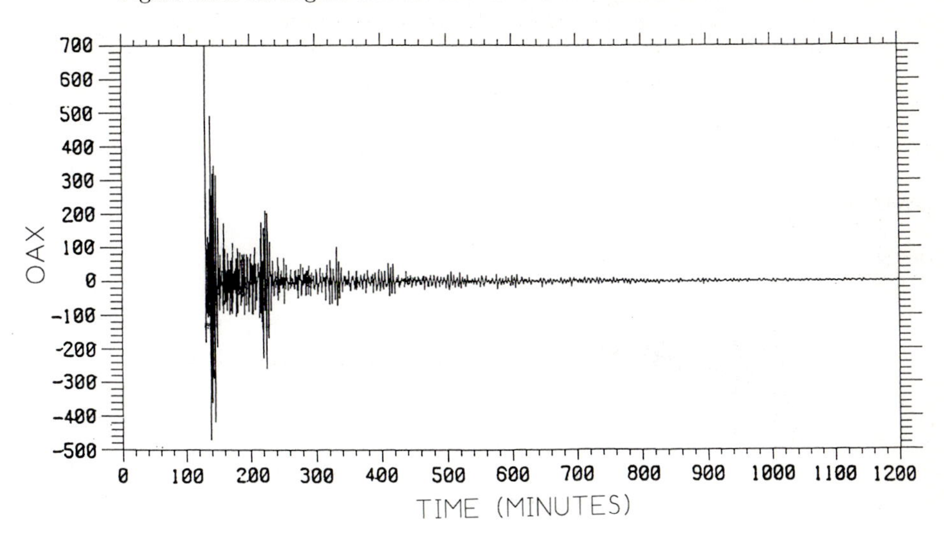

Figure 5.17: Yet again the record from CTAO but this time the full plot of 4 hours that was used to generate the spectra.

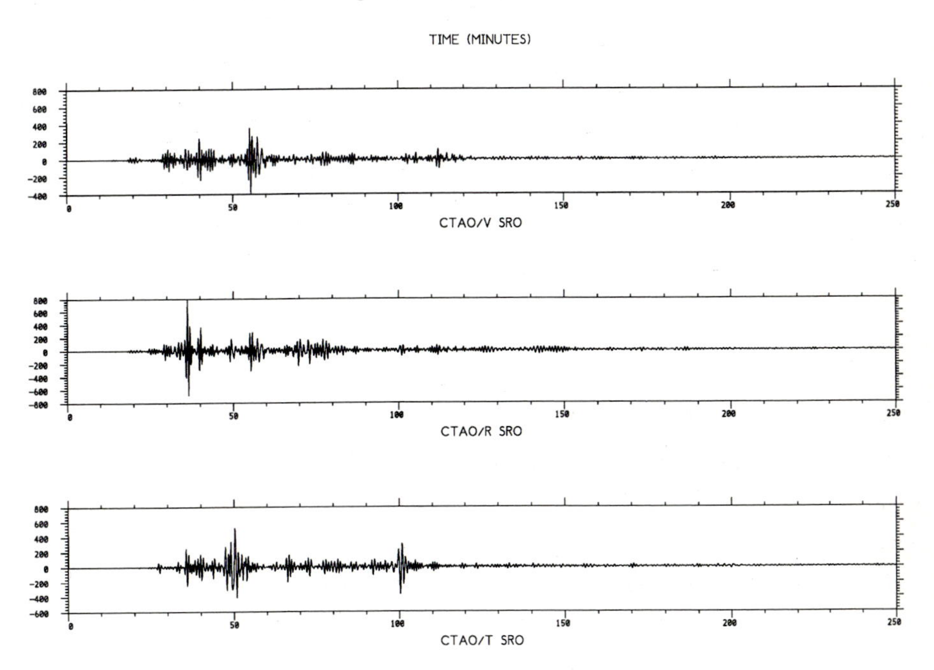

Figure 5.18: Power spectra for the three components of long period at CTAO. The fall-off at very low frequency is due to the instrument response of the *SRO*.

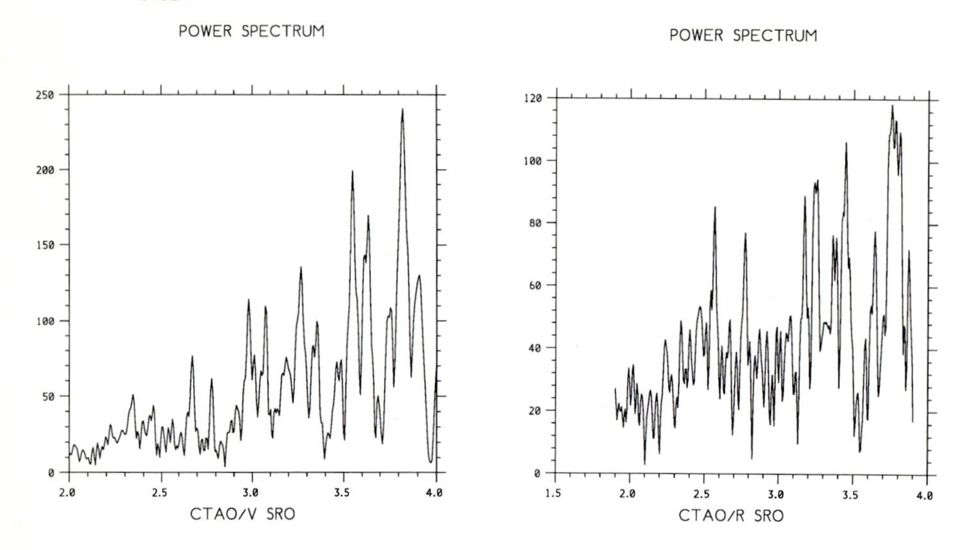

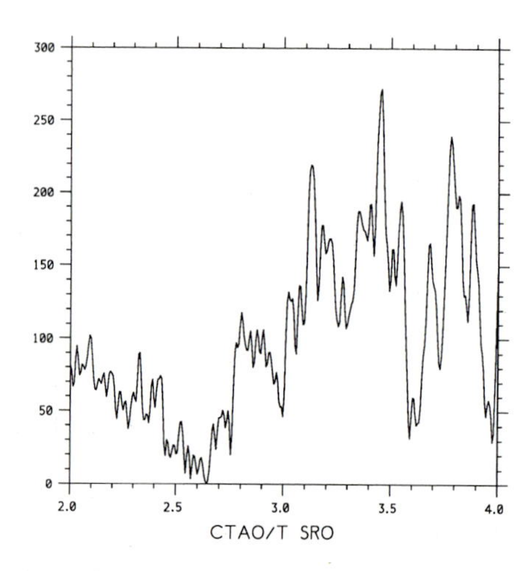

Figure 5.19: Power spectra for *IDA* station CMO. 6 is for higher frequencies; 7 for low frequencies. Frequencies are in mHz in both cases.

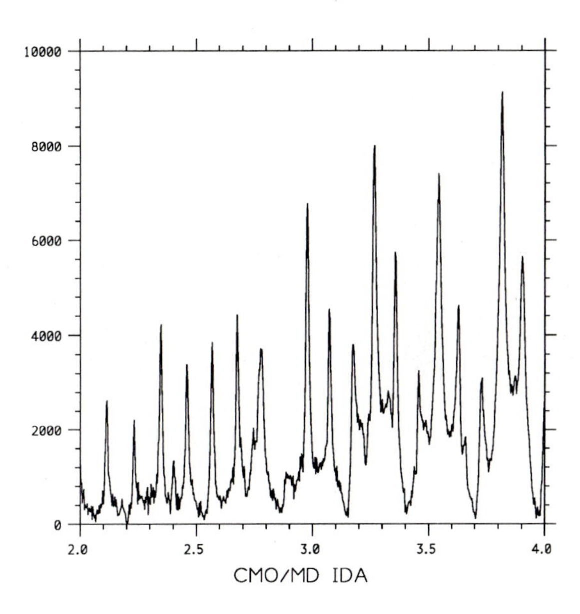

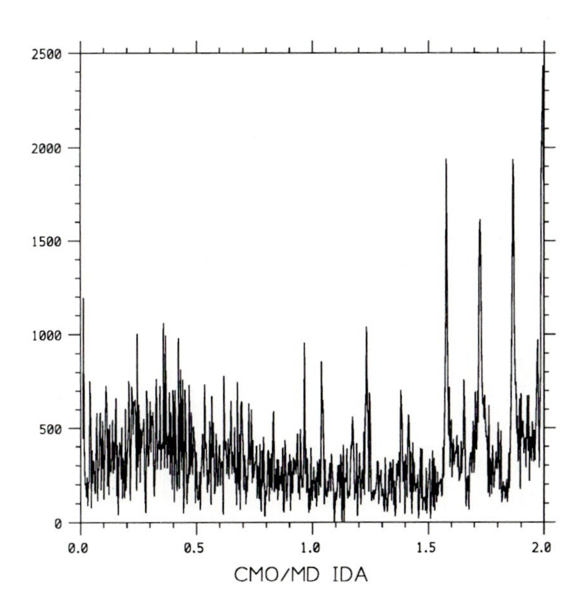

n	mode	ℓ	mHz
0	S	2	0.30943
0	T	2	0.38011
2	S	1	0.40386
0	S	3	0.46860
0	T	3	0.58736
0	S	4	0.64780
1	S	2	0.68034
0	T	4	0.76783
1	S	2	0.81296
0	S	5	0.84020
0	T	5	0.92976
1	S	3	0.94062
3	S	1	0.94266
2	S	2	0.95329
0	S	6	1.03783
0	T	6	1.08046
3	S	2	1.10723
1	S	4	1.17393
0	T	7	1.22243
0	S	7	1.23123
1	T	1	1.23886
2	S	3	1.24134
1	T	2	1.32206
0	T	8	1.35793
1	S	5	1.37138
2	S	4	1.37798
0	S	8	1.41278
4	S	1	1.41385
3	S	3	1.44044
1	T	3	1.44186
0	T	9	1.48858
2	S	5	1.51375
1	S	6	1.52242
0	S	9	1.57741
1	T	4	1.58751
0	T	10	1.61523
1	S	0	1.62948
1	S	7	1.65513
2	S	6	1.68034
4	S	2	1.72164
0	S	10	1.72546
5	S	1	1.73324
0	T	11	1.73887
1	T	5	1.75270
1	S	8	1.79829
0	T	12	1.86000
0	S	11	1.86129
3	S	4	1.86236
2	S	7	1.86507
1	T	6	1.92809
1	S	9	1.96213
6	S	1	1.97836
0	T	13	1.97909
0	S	12	1.98917
4	S	3	2.04756
2	S	8	2.04971
5	S	2	2.09219
0	T	14	2.09648
1	T	7	2.10658
0	S	13	2.11170
1	S	10	2.14622
5	S	3	2.17059
2	T	1	2.19159
0	T	15	2.21248
7	S	1	2.22607
2	S	9	2.22955
0	S	14	2.23017
2	T	2	2.23426
3	S	5	2.26899
4	S	4	2.27988
1	T	8	2.28326
2	T	3	2.29785
0	T	16	2.32733
1	S	11	2.34475
0	S	15	2.34521
5	S	4	2.37811
2	T	4	2.38212
2	S	10	2.40396
4	S	9	2.41156
6	S	2	2.43711
0	T	17	2.44122

Table 5.2: Normal mode frequencies tabulated for model 1066B

n	mode	ℓ	mHz
1	T	9	2.45562
0	S	16	2.45712
2	T	5	2.48656
2	S	0	2.50935
7	S	2	2.51882
3	S	6	2.54982
1	S	12	2.55169
0	T	18	2.55433
0	S	17	2.56612
2	S	11	2.57335
2	T	6	2.61067
1	T	10	2.62305
4	S	6	2.66672
0	T	19	2.66679
0	S	18	2.67241
3	S	7	2.68680
5	S	5	2.70222
2	S	12	2.73854
2	T	7	2.75346
1	S	13	2.76246
0	S	19	2.77617
0	T	20	2.77872
1	T	11	2.78608
6	S	3	2.82008
3	S	8	2.82045
8	S	1	2.87095
0	S	20	2.87764
0	T	21	2.89021
2	S	13	2.90088
2	T	8	2.91301
1	T	12	2.94558
3	S	9	2.95265
1	S	14	2.97233
0	S	21	2.97703
0	T	22	3.00134
5	S	6	3.01049
4	S	7	3.05863
2	S	14	3.06344
0	S	22	3.07459
3	S	10	3.08597
2	T	9	3.08616

n	mode	ℓ	mHz
6	S	4	3.09034
1	T	13	3.10234
0	T	23	3.11219
0	S	23	3.17054
9	S	1	3.20194
3	T	1	3.20755
8	S	2	3.21324
3	S	11	3.22240
0	T	24	3.22279
9	S	2	3.23427
2	S	15	3.23707
3	T	2	3.23797
1	T	14	3.25695
0	S	24	3.26511
6	S	5	3.26552
2	T	16	3.26881
3	S	9	3.27066
3	T	3	3.28337
8	S	3	3.28658
5	S	7	3.29203
0	T	25	3.33319
1	S	16	3.34005
3	T	4	3.34348
0	S	25	3.35845
3	S	12	3.36302
6	S	6	3.40152
1	T	15	3.40975
7	S	4	3.41407
3	T	5	3.41800
10	S	1	3.42176
2	S	16	3.43631
0	T	26	3.44344
4	S	8	3.44701
0	S	26	3.45088
2	T	11	3.45655
1	S	17	3.49569
3	T	6	3.50659
3	S	13	3.50812
5	S	8	3.52788
0	S	27	3.54245
6	S	7	3.54842

n	mode	ℓ	mHz
0	T	27	3.55357
9	S	3	3.55419
1	T	16	3.56098
3	T	7	3.60893
0	S	28	3.63336
2	T	12	3.64556
1	S	18	3.64654
2	S	17	3.64866
3	S	14	3.65744
7	S	5	3.65796
0	T	28	3.66360
11	S	1	3.68601
4	S	9	3.71014
1	T	17	3.71073
0	S	29	3.72373
3	T	8	3.72474
6	S	8	3.73225
8	S	4	3.74215
0	T	29	3.77355
1	S	19	3.79519
3	S	15	3.81043
0	S	30	3.81368
5	S	9	3.83293
2	T	13	3.83300
3	T	9	3.85372
1	T	18	3.85905
2	S	18	3.86441
4	S	18	3.86513
9	S	4	3.88137
0	T	30	3.88343
0	S	31	3.90331
1	S	20	3.94251
7	S	6	3.95585
6	S	9	3.96084
3	S	16	3.96643
0	S	32	3.99271
0	T	31	3.99328
3	T	10	3.99555

Summary

After a great earthquake (or a large meteorite impact for that matter) the earth rings like a bell; it has resonant frequencies just like a bell but with periods of many minutes. These are the earth's NORMAL MODES or FREE OSCILLATIONS. When the elastic parameters are spherically symmetric and depend only on radius we find there are two distinct sets of modes: SPHEROIDAL modes analogous to *P-SV* motion and TORSIONAL modes involving only shear motion and equivalent to *SH* motion. Each mode can be labelled with three integers that are analogous to quantum numbers for the orbits of electrons in an atom: the ANGULAR ORDER NUMBER, ℓ, which gives the number of nodal lines on the earth's surface, the OVERTONE NUMBER, n, which gives the number of nodal surfaces in depth within the earth, and the AZIMUTHAL ORDER NUMBER, m, which gives the number of nodal lines in longitude. The frequencies are the same for the same n and ℓ but different m: the modes are DEGENERATE. They are usually written as ${}_nS_\ell$ for a spheroidal mode or ${}_nT_\ell$ for a torsional mode. Modes with $n = 0$ are called FUNDAMENTALS; they are similar to surface waves. At high ℓ torsional modes become standing Love waves and spheroidal modes become standing Rayleigh waves.

Any ground response may be represented as a combination of the doubly-infinite set of normal modes; the seismogram has a dual representation either as a function of time or a set of amplitudes, one for each mode. The normal mode representation is most useful in describing the long period part of the seismogram because the spectrum becomes dense at high frequency and a great many modes must be summed to account for the short period part of the seismogram. Normal mode frequencies can be calculated analytically for a uniform sphere but a numerical computation is needed for real earth models. To observe normal modes you need an instrument that is sensitive to very long period ground motion, a big earthquake, and some data processing. Many hours of data must be used in the Fourier transform to show many succesive passes of the surface wave trains as they travel around and around the world every three hours or so; the Fourier transform shows many peaks at more or less regular intervals in frequency.

The degeneracy of the frequencies is caused by the earth's spherical symmetry: any departure from this spherical symmetry will split the degenerate MULTIPLET into individual SINGLETS. Splitting caused by the earth's rotation has been observed; the ellipticity of figure also causes splitting of the peaks. The width of an observed spectral peak is much larger than would be expected from damping of the mode: it is due mainly to splitting, the collection of split singlets appearing as a single broad peak because of lack of resolution. Splitting of the low-order modes can be observed directly.

The measured frequencies have been used to refine models of the elastic parameters within the earth. A normal mode has ground motion throughout the earth, unlike a body wave which has ground motion confined to a small region in the vicinity of its ray path, and therefore should give an earth model representative of the spherically-averaged earth. Fundamental modes that are equivalent to surface waves sample only the great circle path containing the source and receiver. A consistent geographic pattern to the frequency anomalies suggests a systematic departure from the spherical earth structure deep within the earth; a structure that does not appear to be related to the ocean-continent difference.

Further reading

Free oscillations are treated in BULLEN & BOLT (Capter 5.6) and AKI & RICHARDS (Chapter 8); there is also the book by LAPWOOD & USAMI. None of these books give an adequate account of the interesting processing methods used to extract the peak frequencies from the data, or separate close peaks. Thoroughly recommended, if you can find them, are the lecture notes on "An Introduction to Low-Frequency Seismology" by F. GILBERT, given at the Enrico Fermi School on Physics of the Earth's Interior in 1979, published by the Italian Physical Society; also lectures A. DZIEWONSKI & WOODHOUSE at the same summer school in 1982. The *SRO* network of seismometers is described by Peterson & Orsini (1976) and the *IDA* network by Agnew *et al.* (1976).

6

The seismic source

In this chapter we discuss how seismic waves are excited by earthquakes, how the elastic response can be calculated from a given earthquake source, the mathematical representation of the source, and how measurements of the teleseismic response can be used to determine the source parameters. There are many different models of the seismic source and we shall concentrate on only the simplest and most useful: slip across a small planar fault surface (shear faulting). This source emits a characteristic radiation pattern which, if measured, can be used to obtain the orientation of the fault plane and the direction of slip — the so- called *fault plane solution.* The reader who is interested only in the practical aspects of obtaining a fault plane solution can skip the mathematical derivation and proceed directly to Section 6.10, which contains a simple heuristic explanation of the radiation pattern for shear faulting. Although we shall restrict ourselves to one specific type of seismic source (albeit an important one), the mathematical development that provides the teleseismic response from a given source relies on simple physical principles and is given quite generally: little complexity is added by treating the general problem.

6.1 Uniqueness and reciprocal theorems

Consider an elastic body of either finite or infinite extent that is subject to body forces. We wish to calculate its displacement $\mathbf{u}(\mathbf{x}, t)$ by solving the equation of motion (2.30) subject to initial conditions $\mathbf{u}(\mathbf{x}, 0)$ and suitable boundary conditions (for an infinite medium conditions on $\mathbf{u}$ and the stresses at infinity must be imposed). The problem can be solved uniquely provided the displacement $\mathbf{u}$, or the traction $\mathbf{T}$, or a linear combination of the two, is specified on the boundary. This is the *uniqueness theorem* of elasticity and is proved in AKI & RICHARDS, p. 24. We already made implicit use of the theorem in Chapter 3 when we calculated the reflection coefficients for plane waves incident on a free surface: in that case the free surface had a boundary condition $\mathbf{T} = 0$ and $\mathbf{u}$ was not specified. The free surface boundary condition is the only one we shall make use of. An important corollary to the uniqueness theorem is that, because once **T** is specified on the boundaries then **u** is determined there, it is not possible to specify both **T** and **u** as boundary conditions. That would mean over-constraining the solution to the equation of motion.

Another important fundamental result is called the *reciprocal theorem.* Let **u** be the displacement at one point due to a body force **f** applied at another point. **g** is another body force parallel to **u** and applied at the same point **u** was measured, and **v** the corresponding displacement at the point of application of **f**. The reciprocal theorem states that **v** will be parallel to **f**. Thus **f** and **u** are interchangeable. This result depends only on the symmetry of the stress tensor, which in turn rests on the principle of conservation of angular momentum. It even applies in an anisotropic medium. The result is central to all seismic source theory, and therefore the proof is given here.

The equations of motion for body forces **f** and **g** with displacements **u** and **v** respectively are

$$\rho \ddot{u}_i = f_i + \frac{\partial \sigma^u_{ji}}{\partial x_j} \tag{6.1}$$

$$\rho \ddot{v}_i = g_i + \frac{\partial \sigma^v_{ji}}{\partial x_j} \tag{6.2}$$

where σ^u_{ij} is the stress tensor corresponding to displacement **u** and σ^v_{ij} that corresponding to displacement **v**. Now form the scalar product of (6.1) with **v**, (6.2) with **u**, integrate over the volume of the medium V, and subtract, to give

$$\begin{aligned} &\int_V (\mathbf{f} - \rho\ddot{\mathbf{u}}) \cdot \mathbf{v} dV + \oint_{\partial V} \mathbf{v} \cdot \mathbf{T}(\mathbf{u}, \mathbf{n})\, dS \\ &= \int_V (\mathbf{g} - \rho\ddot{\mathbf{v}}) \cdot \mathbf{u} dV + \oint_{\partial V} \mathbf{u} \cdot \mathbf{T}(\mathbf{v}, \mathbf{n})\, dS \end{aligned} \tag{6.3}$$

where **T** is the traction on the bounding surface. In deriving (6.3) we have used the results

$$\frac{\partial v_i}{\partial x_j}\sigma^u_{ji} = \frac{\partial v_j}{\partial x_i}\sigma^u_{ji} = \sigma^u_{ij} e^v_{ij} \tag{6.4}$$

(which relies on symmetry of the stress tensor), the identity

$$\begin{aligned} \int_V v_i \frac{\partial \sigma^u_{ij}}{\partial x_j} dV &= \int_V \left[\frac{\partial}{\partial x_j}\left(v_i \sigma^u_{ij} \right) - \frac{\partial v_i}{\partial x_j}\sigma^u_{ij} \right] dV \\ &= \oint_{\partial V} v_i \sigma^u_{ij} n_j dS - \int_V e^v_{ij}\sigma^u_{ij} dV \end{aligned} \tag{6.5}$$

(by the divergence theorem), and Hooke's law (2.83) to write the final integral in (6.5) as

$$\int_V C_{ijkl} e^u_{ij} e^v_{kl} dV$$

This integral appears in both expressions obtained from (6.1) and (6.2) and cancels on subtraction leaving (6.3).

Note that (6.3) holds even if $\mathbf{u}$ and $\mathbf{v}$ are expressed at different times. Let $\mathbf{u}$ be expressed at time t and $\mathbf{v}$ at time $\tau - t$, and integrate (6.3) over all time. The time integral can be taken inside the volume integrals (provided displacements are small) and the acceleration terms combine on integration by parts to give

$$\begin{aligned} &\int_{-\infty}^{\infty} \rho \left[\ddot{u}(t)\, v(\tau - t) - u(t)\, \ddot{v}(\tau - t)\right] dt \\ &= \rho \left[\dot{u}(t)\, v(\tau - t) - u(t)\, \dot{v}(\tau - t)\right]_{-\infty}^{\infty} \end{aligned} \tag{6.6}$$

which is zero if our medium is in an undisturbed initial state of rest. Define the convolution of two vector functions as

$$\begin{aligned} \mathbf{u} * \mathbf{g} &= \int_{-\infty}^{\infty} \mathbf{u}(\tau - t) \cdot \mathbf{g}(t)\, dt \\ &= \int_{-\infty}^{\infty} \mathbf{u}(t) \cdot \mathbf{g}(\tau - t)\, dt \end{aligned} \tag{6.7}$$

Then (6.3) gives

$$\int_V (\mathbf{u} * \mathbf{g} - \mathbf{v} * \mathbf{f})\, dV = \oint_{\partial V} [\mathbf{v} * \mathbf{T}(\mathbf{u}) - \mathbf{u} * \mathbf{T}(\mathbf{v})]\, dS \tag{6.8}$$

Equation (6.8) is a form of the reciprocal theorem; we shall see how it corresponds to the verbal statement of the theorem in a moment. Here we note the symmetry of $\mathbf{u}$ and $\mathbf{v}$ in (6.8). (6.8) applies to the elastic equation of motion; its counterpart for Laplace's equation is known as Green's theorem:

$$\int_V \left(\phi \nabla^2 \psi - \psi \nabla^2 \phi\right) dV = \oint_{\partial V} (\phi \nabla \psi - \psi \nabla \phi) \cdot \mathbf{n} dS \tag{6.9}$$

Green's theorem is used in potential theory to develop a representation of the solution to Laplace's equation as an integral over the boundaries; we shall use (6.8) to do the same thing for the elastodynamic equations.

6.2 The Green's tensor

A Green's function is the response of a system to an impulsive input. The counterpart in elastodynamics is the elastic response to an impulsive point force of unit strength. Both force and displacement are vectors; therefore a complete description of the response to a general point impulsive force is a tensor. We shall refer to it as Green's tensor. A unit point force in the m direction applied at the point $\mathbf{x} = \boldsymbol{\xi}_1$ and time $t = \tau_1$, is written

$$f_i = \delta\left(\mathbf{x} - \boldsymbol{\xi}_1\right) \delta\left(t - \tau_1\right) \delta_{im} \tag{6.10}$$

The corresponding elastic response is $G_{ij}\left(\mathbf{x}, t; \boldsymbol{\xi}_1, \tau_1\right)$, which satisfies the equation of motion

$$\begin{aligned} &\rho \frac{\partial^2}{\partial t^2} G_{im}\left(\mathbf{x}, t; \boldsymbol{\xi}_1, \tau_1\right) \\ &\quad = \delta\left(\mathbf{x} - \boldsymbol{\xi}_1\right) \delta\left(t - \tau_1\right) \delta_{im} + \frac{\partial}{\partial x_j}\left(C_{ijkl} \frac{\partial G_{km}}{\partial x_l}\right) \end{aligned} \tag{6.11}$$

and initial conditions:

$$G_{im}\left(\mathbf{x}, t; \boldsymbol{\xi}_1, \tau_1\right) = \frac{\partial G_{im}}{\partial t} = 0 \text{ if } t \leq \tau_1; \mathbf{x} \neq \boldsymbol{\xi}_1 \tag{6.12}$$

and homogeneous boundary conditions: either displacement G_{im} or traction $C_{ijkl}(\partial G_{km}/\partial x_l) n_j$ is zero on ∂V.

We apply the reciprocal theorem (6.8) with $\mathbf{f}$ given by (6.10) and $\mathbf{g}$ by

$$g_i = \delta\left(\mathbf{x} - \boldsymbol{\xi}_2\right) \delta\left(t + \tau_2\right) \delta_{in} \tag{6.13}$$

so $u_i(\mathbf{x}, t)$ becomes $G_{im}(\mathbf{x}, t; \boldsymbol{\xi}_1, \tau_1)$ and $v_i(\mathbf{x}, t)$ becomes $G_{in}(\mathbf{x}, t; \boldsymbol{\xi}_2, -\tau_2)$. The right hand side of (6.8) vanishes, for homogeneous boundary conditions, leaving

$$\begin{aligned} \int_{-\infty}^{\infty} dt \int_V dV \quad &\left[G_{im}\left(\mathbf{x}, \tau - t; \boldsymbol{\xi}_1, \tau_1\right) \delta\left(\mathbf{x} - \boldsymbol{\xi}_2\right) \delta\left(t + \tau_2\right) \delta_{in}\right. \\ &\left. - G_{in}\left(\mathbf{x}, \tau - t; \boldsymbol{\xi}_2, -\tau_2\right) \delta\left(\mathbf{x} - \boldsymbol{\xi}_1\right) \delta\left(t - \tau_1\right) \delta_{im}\right] \\ &= 0 \end{aligned} \tag{6.14}$$

Performing the integrations leaves

$$G_{nm}\left(\boldsymbol{\xi}_2, \tau + \tau_2; \boldsymbol{\xi}_1, \tau_1\right) = G_{mn}\left(\boldsymbol{\xi}_1, \tau - \tau_1; \boldsymbol{\xi}_2, -\tau_2\right) \tag{6.15}$$

and setting $\tau = 0$ leaves the desired relationship satisfied by the Green's tensor

$$G_{nm}(\boldsymbol{\xi}_2, \tau_2; \boldsymbol{\xi}_1, \tau_1) = G_{mn}(\boldsymbol{\xi}_1, -\tau_1; \boldsymbol{\xi}_2, -\tau_2) \tag{6.16}$$

This is a mathematical formulation of the reciprocal theorem as stated above. The left hand side gives the displacement in direction n in response to a force in direction m; the force is at position $\boldsymbol{\xi}_1$ and time τ_1 while the displacement is measured at $(\boldsymbol{\xi}_2, \tau_2)$. The right hand side gives the displacement in direction m (that of the force on the left hand side) in response to a force in direction n (that of the displacement on the left hand side); the force is applied at $(\boldsymbol{\xi}_2, -\tau_2)$ and the displacement at $(\boldsymbol{\xi}_1, -\tau_1)$. The minus signs arise from causality — displacements must follow the force. In fact for homogeneous boundary conditions the Green's tensor is a function only of $\tau_1 - \tau_2$, the time interval between measurement of the displacement and application of the force. This result is obvious physically, and evident mathematically since the governing equation (6.11) and the boundary conditions (6.12) depend only on the time difference $t - \tau$. We can therefore shift the time origin in (6.16) to give

$$G_{nm}(\boldsymbol{\xi}_2, \tau_2; \boldsymbol{\xi}_1, 0) = G_{mn}(\boldsymbol{\xi}_1, 0; \boldsymbol{\xi}_2, -\tau_2) \tag{6.17}$$

The nature of the reciprocity is illustrated in Figure 6.1.

We have introduced the concept of the Green's tensor and made use of it without evaluating it for a specific medium. Determining Green's tensors for the earth is a very complicated problem that lies at the heart of

Figure 6.1: The reciprocal theorem as given by equation (6.17). Suppose we know the displacement $\mathbf{u}_\mathrm{B}$ at point B due to a point force of unit strength at point A, $\mathbf{f}_\mathrm{A}$. Then (6.17) shows that the displacement at A, $\mathbf{u}_\mathrm{A}$, in response to a point unit force, $\mathbf{f}_\mathrm{B}$, applied at B in the direction of $\mathbf{u}_\mathrm{B}$ will be of strength $\mathbf{u}_\mathrm{B}$ in the direction of $\mathbf{f}_\mathrm{A}$, the time delay between $\mathbf{u}_\mathrm{A}, \mathbf{f}_\mathrm{B}$ being the same as that between $\mathbf{u}_\mathrm{B}, \mathbf{f}_\mathrm{A}$.

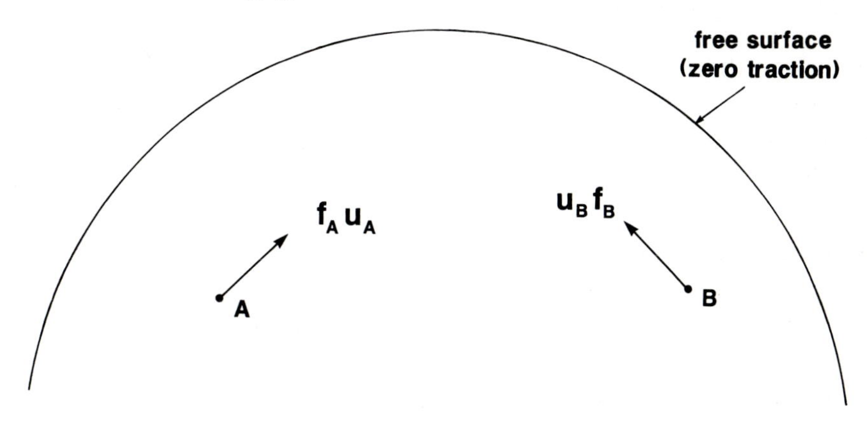

modern theoretical seismology. Later in this chapter we give the Green's tensor for a uniform isotropic medium. Before that, however, we make use of the general Green's tensor to separate the elastic response into a part due to the source and one due to propagation through the elastic medium — the latter effect being determined by the Green's tensor.

6.3 Representation theorems

The reciprocal theorem in the form (6.8) can be used to express the displacement as a volume integral over body forces plus surface integrals over displacement and traction. In (6.8) we take $\mathbf{g}$ to be a unit point force so that $\mathbf{v}$ is the appropriate Green's tensor:

$$g_i(\boldsymbol{\xi}, t) = \delta_{in}\delta(\boldsymbol{\xi} - \mathbf{x})\delta(t) \tag{6.18}$$

$$v_i(\boldsymbol{\xi}, t) = G_{in}(\boldsymbol{\xi}, t; \mathbf{x}, 0) \tag{6.19}$$

and corresponding tractions are

$$T_i(\mathbf{v}, \mathbf{n}) = n_j C_{ijkl} \frac{\partial G_{kn}}{\partial x_l} \tag{6.20}$$

$\mathbf{f}$ and $\mathbf{u}$ remain quite general in (6.8), which may be rearranged to give

$$\begin{aligned} u_n(\mathbf{x}, t) \quad = \quad & \int_{-\infty}^{\infty} d\tau \, \{ \int f_i(\boldsymbol{\xi}, \tau) G_{in}(\boldsymbol{\xi}, t-\tau; \mathbf{x}, 0) dV_\xi \\ & + \oint T_i \left[\mathbf{u}(\boldsymbol{\xi}, \tau), \mathbf{n}\right] G_{in}(\boldsymbol{\xi}, t-\tau; \mathbf{x}, 0) dS_\xi \\ & - \oint n_j C_{ijkl} u_i(\boldsymbol{\xi}, \tau) \frac{\partial}{\partial \xi_l} G_{np}\left(\boldsymbol{\xi}, t-\tau; \mathbf{x}, 0\right) dS_\xi \, \} \end{aligned} \tag{6.21}$$

The convolution has been made explicit in (6.21) for clarity; $\mathbf{x}$ and $\boldsymbol{\xi}$ have been exchanged to give the displacement at point $\mathbf{x}$.

Equation (6.21) enables $\mathbf{u}$ to be determined from body forces and surface displacement and tractions. At first sight it might seem that both displacement and traction are needed everywhere on the surface in order to evaluate $\mathbf{u}$, in contradiction to our earlier statement of the uniqueness theorem; but the Green's tensor also satisfies suitable boundary conditions which render knowledge of both $\mathbf{u}$ and $\mathbf{T}$ unnecessary. For example, if G_{in} were chosen to satisfy rigid (zero displacement) boundary conditions the first surface integral in (6.21) would vanish, leaving only the surface integral over displacement. Likewise, if the Green's function is chosen to satisfy stress-free boundary conditions the second surface integral vanishes, leaving only a surface integral over applied tractions.

6.4 Mathematical models of the earthquake source

We need a useful working model of the source of seismic waves. In constructing such a model it is important that details of the distant elastic radiation be reproduced; it is less important that the model match the actual physical processes governing faulting near the source itself. In fact most of our knowledge about the seismic source comes from observations of the elastic radiation some distance away. Only occasionally are fault breaks seen on the earth's surface, and even when they are seen the major part of the faulting takes place deep underground. The orientation of a fault may also vary with depth, making surface breaks misleading. We use a simple kinematic model of faulting. It is important to distinguish kinematics, which involve only the geometry of the fault and associated displacement, with the dynamics, which involves the forces at play. Dynamics of faulting is an important and difficult area of study that deals with questions such as how the material fractures under stress and how cracks propagate: it is not needed for the applications to seismology in this book.

Referring to the representation theorem (6.21) we note that both surface integrals vanish when the surfaces are stress free (or rigid) and the Green's tensor is chosen to satisfy the same boundary condition. We shall ignore the effect of the gravitational and all other body forces (see Chapter 2). The only contribution to the right hand side of (6.21) therefore comes from surfaces that do not have the same stress-free (or rigid) boundary condition as the Green's function.

A fault is now defined as a break in the elastic medium. There are two surfaces in contact across which tractions are continuous when the fault breaks. The displacement across the fault surface jumps giving a discontinuity $[\mathbf{u}(\mathbf{x},t)]$ across some area Σ drawn in the medium that occurs over the duration of the event. Later we shall simplify to events that are small in area and short in duration, but there is little to be gained by simplification at this stage.

Equation (6.21) becomes

$$u_n(\mathbf{x},t) = \int_{-\infty}^{\infty} d\tau \int_{\Sigma} \left[u_i\left(\boldsymbol{\xi},\tau\right)\right] C_{ijpq}\nu_j \frac{\partial}{\partial \xi_q} G_{pn}\left(\boldsymbol{\xi}, t-\tau; \mathbf{x}, 0\right) d\Sigma \tag{6.22}$$

where Σ represents the two faces of the crack, Σ^+ and Σ^-, $[\mathbf{u}]$ is taken to mean $\mathbf{u}^+ - \mathbf{u}^-$, and $\boldsymbol{\nu}$ is normal to the surface and points from Σ^+ towards Σ^-. The seismic source is therefore represented in terms of the fault surface Σ and the discontinuity in displacement or "throw" across it. No forces are involved in this description at all; it is purely kinematic.

6.5 Equivalent forces

By mathematical manipulation we may convert the surface integral in (6.22) to a volume integral that takes the form of the body force integral on the right side of (6.21). This gives a mathematical representation of the seismic source in terms of fictitious forces — called the *equivalent force system.* These forces could be applied to give the same elastic response as the faulting mechanism.

The easiest way to demonstrate an equivalent force system to (6.22) is to write down the force and verify its equivalence. The force is

$$f_i(\boldsymbol{\eta},\tau) = -\int_\Sigma [u_p(\boldsymbol{\xi},\tau)]\, C_{ijpq}\nu_j \frac{\partial}{\partial \eta_q}\delta(\boldsymbol{\eta}-\boldsymbol{\xi})\, d\Sigma_\xi \tag{6.23}$$

Substituting into the volume integral in (6.21) with β replacing $\boldsymbol{\xi}$ gives

$$-\int_{-\infty}^{\infty} d\tau \int_V dV_\eta G_{in}(\boldsymbol{\eta},t-\tau;\mathbf{x},0) \int_\Sigma [u_p(\boldsymbol{\xi},\tau)]\, C_{ijpq}\nu_j \frac{\partial}{\partial \eta_q}\delta(\boldsymbol{\eta}-\boldsymbol{\xi})\, d\Sigma_\xi \tag{6.24}$$

We must now do the η integration. The relevant part is

$$\int_V G_{in}(\boldsymbol{\eta},t-\tau;\mathbf{x},0)\frac{\partial}{\partial \eta_q}\delta(\boldsymbol{\eta}-\boldsymbol{\xi})\, dV_\eta \tag{6.25}$$

Writing the integrand as

$$\frac{\partial}{\partial \eta_q}[G_{in}\delta(\boldsymbol{\eta}-\boldsymbol{\xi})] - \frac{\partial G_{in}}{\partial \eta_q}\delta(\boldsymbol{\eta}-\boldsymbol{\xi}) \tag{6.26}$$

the first part gives a surface integral, by the divergence theorem, which is zero because of the boundary condition on G_{in}, while the second part gives the derivative of the Green's tensor evaluated at $\boldsymbol{\xi}$ by the substitution property of the delta function: (6.25) becomes

$$-\frac{\partial}{\partial \xi_q}G_{in}(\boldsymbol{\xi},t-\tau;\mathbf{x},0)$$

and (6.24) reduces to the right hand side of (6.22), as required.

We have shown that displacement across the fault surface, and the fault surface itself, may be replaced by a body force distribution that is concentrated onto the surface itself: the elastic displacement remains unchanged. This has been accomplished by a mathematical transformation: we have not attributed any physical reality or respectability to the equivalent forces. We return later to discuss equivalent forces in the context of shear faulting and a small source.

6.6 The moment tensor

A useful quantity that fully describes the faulting is the *moment density tensor* m, defined by

$$m_{pq}(\boldsymbol{\xi}, t) = [u_i(\boldsymbol{\xi}, t)]\, \nu_j C_{ijpq} \tag{6.27}$$

Then (6.22) becomes

$$u_n(\mathbf{x}, t) = \int_{-\infty}^{\infty} d\tau \int_{\Sigma} m_{pq}(\boldsymbol{\xi}, \tau) \frac{\partial G_{pn}}{\partial \xi_q}(\boldsymbol{\xi}, t-\tau; \mathbf{x}, 0)\, d\Sigma_{\xi} \tag{6.28}$$

The displacement is therefore an integral over the fault surface of the convolution of moment density tensor and gradient of the Green's tensor. We have reached our desired aim of separating the effects of source and propagation: the source part being contained in m and propagation effects in G.

It is now time to specialise to small impulsive sources, which is the only type of source that will concern us. If the fault surface Σ is sufficiently small then $\partial G_{pn}/\partial \xi_q$ in (6.28) will be constant throughout the integral. (6.28) then becomes

$$u_n(\mathbf{x}, t) = M'_{pq} * \frac{\partial}{\partial \xi_q} G_{np}(\boldsymbol{\xi}, t; \mathbf{x}, 0) \tag{6.29}$$

where $\boldsymbol{\xi}$ now denotes the point of occurrence of the source, and

$$M'_{pq}(t) = \int_{\Sigma} m_{pq}(\boldsymbol{\xi}, t)\, d\Sigma_{\xi} \tag{6.30}$$

is called the *moment tensor* for the small source. For a point source we need only specify the moment tensor and not the distribution of the moment density tensor over the fault surface. We can be quite specific of the term "small" as applied to the source: the Green's tensor must not vary significantly across the fault surface Σ, otherwise it cannot be regarded as a constant and be removed from the integral in (6.28). G describes the waves that propagate in the elastic medium and its variability is determined by the shortest wavelength. The source must therefore be smaller than the shortest wavelength considered.

We shall also consider exclusively impulsive sources; these have a step-function time dependence. With

$$M'_{pq}(t) = M_{pq} H(t) \tag{6.31}$$

the convolution in (6.29) becomes a multiplication

$$u_n(\mathbf{x}, t) = M_{pq} \int_0^\infty \frac{\partial}{\partial \xi_q} G_{pn}(\xi, t-\tau; \mathbf{x}, 0)\, d\tau \tag{6.32}$$

Note, from the definition of m (6.27) and the symmetry properties of the elasticity tensor, that M is symmetric. M has six independent elements which determine the source completely. The assumption of a small impulsive source is inadequate for many real earthquakes. Fault breaks from the great Alaskan earthquake of 1964 ran for about 1000 km; the source region is clearly many times larger than the wavelength of one second P waves. Larger earthquakes are frequently observed to emit complicated waves as if they were in fact multiple events with several fault breaks occurring at once. Still another common observation is the "stopping phase" which appears as a second pulse a short time after the first onset: it is interpreted as emission of radiation when fault slip ceases. Little has been done with distant observations of seismic radiations in elucidating these more complex sources. It is a fascinating subject but beyond the scope of this book.

6.7 Shear faulting

Most earthquake source mechanisms have no measurable change in volume associated with them: the displacement jump across the fault surface then lies in the fault surface, or

$$[\mathbf{u}] \cdot \boldsymbol{\nu} = 0 \tag{6.33}$$

This condition defines *shear faulting.* We also specialise to isotropic media; substituting (2.82) into equation (6.27) for the moment density tensor gives

$$m_{pq} = \mu\,(\nu_p [u_q] + \nu_q [u_p])$$

Applying (6.33) for shear faulting eliminates the first term to leave

$$m_{pq} = \mu\,(\nu_q [u_q] + \nu_q [u_p]) \tag{6.35}$$

Integrating over the planar fault surface and assuming μ and $[\mathbf{u}]$ to be uniform across it gives the moment tensor

$$M_{pq} = \mu U A\,(\nu_p [\hat{u}_q] + \nu_q [\hat{u}_p]) \tag{6.36}$$

where $[\hat{\mathbf{u}}]$ is a unit vector parallel to $[\mathbf{u}]$, which we shall call the *slip vector*, and U is the magnitude of $[\mathbf{u}]$, the *throw* across the fault, and A its area.

This seismic source is specified completely by two unit vectors describing the fault and slip geometry, $\boldsymbol{\nu}$ and $[\hat{\mathbf{u}}]$, and a single scalar describing the "size" of the fault, μUA, called the *seismic moment.* The fault geometry is alternatively specified by three angles: the *dip*, *strike* and *rake* (or *plunge*). The dip is the angle between the fault plane and the horizontal. The strike is the angle between true north and the intersection of the fault plane with the horizontal, measured clockwise from north (Figure 6.2). If the fault breaks the surface the strike gives the direction of the break. These two angles specify the fault plane. For shear faulting the slip vector lies in the fault plane and needs just a single angle for its specification: either the rake measured upwards from the horizontal, or plunge if downwards from the horizontal.

The seismic moment determines the intensity of the emitted seismic radiation and is therefore the best measure of the "size" of an earthquake, at least as far as the elastic radiation is concerned. It is a more logical indication of the size of an event than the earthquake magnitude, which is based on an arbitrary measure of the radiated elastic energy (either one second body waves, as described in Section 3.1, which give a poor indication of the size of a large earthquake, or surface waves)

A simple example of a shear fault is shown in Figure 6.3. The normal to the fault plane is in the x_3 direction and the slip vector is in the x_1 direction. If x_3 lies along the vertical and x_1 points north this fault will

Figure 6.2: Angles used in specifying an arbitrary fault geometry. $ABCD$ is the fault plane. $PQRS$ is a horizontal plane. AE lies in the horizontal plane perpendicular to AB. $\boldsymbol{\nu}$ is normal to the fault plane; $[\mathbf{u}]$ lies in the fault plane. ϕ is the strike, δ the dip, λ the rake, measured upwards from the horizontal.

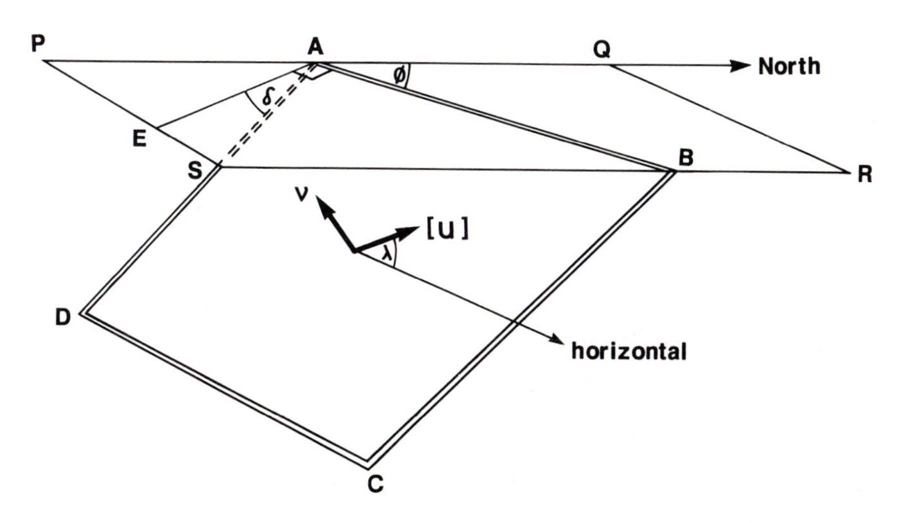

have zero dip and zero rake (the strike in this case is undefined). The corresponding moment tensor is, from (6.36),

$$\mathrm{M} = M_0 \begin{pmatrix} 0 & 0 & 1 \\ 0 & 0 & 0 \\ 1 & 0 & 0 \end{pmatrix} \tag{6.37}$$

where $M_0 = \mu UA$, the seismic moment. The displacement response follows from (6.32)

$$u_n(\mathbf{x}, t) = M_0 \left(\frac{\partial \overline{G}_{n1}}{\partial x_3} + \frac{\partial \overline{G}_{n3}}{\partial x_1} \right) \tag{6.38}$$

where the bar denotes time integration as in (6.32). The equivalent force system is defined in (6.23). Specialising to small sources, (6.23) may be written

$$F_p(\mathbf{x}, t) = -M_{pq} \frac{\partial}{\partial x_q} \delta(\mathbf{x} - \boldsymbol{\xi})\, \delta(t) \tag{6.39}$$

which for the moment tensor (6.37) with source at the origin ($\boldsymbol{\xi} = 0, t = 0$) becomes

$$\begin{aligned} F_1 &= -M_0 \frac{\partial}{\partial x_3} \delta(\mathbf{x})\, \delta(t) \\ F_2 &= 0 \\ F_3 &= -M_0 \frac{\partial}{\partial x_1} \delta(\mathbf{x})\, \delta(t) \end{aligned} \tag{6.40}$$

Figure 6.3: The geometry for simple shear faulting

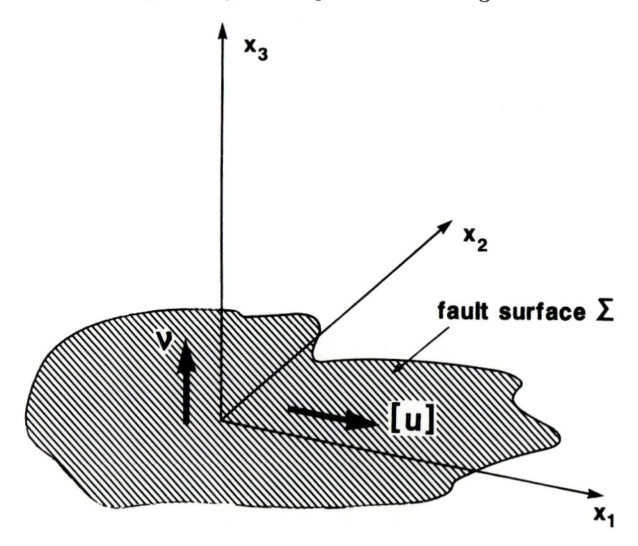

Each component of force is zero on average. The derivative of a delta function is a double spike and each of F_1, F_3 represents a couple as shown in Figure 6.4. The two couples act in opposite senses and are equal in magnitude: they cancel, leaving zero net moment. This force distribution is called a *double couple.*

The *double couple* source is the simplest possible equivalent force system for an intrinsic source and it is worth reviewing to examine the underlying principles. The earth must obey the principle of conservation of momentum: an extrinsic source like a meteorite impact will change the earth's momentum but an intrinsic source like earthquake faulting cannot. Its equivalent force system must therefore have zero net force component. Likewise an intrinsic source cannot change the earth's angular momentum; the equivalent force distribution must therefore have zero net moment. The simplest force distributions satisfying both these conditions is a pair of couples, oppositely directed. The principle of conservation of momentum entered our mathematical derivation of (6.40) via the equation of motion. Conservation of angular momentum entered the derivation through symmetry of the stress tensor, which is at the heart of the reciprocal theorem (6.16).

6.8 Green's tensors for a uniform medium

It should be clear from the preceding sections that knowledge of the Green's tensor allows calculation of the displacement response for an arbitrary source. Calculation of the Green's tensor for general elastic media is a very difficult problem: we shall restrict ourselves to uniform media. The full derivation is given in AKI & RICHARDS (Section 4.1) and

Figure 6.4: The double couple equivalent force system.

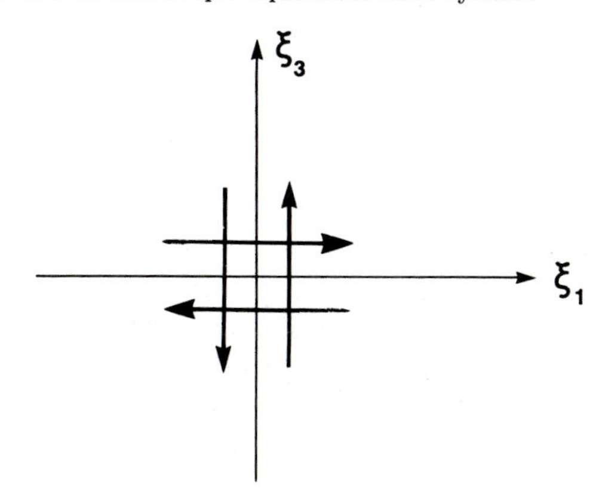

we do not repeat it here. There is one complication worth mentioning, the existence of *near-field* and *far-field* radiation. Consider a scalar spherical wave emitted by a point source. The potential for a spherical wave has the general form

$$\Phi = \frac{f\left(t - r/c\right)}{r} \tag{6.41}$$

where c is the wave speed. The elastodynamic potential for P waves would have the form (6.41) in a liquid. The displacement is the gradient of the potential

$$u_i = \frac{\partial \Phi}{\partial x_i} = -\frac{x_i f}{r^3} - \frac{x_i f'}{c r^2} \tag{6.42}$$

The first term on the right hand side of (6.42) falls off as r^{-2}, the second as r^{-1}. Therefore the second term will dominate the radiated energy at large distances. Most observations in seismology, and all the observations we shall deal with, are in the range where the second term dominates, the far-field radiation, and the near-field radiation can be neglected. We shall ignore all but the far-field terms henceforth.

The Green's tensor for a source at the origin is, with the far-field approximation,

$$\begin{aligned} G_{ij}\left(\mathbf{x}, t; 0, 0\right) &= \frac{1}{4\pi\rho\alpha^2} \frac{x_i x_j}{r^3} \, \delta\left(t - r/\alpha\right) \\ &+ \frac{1}{4\pi\rho\beta^2} \left(-\frac{x_i x_j}{r^3} + \frac{\delta_{ij}}{r}\right) \delta\left(t - r/\beta\right) \end{aligned} \tag{6.43}$$

The corresponding form for a source at $\boldsymbol{\xi}, \tau$ is obtained simply by replacing $\mathbf{x}$ with $\mathbf{x} - \boldsymbol{\xi}$ and t with $t - \tau$. The first term on the right hand side of (6.43) gives the P wave response. It gives an expanding spherical wave travelling with the P wave velocity α. The dependence on $x_i x_j$ ensures the motion is longitudinal and satisfies the reciprocal theorem (6.17). The normalising constant $\left(4\pi\rho\alpha^2\right)^{-1}$ is needed for a unit point force.

The second term gives the S wave response, which is again an expanding spherical wavefront. It travels with the S wave velocity β. The dependence on $\left(x_i x_j / r^2 - \delta_{ij}\right)$ ensures the motion is transverse. The normalising constant $\left(4\pi\rho\beta^2\right)^{-1}$ corresponds to a point force.

The form (6.43) is derived in full using elastodynamic potentials by AKI & RICHARDS. Unfortunately it is not possible to substitute (6.43) into the equation of motion to verify their validity because we have neglected near-field terms in deriving it. An alternative derivation, simpler than that of AKI & RICHARDS, is given by HUDSON.

6.9 P wave radiation from a shear fault

Consider P wave radiation from the shear fault shown in Figure 6.3, with normal to the fault plane along x_3 and slip along x_1. Radiation from a shear fault of arbitrary orientation can be deduced subsequently by rotation of axes. The displacement is given by (6.38) with Green's tensor given by the first part of (6.43), the part associated with the P wave. Again we must apply the far-field when differentiating (6.43): only the derivative of the delta function contributes.

Since from (6.43), for the P wave

$$G_{in}^P = \frac{1}{4\pi\rho\alpha^2}\frac{x_i x_n}{r^3}\,\delta\left(t-\frac{r}{\alpha}\right) \tag{6.44}$$

then

$$\frac{\partial G_{1n}}{\partial x_3} = -\frac{1}{4\pi\rho\alpha^3}\frac{x_1 x_n}{r^3}\frac{x_3}{r}\delta'\left(t-\frac{r}{\alpha}\right) \tag{6.45}$$

where δ' denotes the derivative of the delta function, a double pulse. The same expression holds for the other derivative, $\partial G_{3n}/\partial x_1$, appearing in (6.38). Integrating with respect to time and substituting into (6.38) gives

$$u_n^P = \frac{2M_0}{4\pi\rho\alpha^3}\frac{x_1}{r}\frac{x_3}{r}\frac{x_n}{r}\frac{1}{r}\delta\left(t-\frac{r}{\alpha}\right) \tag{6.46}$$

which is more conveniently expressed in spherical polar coordinates as shown in Figure 6.3:

$$\begin{aligned} \frac{x_3}{r} &= \cos\theta \\ \frac{x_1}{r} &= \sin\theta\cos\phi \\ \frac{x_n}{r} &= (\hat{\mathbf{r}})_n \end{aligned} \tag{6.47}$$

where $\hat{\mathbf{r}}$ is the unit vector in the radial direction. (6.46) becomes

$$\mathbf{u}^P = \frac{M_0\delta(t-r/\alpha)}{4\pi\rho\alpha^2}\frac{1}{r}\sin 2\theta\cos\phi\,\hat{\mathbf{r}} \tag{6.48}$$

Equation (6.48) shows that P wave radiation is emitted in four equal quadrants delineated by *nodal surfaces* where displacement is zero: at $\theta = \pi/2$ and $\phi = \pi/2$ (and $\phi = 3\pi/2$). The quadrants $\theta \le \pi/2,\ |\phi| \le \pi/2$ and $\theta > \pi/2,\ |\phi| > \pi/2$ have positive initial motion; the other two have negative initial motion. Note that one of the nodal surfaces is an extension of the fault surface: the plane with normal $\boldsymbol{\nu}$. It is called the *fault plane.* The other nodal surface is a plane with normal along x_1, the direction of slip. It is called the *auxiliary plane.*

6.10 The fault plane solution

A simple explanation of the P wave radiation pattern was given for a thrust fault in Chapter 1 (see also Figure 1.6). The fault is drawn in section and the fault plane extends perpendicular to the paper. The dashed line represents a second plane perpendicular to the fault plane and normal to the direction of slip across the fault, which is indicated by arrows. This second plane is the auxiliary plane. An observer in one of the two quadrants marked with plus signs sees first motion away from him, while an observer in one of the two quadrants marked with minus signs will see motion away from him; an observer on either the fault or the auxiliary plane will see no motion. The results derived earlier in the section show that propagation of ground displacement by elastic radiation preserves this sense of initial motion: a seismometer in either of the plus quadrants will record initial motion vertically upwards; while one in either of the negative quadrants will record downward vertical motion.

The polarity of the initial displacement at a seismometer is called its *first motion.* By recording the polarities of the first motions on a global distribution of seismometers we can separate the two nodal lines and define the fault and auxiliary planes. This is the basic principle by which we find the source mechanism of an earthquake using teleseismic data. Any phase can be used in principle: the S waves, pP or PP reflections etc., but we shall use only the first arrival P because first motions of later phases are more difficult to read.

Consider an earthquake at some depth and draw a small sphere around its focus. This sphere is called the *focal sphere.* Seismic rays emanating from the focus will cut the focal sphere on their way to their receiving stations as shown in Figure 6.5. The focal sphere is drawn so small that the refraction suffered by the rays before they cut the sphere can be neglected: they can be regarded as straight line segments within the focal sphere. The point on the sphere where the ray crosses can be calculated from the angle of incidence and the azimuth of the receiver from the focus; the angle of incidence can be found from Δ, the angular distance of the receiver from the focus, and a travel time table using the methods described in Chapter 4.

By reading the vertical component instrument of a receiving station we can determine whether the first motion is *compressional* (up) or *dilatational* (down) and mark it on the corresponding point on the focal sphere. Provided we can cover the focal sphere with a sufficient density of points we will then be able to divide the surface of the focal sphere into four equal quadrants, separating dilatational and compressional first motions. Receiving stations that happen to lie on or near nodal surfaces will register zero or very small first motions; these stations are particularly helpful in determining the precise location of the nodal lines separating our quadrants because we know the lines must pass somewhere nearby.

In practice we do not draw points on an actual sphere; we project the sphere onto a plane surface and carry out all the measurements on a plane sheet of paper. Even for deep events almost all the rays travel downwards initially, and are refracted back upwards to their receiving stations by the increase of seismic velocity with depth, and so virtually no points plot on the upper half of the focal sphere. Only the lower hemisphere need be projected. The projection used varies; two popular ones are the stereographic projection or Wulff net used in the practical in Chapter 4 for locating earthquakes, or the Schmidt net or Lambert equal area projection used in the practical in this chapter. Projections are discussed in the Appendix.

The geometry of the nodal planes (the fault and auxiliary planes) and associated vectors are drawn in Figure 6.6. The slip vector lies in the fault plane and is normal to the auxiliary plane; likewise the normal to the fault plane must lie in the auxiliary plane. The tension axis points

Figure 6.5: H is the focus at the centre of a small imaginary sphere called the *focal sphere*. HS is a ray path and S' the projection of S onto the focal sphere. S'' is the *apparent position* of the station S as viewed from the focus; it is at greater distance than S because of refraction.

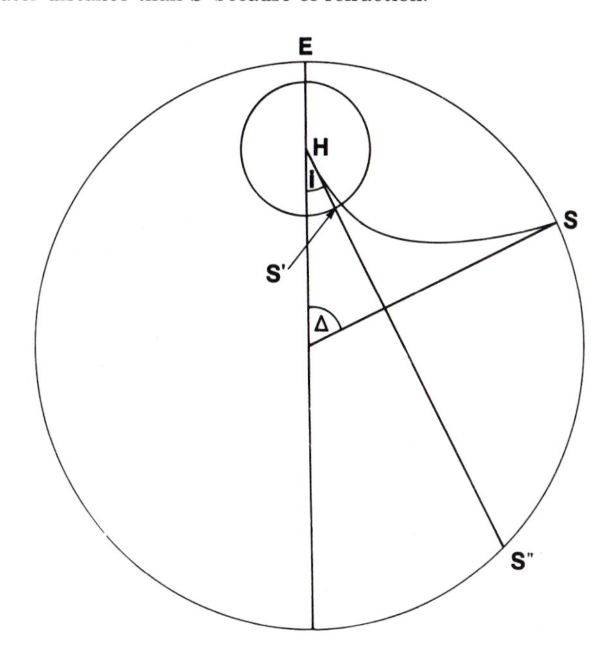

in the direction of maximum compression; it is orthogonal to the line of intersection of the two planes and at 45° to both slip vector and normal to the fault plane. The tension axis points in the direction of maximum dilatation.

Now consider the projection of Figure 6.5. The nodal planes cut the focal sphere in great circles which map into arcs marked on the nets in the projections. The slip vector and normal to the fault plane cut the focal sphere at points. The slip vector lies in the fault plane and therefore in projection it lies on the arc representing the fault plane. Likewise the normal to the fault plane lies in the auxiliary plane and in projection lies on the arc representing the auxiliary plane (Figure 6.6). P and T axes lie on the great circle containing slip vector and normal to the fault plane, and are 90° from the points denoting the intersection of the two nodal planes.

Fault plane solutions are by convention drawn with the compressional quadrants shaded. Focal mechanisms for the three common types of fault are shown in Figure 6.7. Note yet again that we cannot determine which of the two planes is the fault plane; this can only be found by observation of the fault break, correlation with the known tectonics of

Figure 6.6: Elements of the fault plane solution in stereographic projection. Great circles represent the nodal planes and plot as circles in the projection. Slip vector (S) and normal to the fault plane (N) lie on the nodal lines; Pressure (P) and Tension (T) axes lie 45° away along the great circle path through S and N.

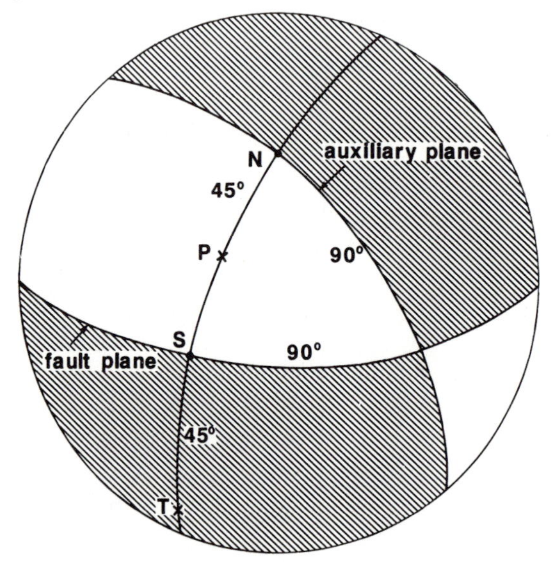

the area, or correlation with fault planes from aftershocks or other events in the locality. Another method will be given in the next chapter.

6.11 Shear wave radiation from a fault

Until recently only P wave first motions were read in finding the fault plane solution. Poor station distribution often led to inadequate determinations of the nodal surfaces. Often this problem can be resolved by reading the first motion of S waves, which have a far-field radiation pattern that is quite different from P waves. Again we use (6.38) with, this time, the second term in (6.43) for the Green's tensor. (6.38) gives

$$u_n = \frac{M_0\delta\,(t-r/\beta)}{4\pi\rho\beta^3}\frac{1}{r}\,(\delta_{n1}\gamma_3 + \delta_{n3}\gamma_1 - 2\gamma_1\gamma_3\gamma_n) \tag{6.49}$$

where $\gamma_n = x_n/r$, the direction cosines of $\mathbf{x}$. It is easily verified that $u_n\gamma_n = 0$, i.e. the radial component in the spherical coordinate system centred on the fault with axis along $\boldsymbol{\nu}$ in Figure 6.3 is zero. The θ component becomes

$$u_\theta = -\frac{u_3}{\sin\theta} = \frac{M_0\delta\,(t-r/\beta)}{4\pi\rho\beta^3}\frac{1}{r}\cos 2\theta\cos\phi \tag{6.50}$$

and the ϕ component

$$\begin{aligned} u_\phi &= (u_\theta\cos\theta\cos\phi - u_1)\,/\sin\phi \\ &= -\frac{M_0\delta\,(t-r/\beta)}{4\pi\rho\beta^3}\frac{1}{r}\cos\theta\sin\phi \end{aligned} \tag{6.51}$$

The radiation pattern is illustrated in Figure 6.8. Note that there are no nodal lines for S motion.

6.12 Synthetic seismograms

A large part of modern theoretical seismology is devoted to the calculation of seismograms from models of the source and elastic constants. Most of the early seismological discoveries of the structure of the earth's interior were made using only travel time data. It seems wasteful to discard most of the wavetrain — indeed, its very complexity indicates that the waves have sampled some complicated, and therefore interesting, structure somewhere along their path. The new digital data can be processed to give more information about the wavetrain in its entirety, which has accelerated theoretical efforts to account for more and more detail in the seismogram.

Figure 6.7: Three types of focal mechanism

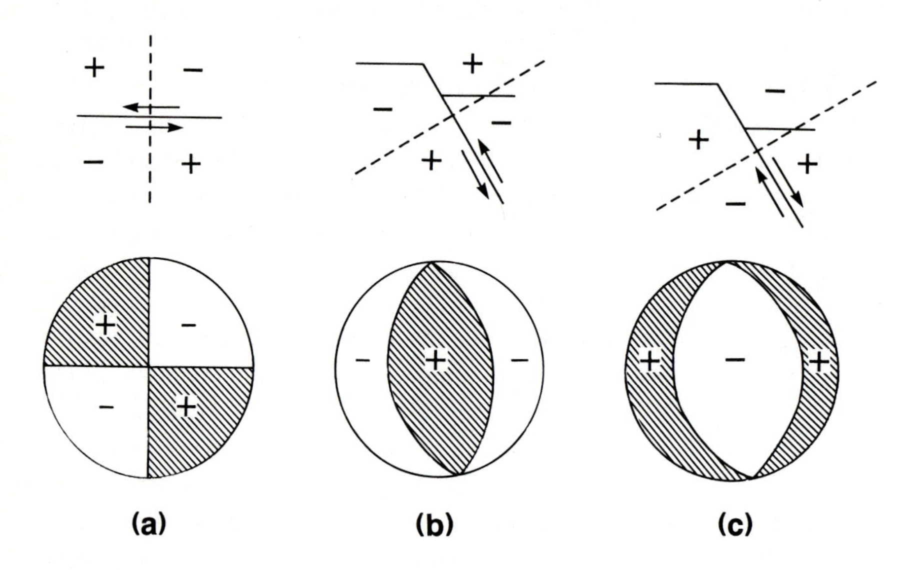

Figure 6.8: Radiation pattern for shear waves

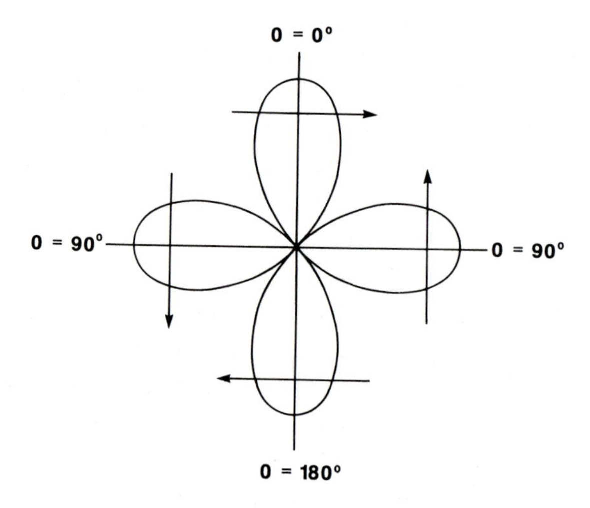

Equation (6.28) shows that the ground motion is the convolution of a source term, described by the moment density tensor, and a propagation term, contained in the Green's tensor. The seismogram is the convolution of the ground motion with the instrument response of the seismometer (Section 3.1). There are therefore three component parts to a synthetic seismogram. The instrument response is the easiest of these to account for: it is either available from the manufacturer or (better) it can be measured directly by giving an impulse to the instrument and recording the response, as is done routinely for the *WWSSN* instruments.

The source effects are described by the six elements of the moment density tensor, which are functions of time and position on the fault plane. Often seismograms from several different sites show similar features which can be attributed to a source effect — a double event for example, very common for larger events. Usually we are interested in either the source or some peculiarity of the propagation path, and it is uncommon to study complex sources and complex Green's functions at the same time.

There are three different approaches to calculating the Green's function; each one is an extension of one of the simple cases discussed in earlier chapters.

1 Ray theory provides the simplest and most effective means of calculating a synthetic seismogram. Geometrical optics (Section 3.13) can account for refraction and determine where nodal planes are located on the earth's surface; it can therefore be used to predict whether a certain phase will be large at a particular receiver. It can also be used to combine several phases at a receiver to produce a composite seismogram. Geometrical optics preserves the shape of the input pulse, and the synthetic seismogram for an impulsive source will be the appropriate combination of arrivals (including reflections, for example) convolved with the instrument response. This simple method can represent simple teleseismic responses very well indeed, and has been useful in separating close phases.

   Ray theory can be developed beyond the infinite frequency limit to include effects of physical optics, diffraction, and more complex propagation effects. This is an active area of research, but already a number of observations have been well matched. The advantage of ray methods is their speed of computation for long ray paths, and that the final seismogram will contain only the rays that were put into it. With other, more complete, methods of calculating synthetic seismograms the interpretation of the synthetic can be as difficult as interpreting the real thing!

2 When the medium contains discontinuities the waves reverberate and quite complex seismograms result. This is very common in the crust where there are often many layers with thicknesses comparable with the wavelengths of the waves. The *reflectivity method* is based on the formulae for transmission and reflection of waves from interfaces developed in Sections 3.4–6. A monochromatic wave will be reflected and transmitted from every interface. All contributions to the final waveform are summed, and all frequencies are summed over, within restricted time and frequency windows, to produce the final seismogram. This is a popular method in interpreting crustal seismic refraction surveys, where the propagation is over a limited distance and the layers have a restricted depth. It is not so efficient for global problems when a great many more reverberations may need to be taken into account. The propagation through a stack of layers can be followed by repeatedly applying scattering matrices of the type (3.51) and (3.65) to the input wave. The formalism is developed in KENNETT.

3 For periods greater than about 40 s we can sum all the normal modes for a particular model of the elastic parameters for the whole earth. This is the best method for long period synthetic seismograms. Modal summation is quite fast but the initial computation of the modes is not. The method is therefore rather clumsy when several different earth models are studied. The number of modes with periods less than 40 s increases very rapidly indeed, limiting the use of the method to long period. Further developments are likely to come from using asymptotic forms for the modes, where specific short period modes are selected as relevant to the particular observation on the seismogram that is to be synthesised. It is possible to reproduce long period seismograms very accurately indeed. The approach is now used routinely to match long period waveforms and determine moment tensors and hypocentral coordinates.

Practical 4. Making a fault plane solution

The records written at a large number of stations by the earthquake in the north Aegean sea, which was located in the practical in Chapter 4, are needed to determine the fault plane solution. The stations have been plotted on the focal sphere in Schmidt projection in Figure 6.9.

1 Read each record and find if the sign of the initial P wave is up (compressional — C) or down (dilatational — D) and mark them in red (C) or blue (D) on the station plot. Use only the long period

records to read the polarities because the short period arrivals suffer from scattering from local anomalies and give unreliable results, but read the arrival times on the short period records to make sure you have identified the first motion on the long period. If you have difficulty about a sign put a question mark on the plot — it may be close to a nodal plane.

2 Draw in one of the nodal planes that separates the blue and red points using the grid provided. It may help to sketch some likely candidates first before deciding on the best arc.
3 Draw in the normal to this first nodal plane by aligning it on the grid and measuring 90° along the axis of the grid. The second, presumably less well determined plane, must pass through this point, so it will be helpful in finding the second plane.
4 Draw in the second nodal plane. Adjust the position of the first one if necessary to obtain a better fit, but ensure that it passes through the normal to the first plane.
5 Draw in the normal to the second plane.
6 Plot the P and T axes by measuring 45° along the great circle joining the two normals.
7 Re-examine the waveforms of stations close to the nodal surfaces: they often have a characteristic shape.
8 Measure the dip and strike of both possible fault planes, and the direction of both possible slip vectors.

Figure 6.9: Focal sphere for Practical 4

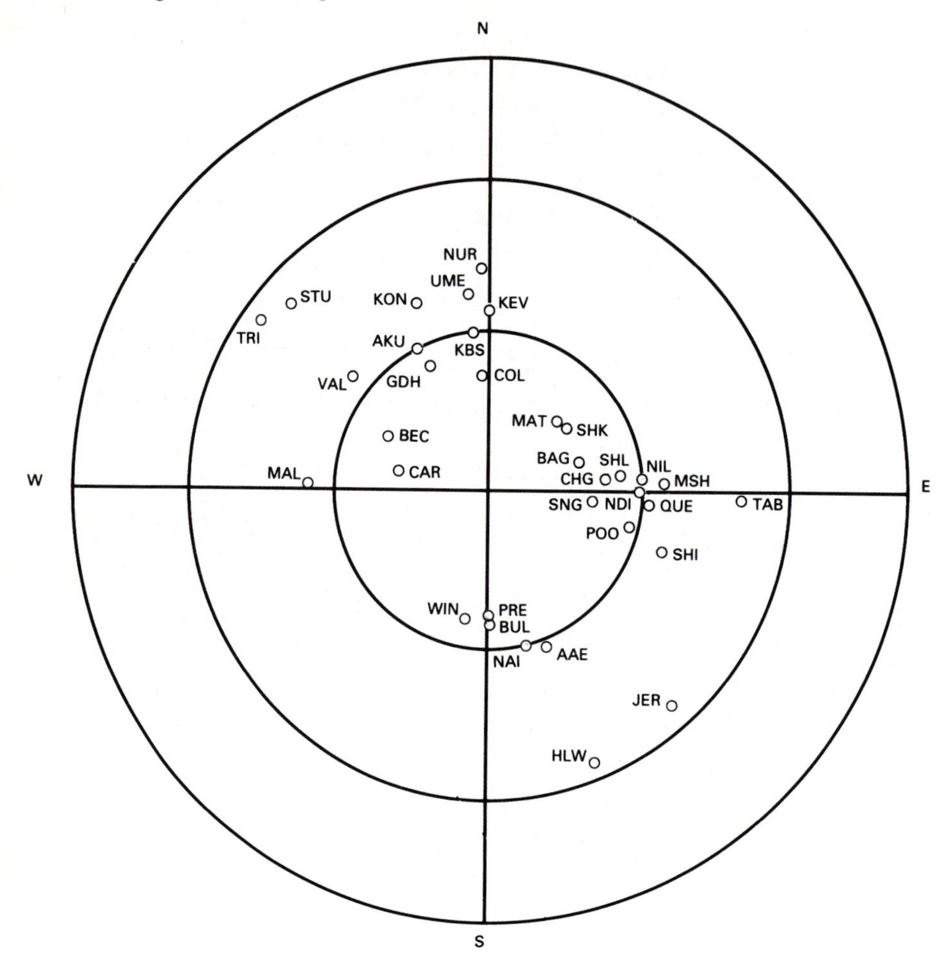

Figure 6.10: Schmidt net for Practical 4

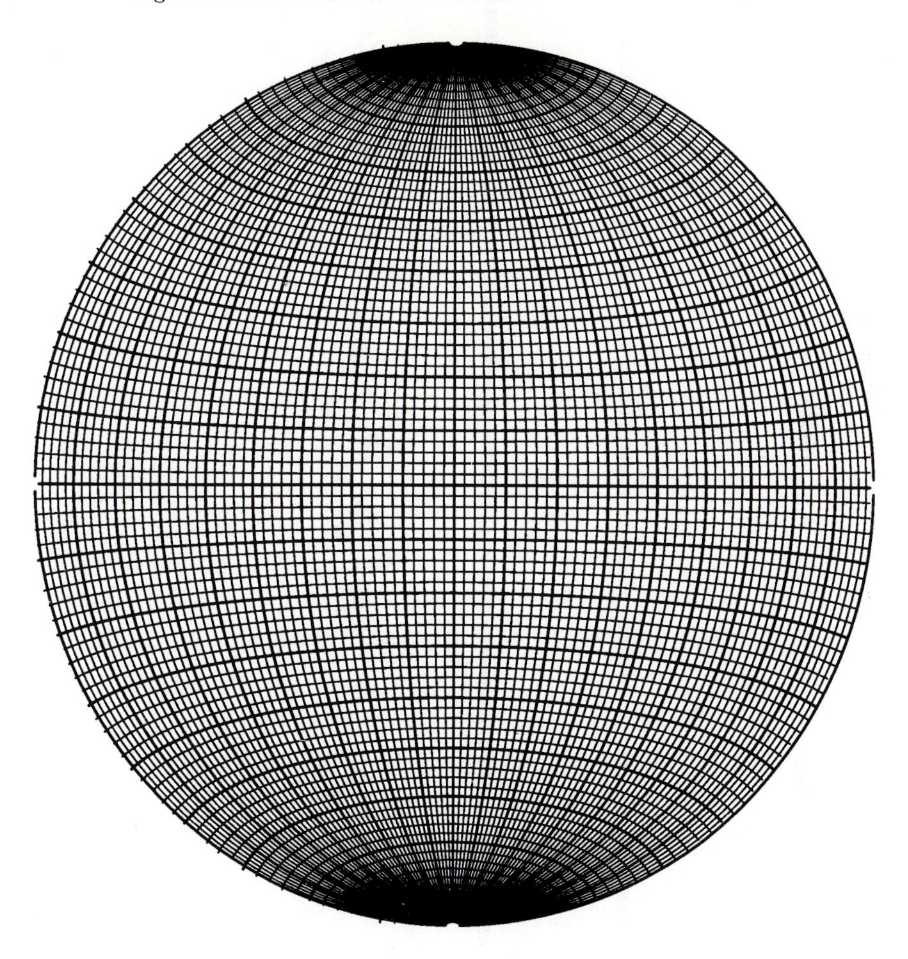

Figure 6.11: Seismogram for Practical 4. AAE(4). All seismograms in this practical have been reduced to half linear dimension.

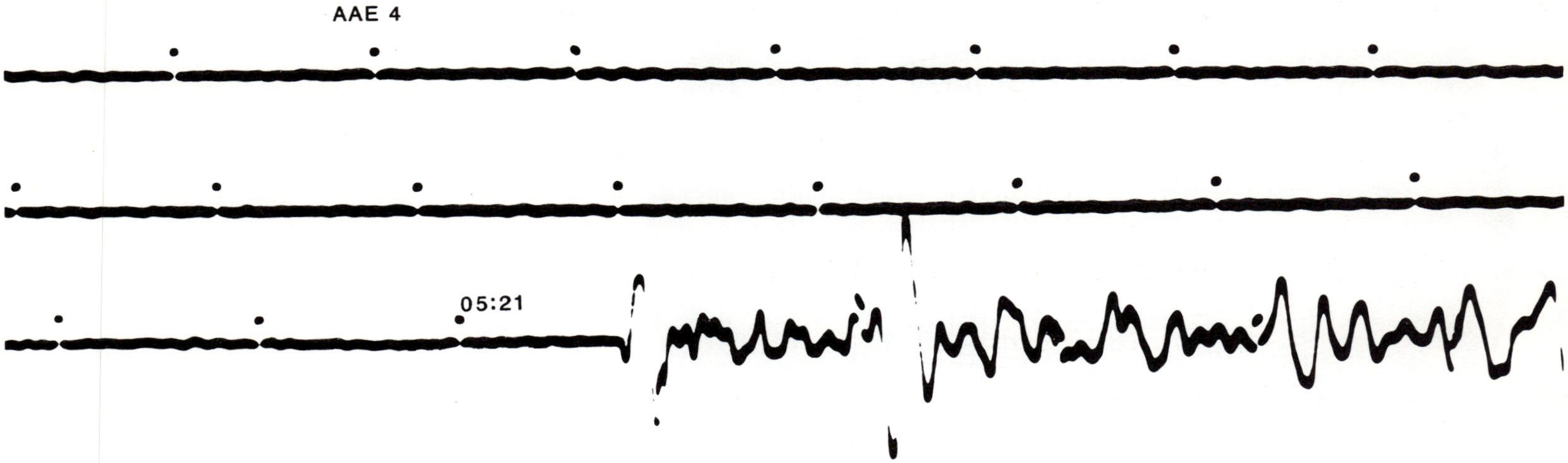

AAE 4

05:21

Figure 6.12: Seismogram for Practical 4. AKU(4)

Figure 6.13: Seismogram for Practical 4. BUL(1,4)

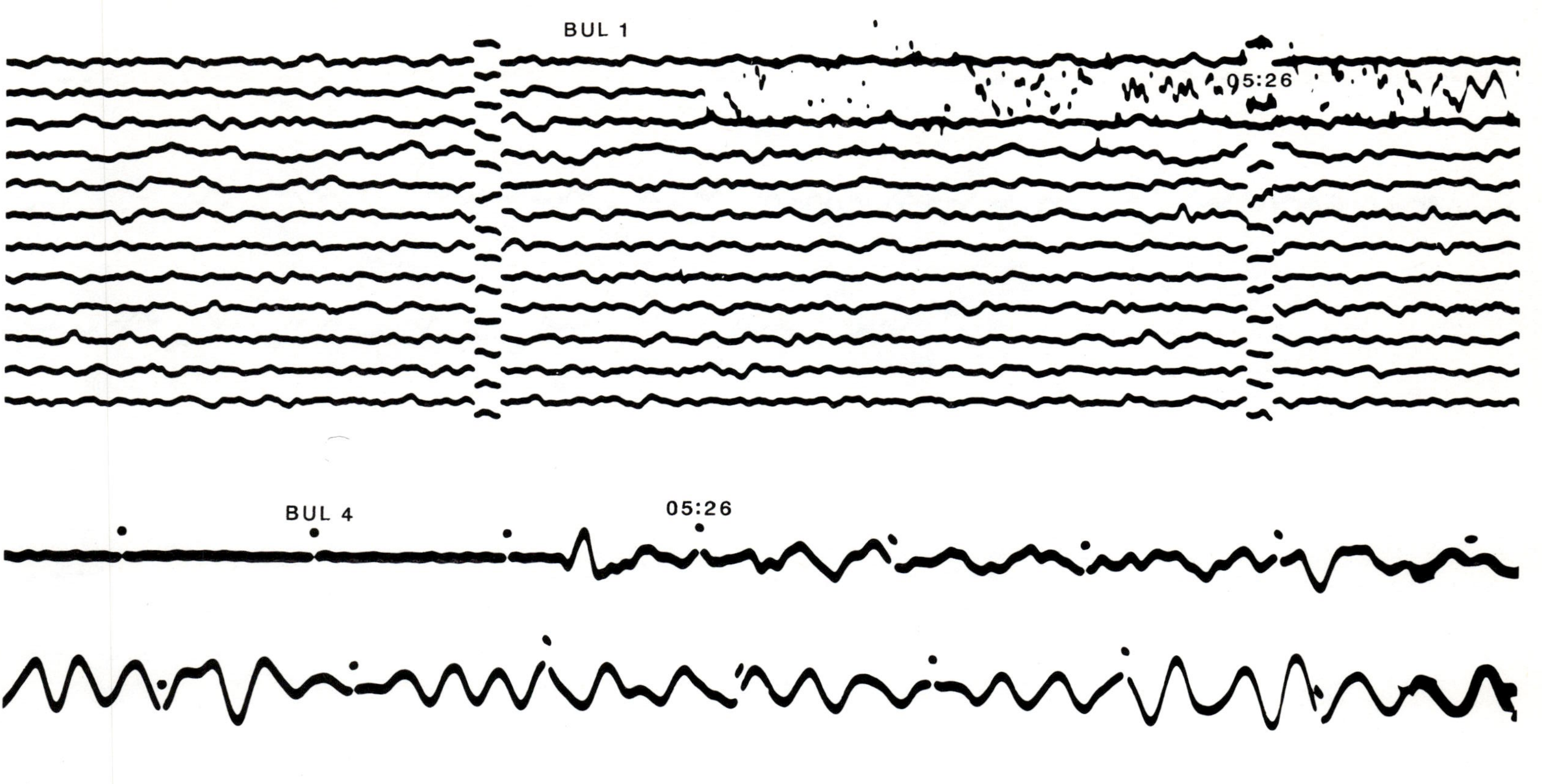

Figure 6.14: Seismogram for Practical 4. BAG(1,4)

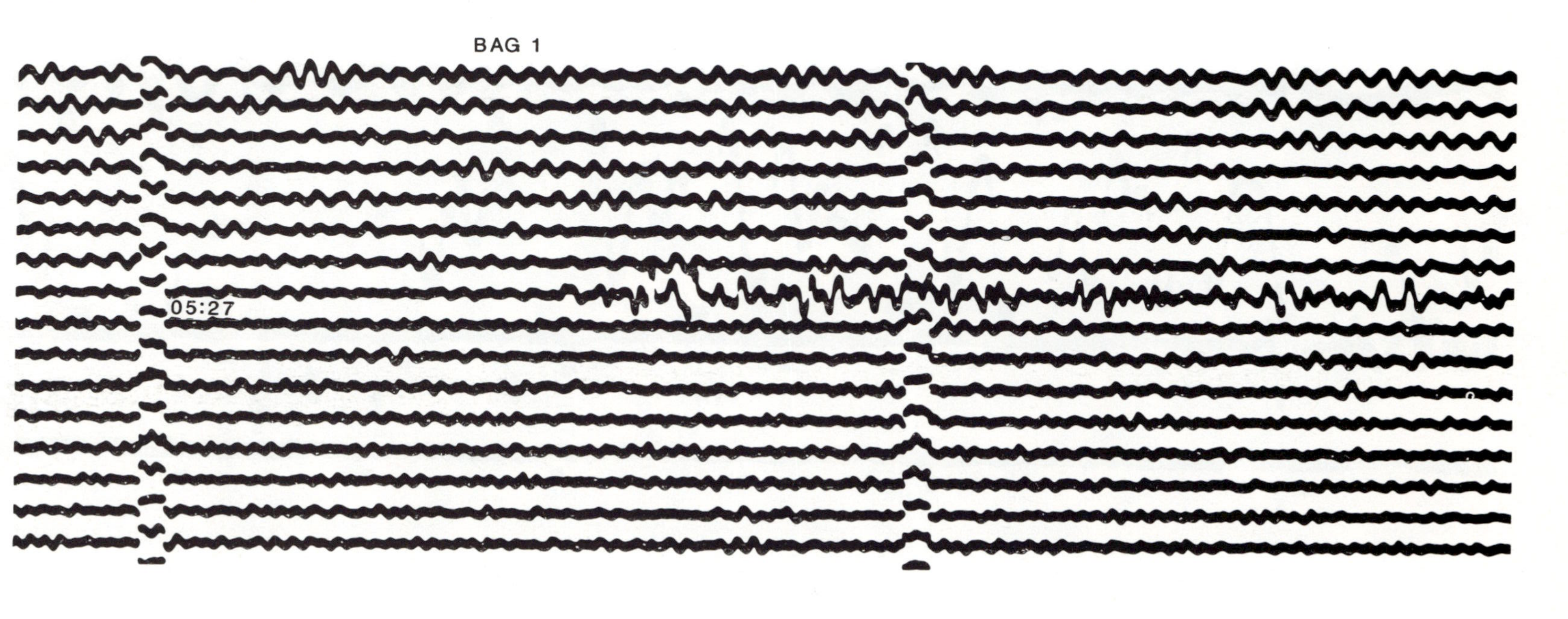

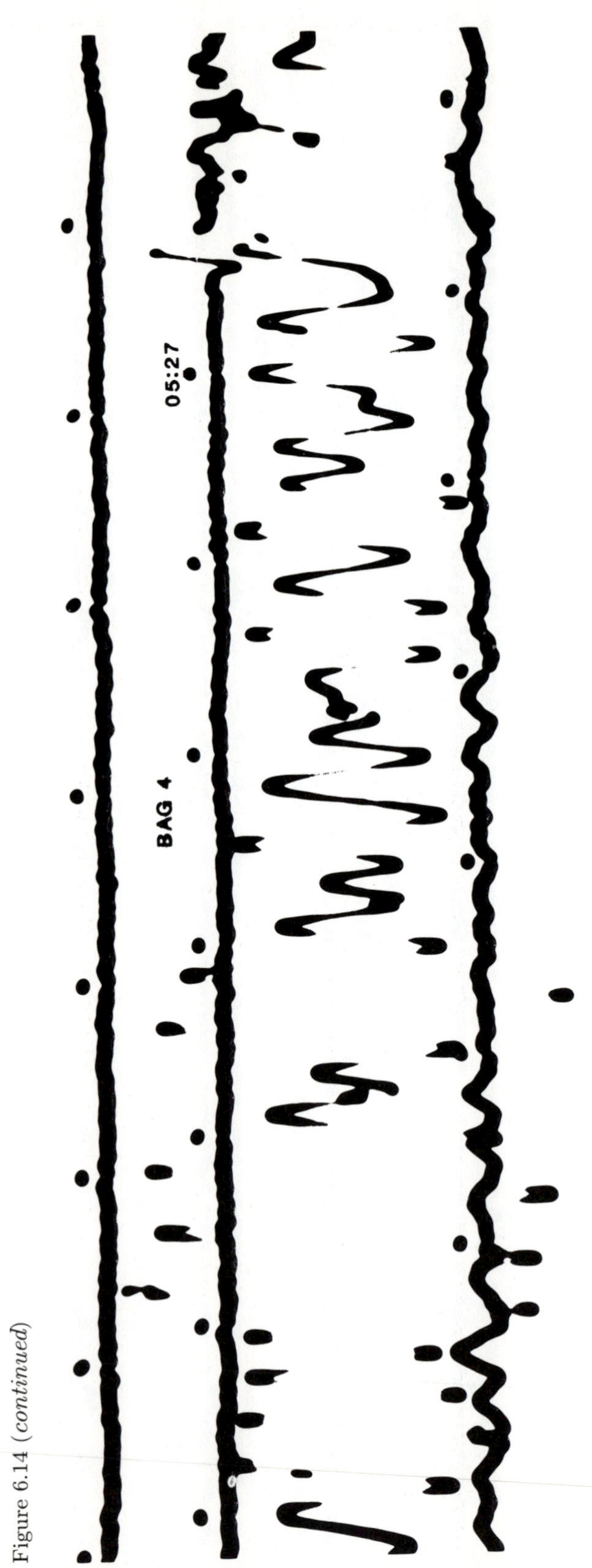

Figure 6.14 (*continued*)

Figure 6.15: Seismogram for Practical 4. BEC(4)

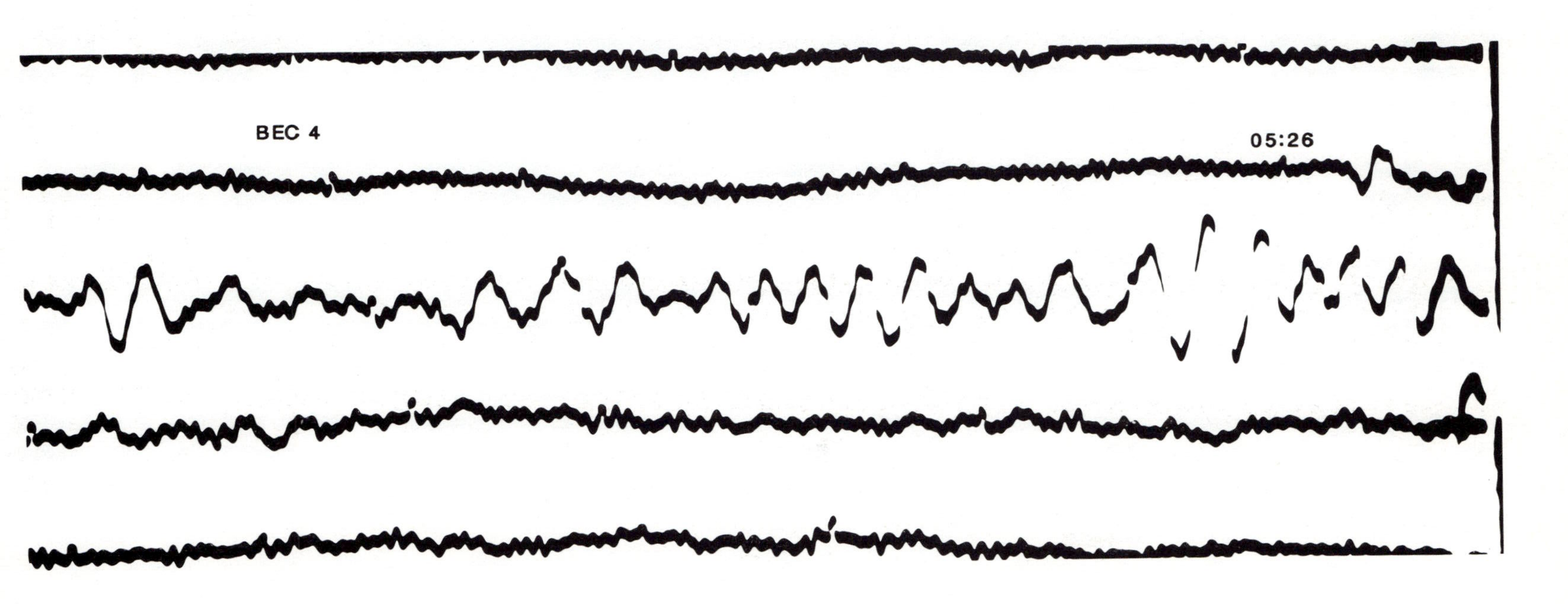

Figure 6.16: Seismogram for Practical 4. CAR(1,4)

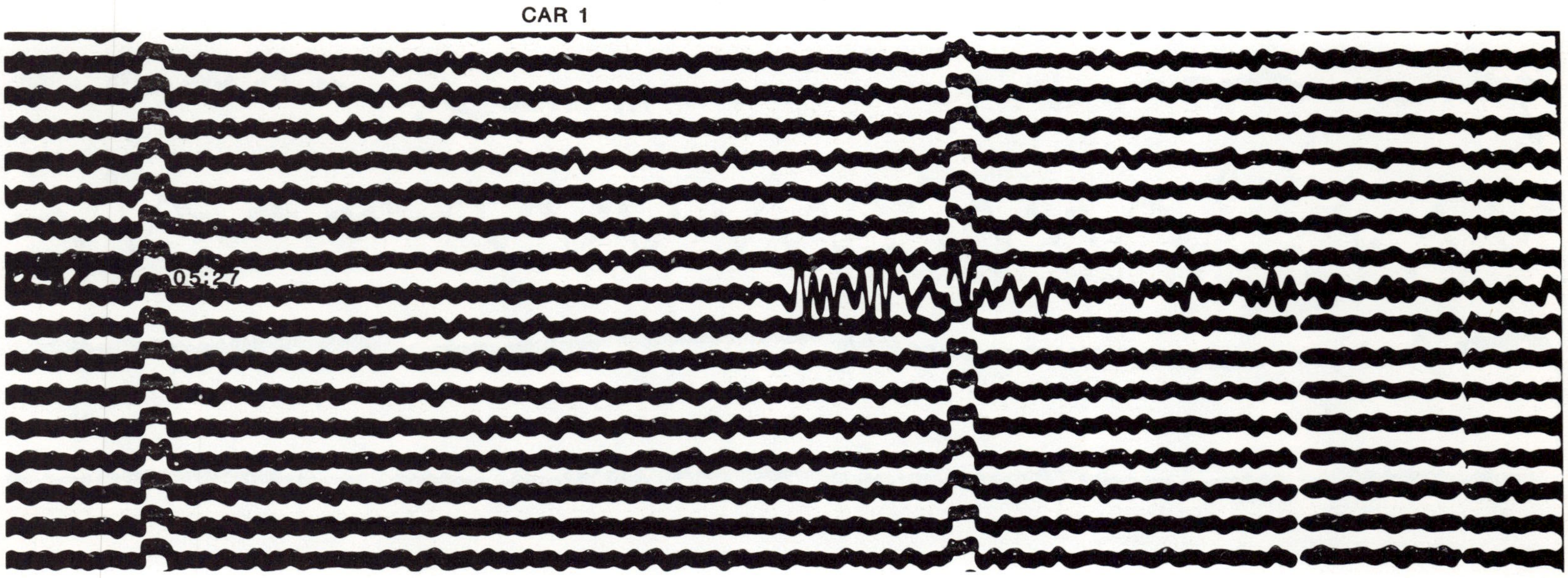

CAR 4
05.27

CHG 1

05:25

Figure 6.17: Seismogram for Practical 4. CHG(1,4)

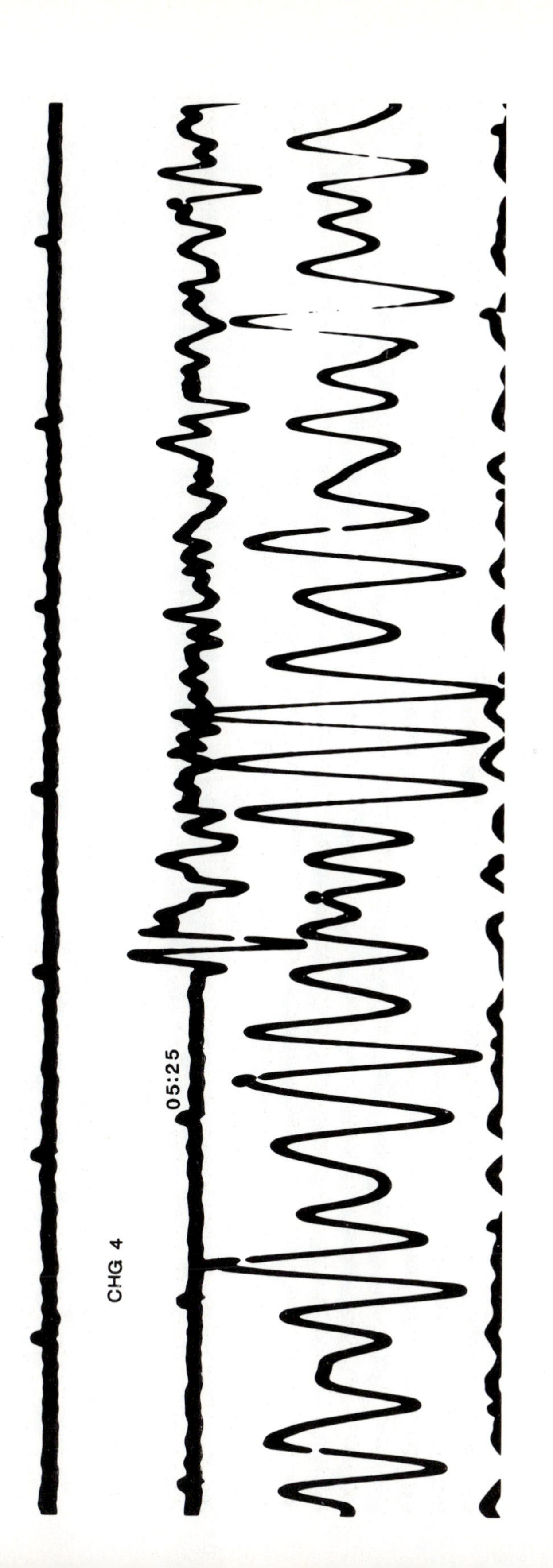
CHG 4
05:25

COL 1

05:26

Figure 6.18: Seismogram for Practical 4. COL(1,4)

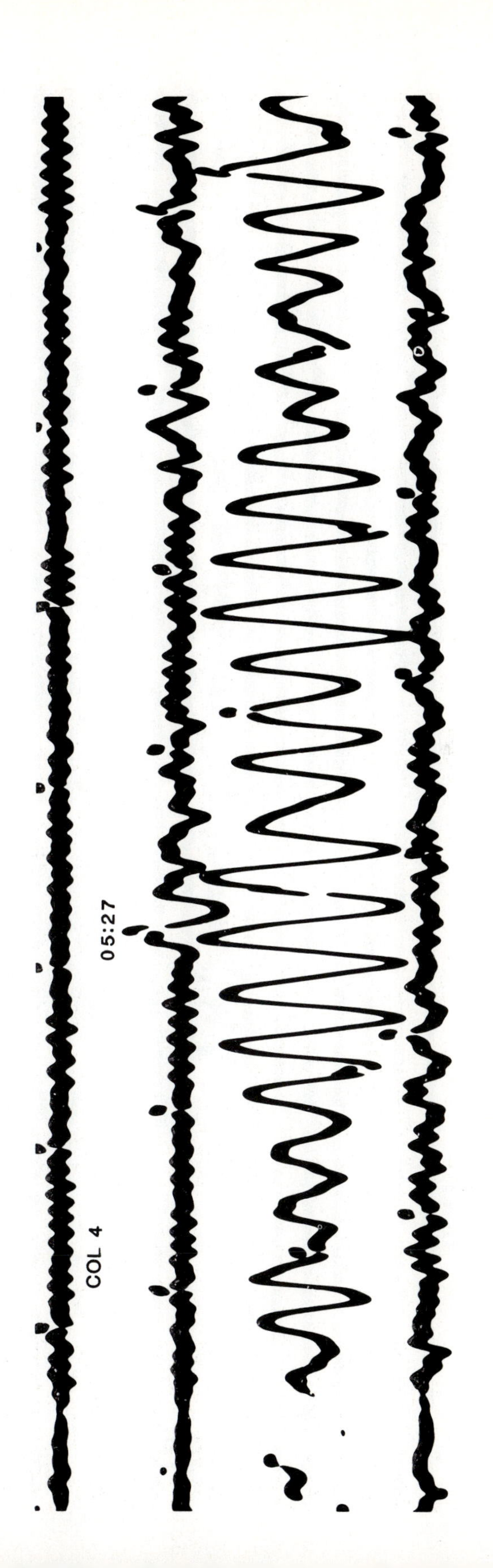
COL 4
05:27

Figure 6.19: Seismogram for Practical 4. GDH(1,4)

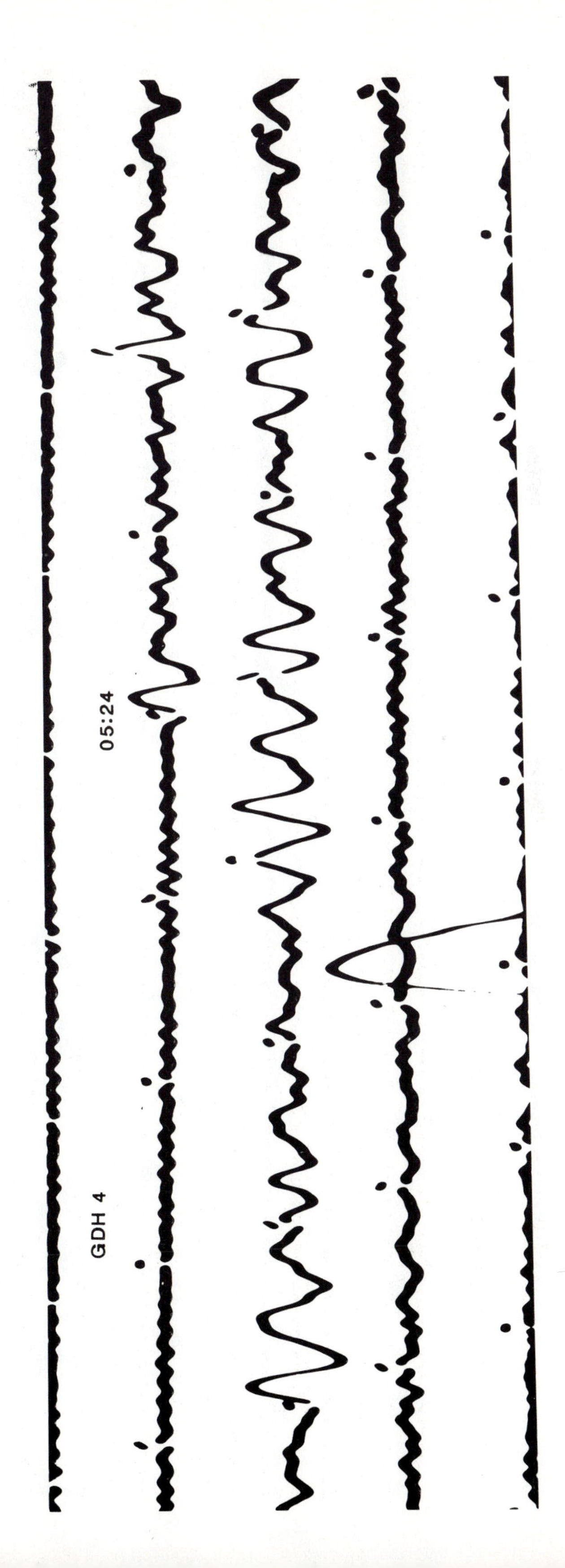
05:24
GDH 4

Figure 6.20: Seismogram for Practical 4. HLW(4)

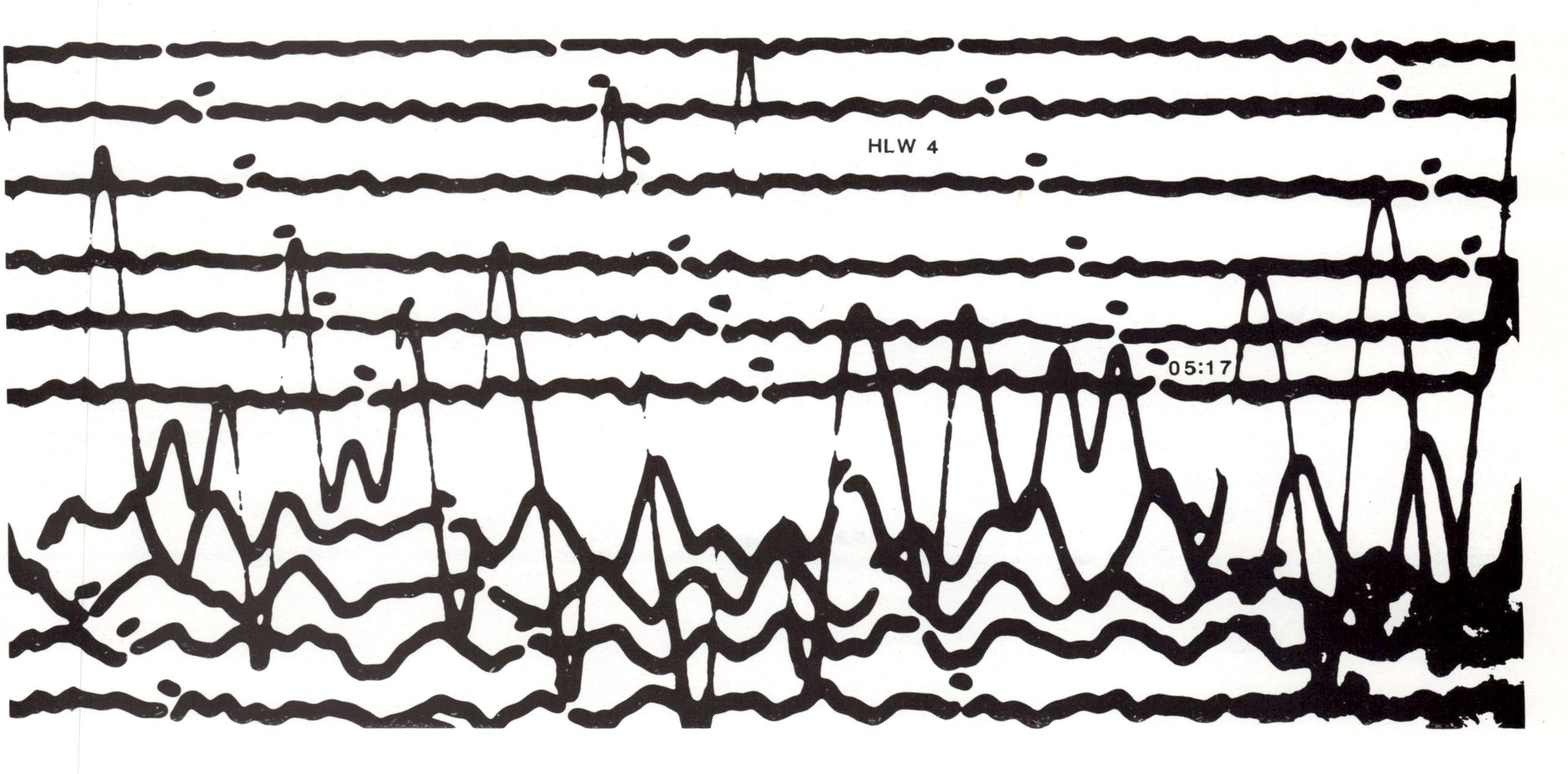

Figure 6.21: Seismogram for Practical 4. JER(1,5,6)

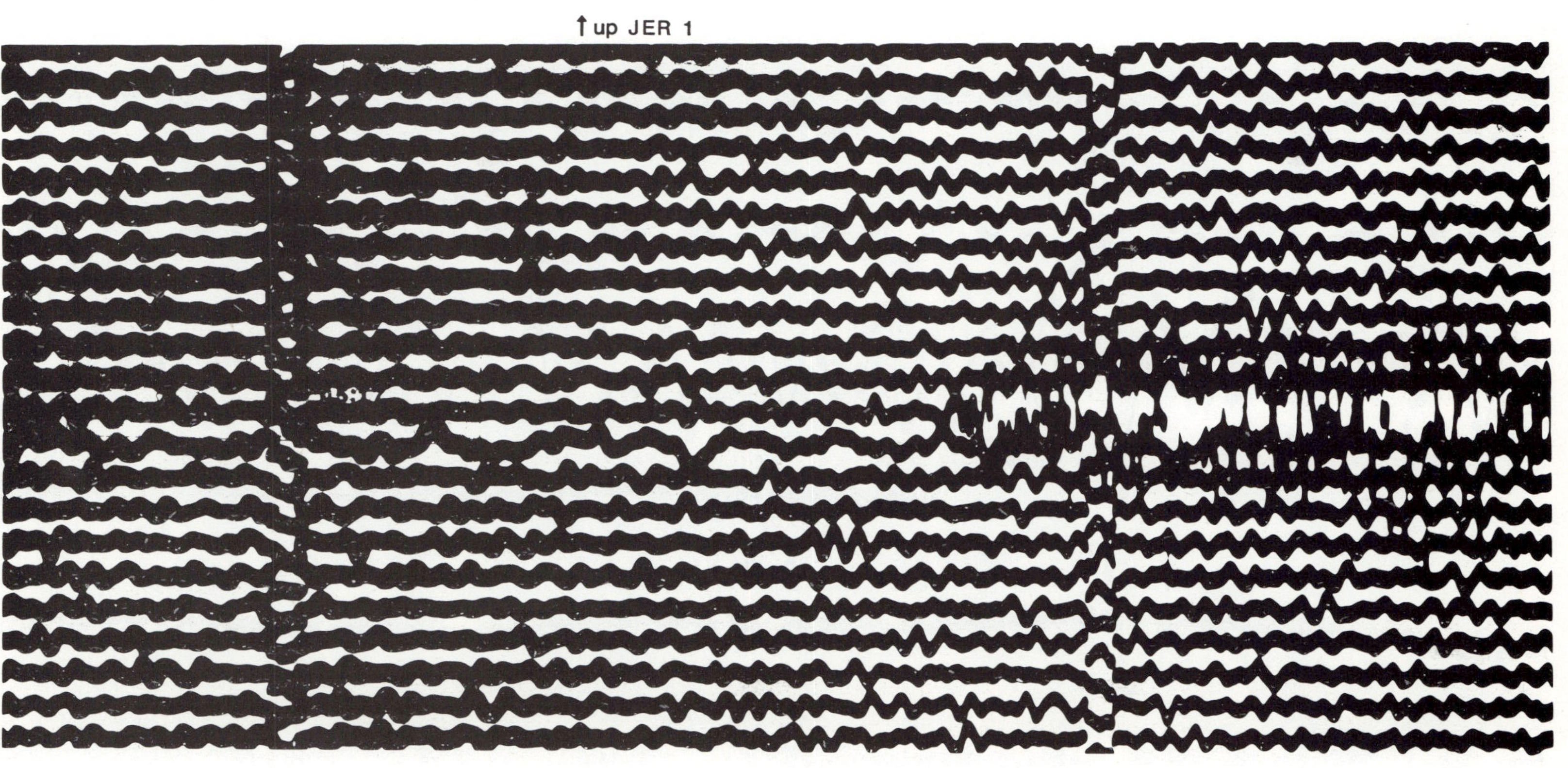

Figure 6.21 (*continued*)

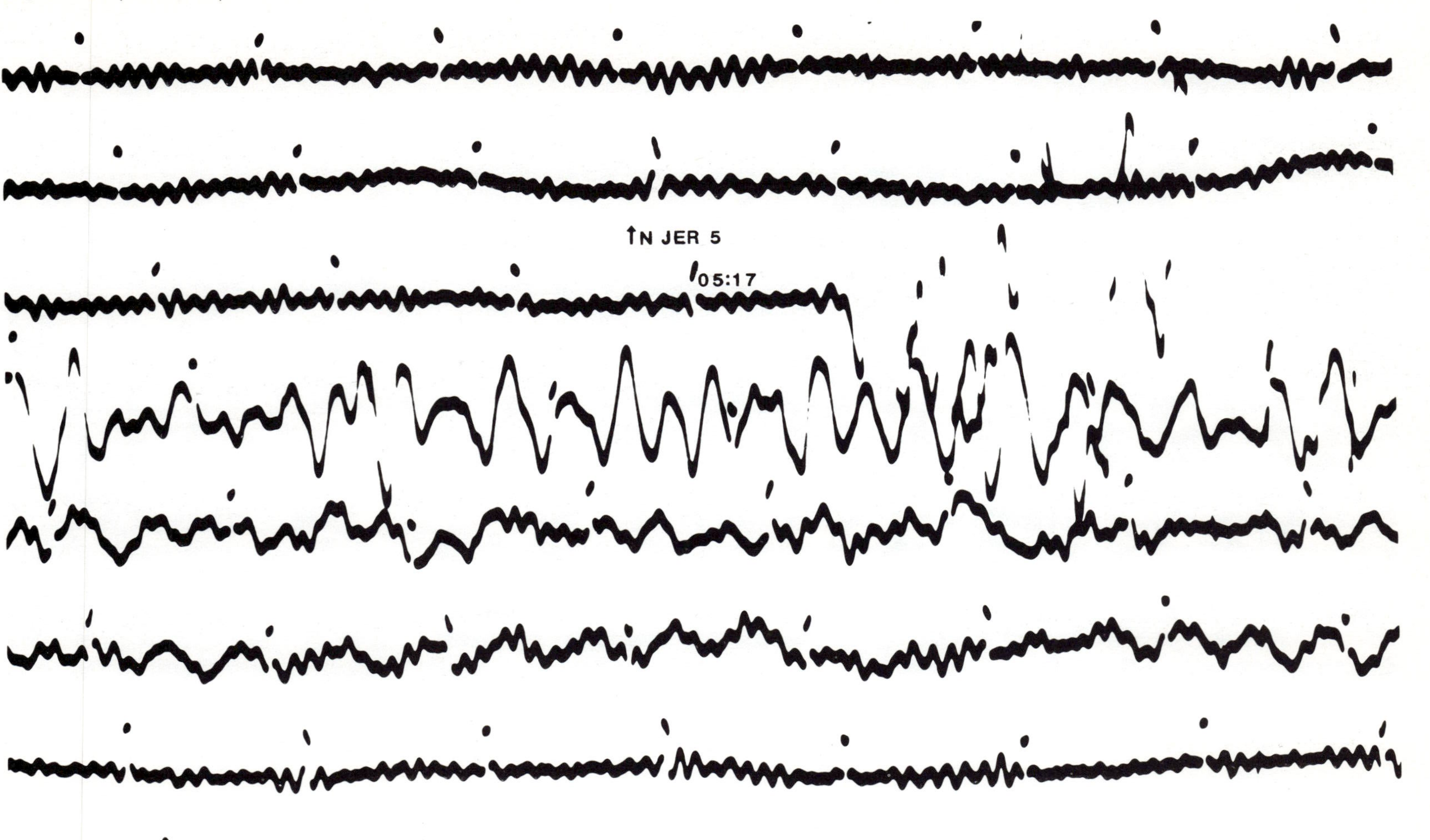

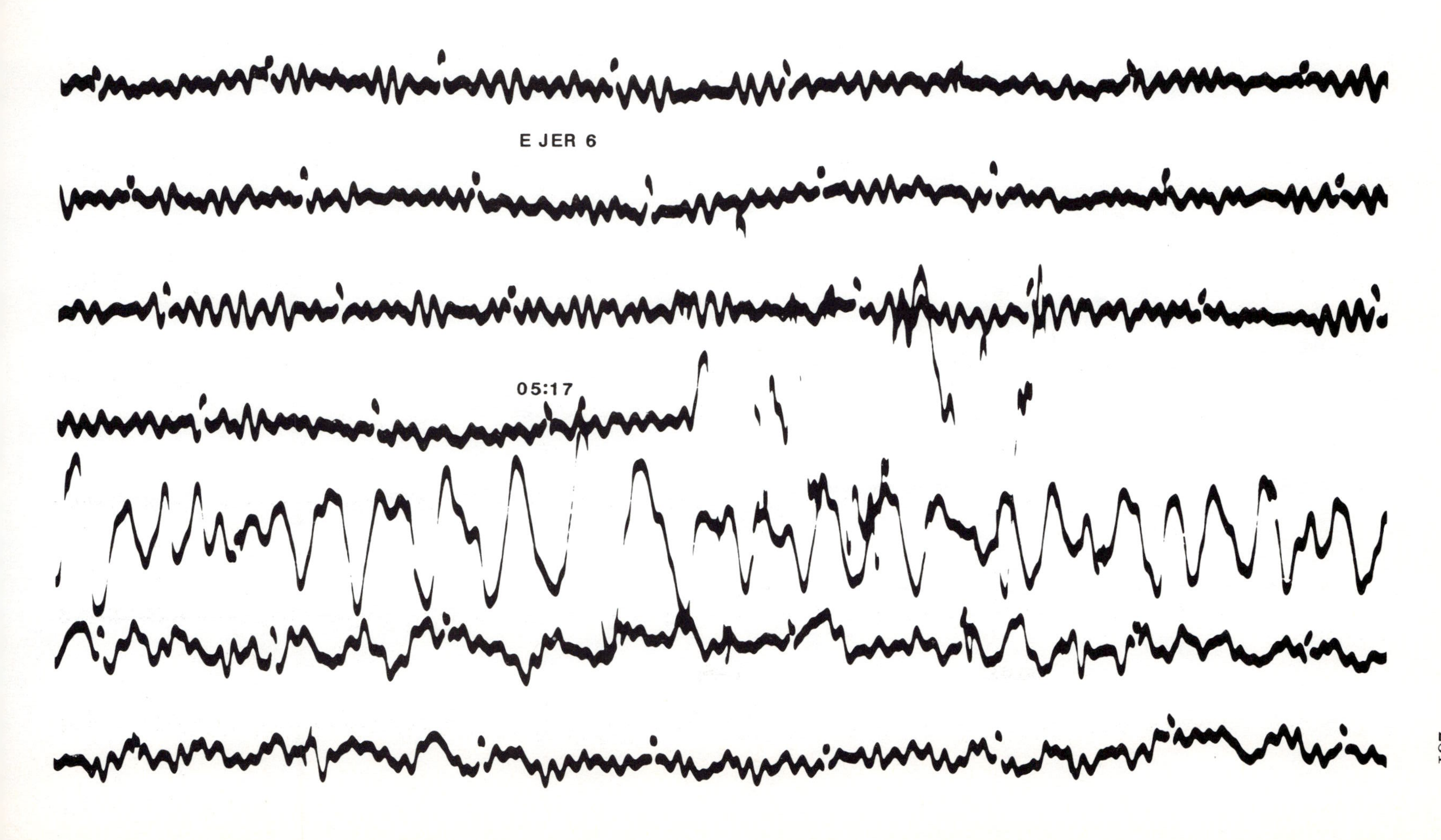
E JER 6
05:17

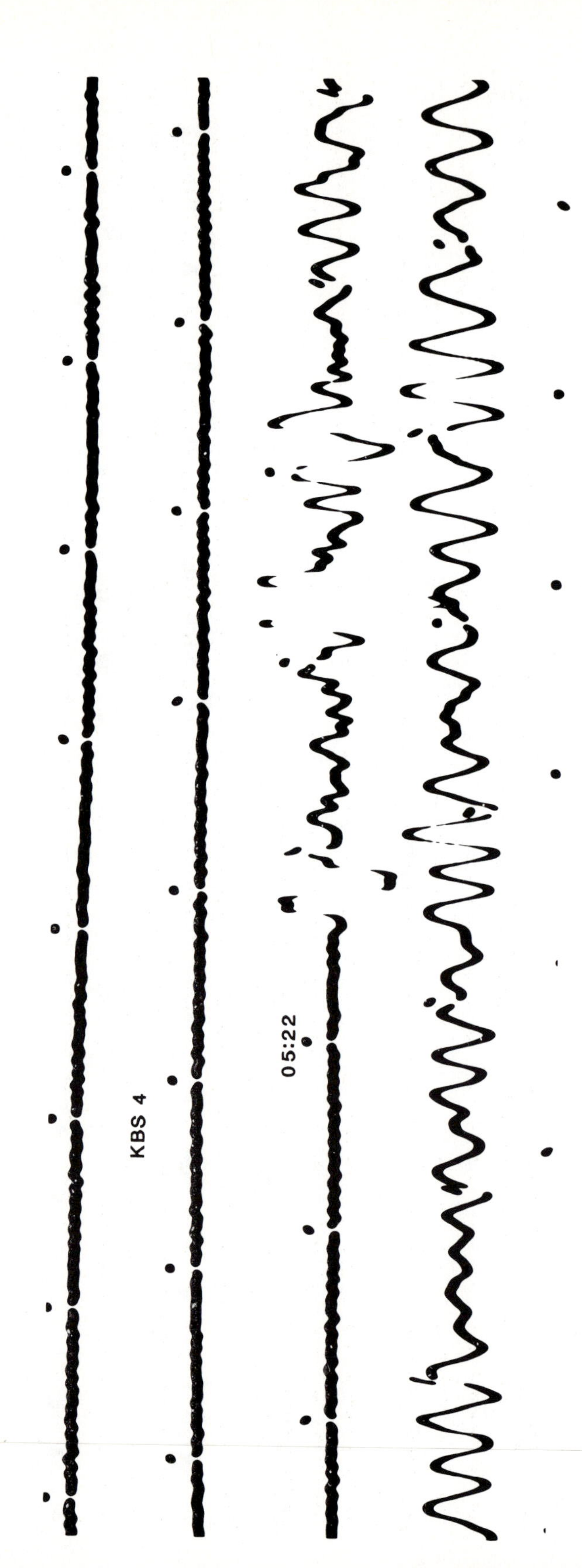

Figure 6.22: Seismogram for Practical 4. KBS(4)

Figure 6.23: Seismogram for Practical 4. KEV(1,4)

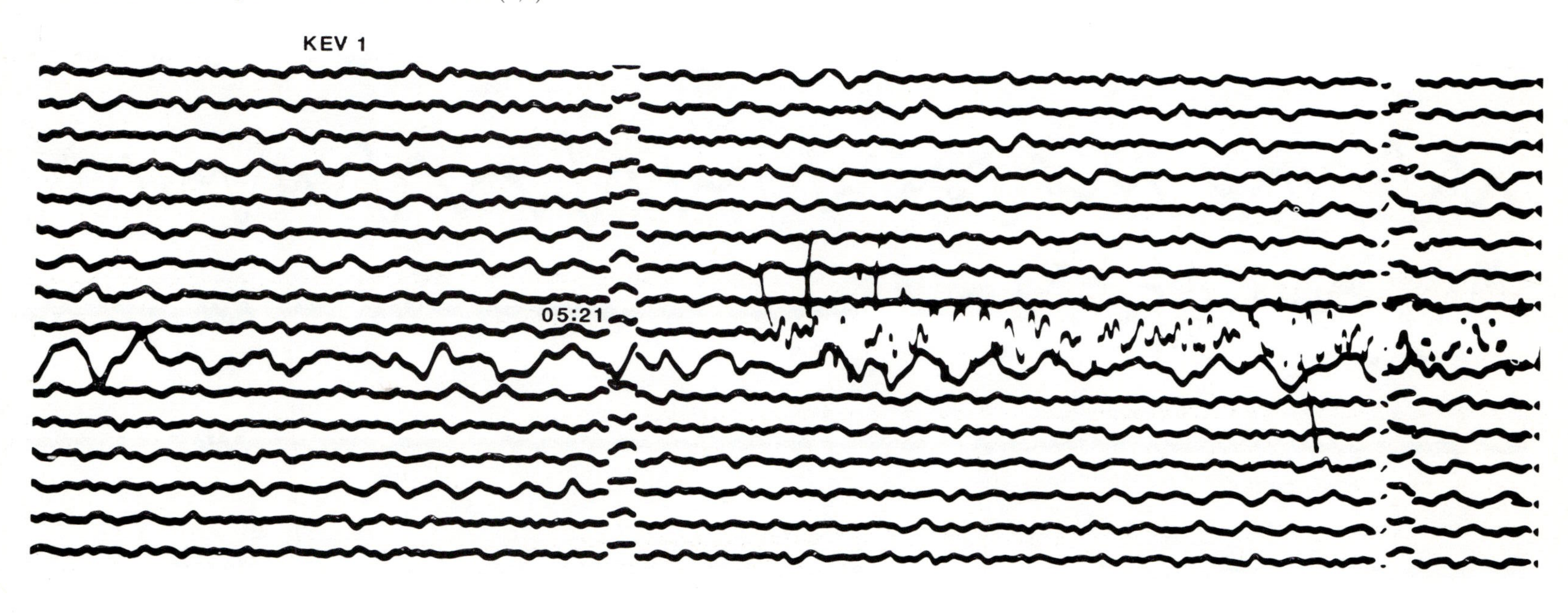

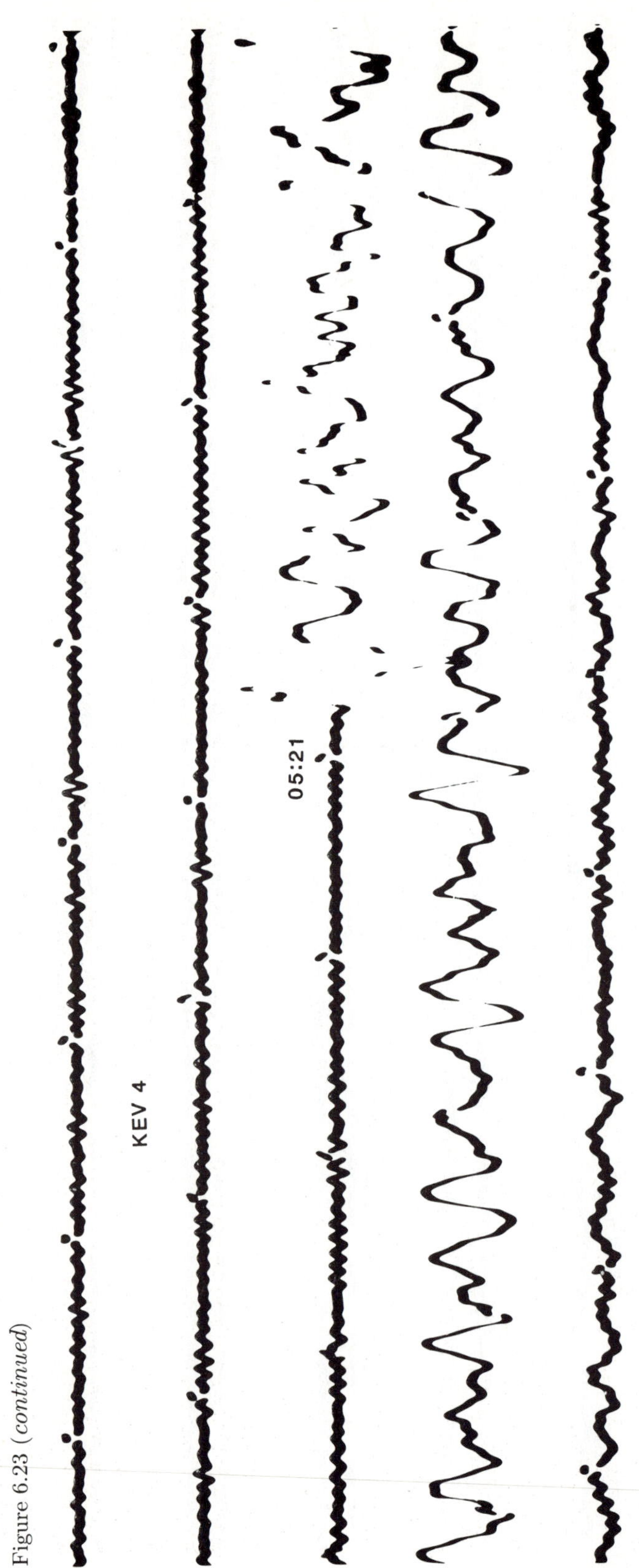

Figure 6.23 (*continued*)

Figure 6.24: Seismogram for Practical 4. KON(1,4)

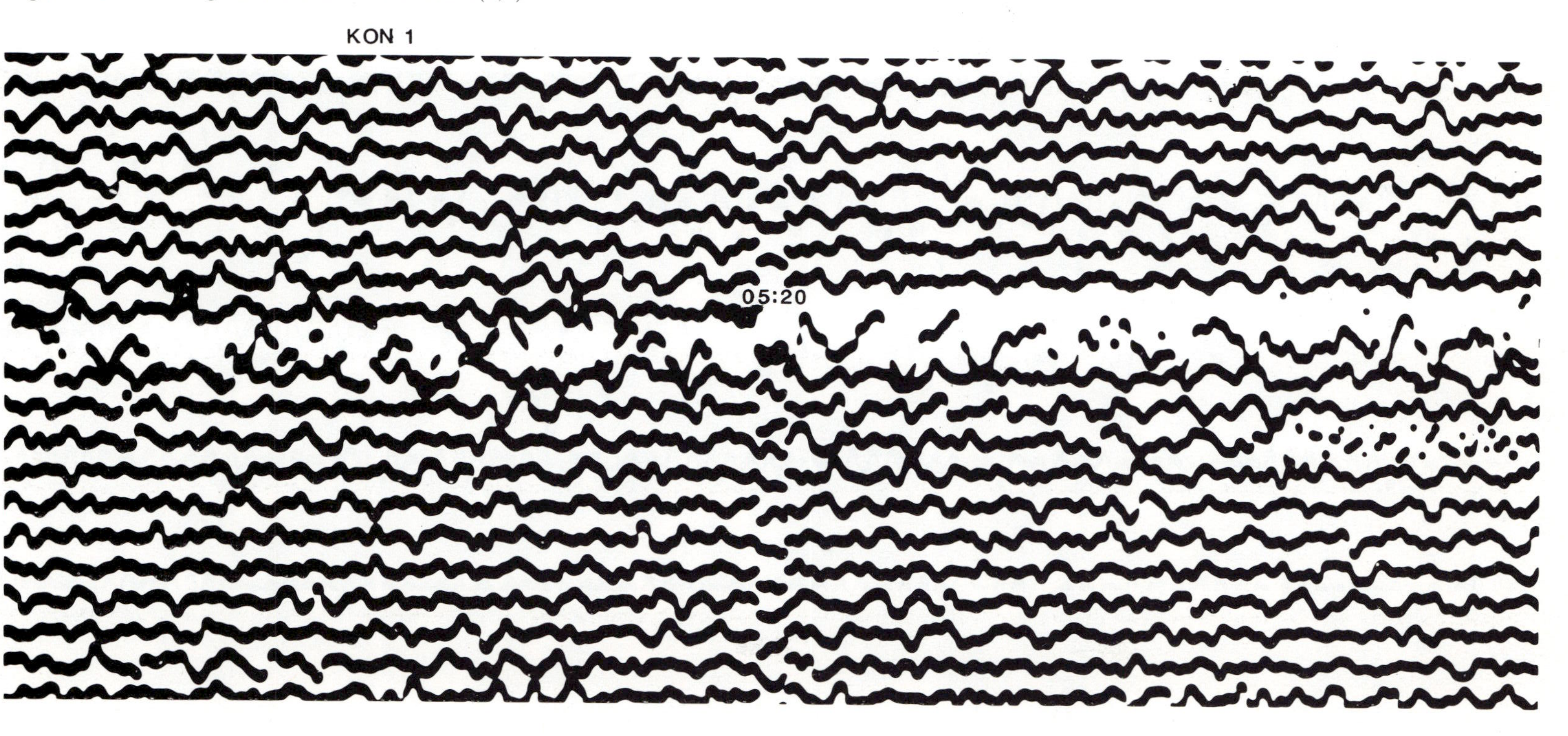

Figure 6.24 (*continued*)

Figure 6.25: Scismogram for Practical 4. MAL(1,4)

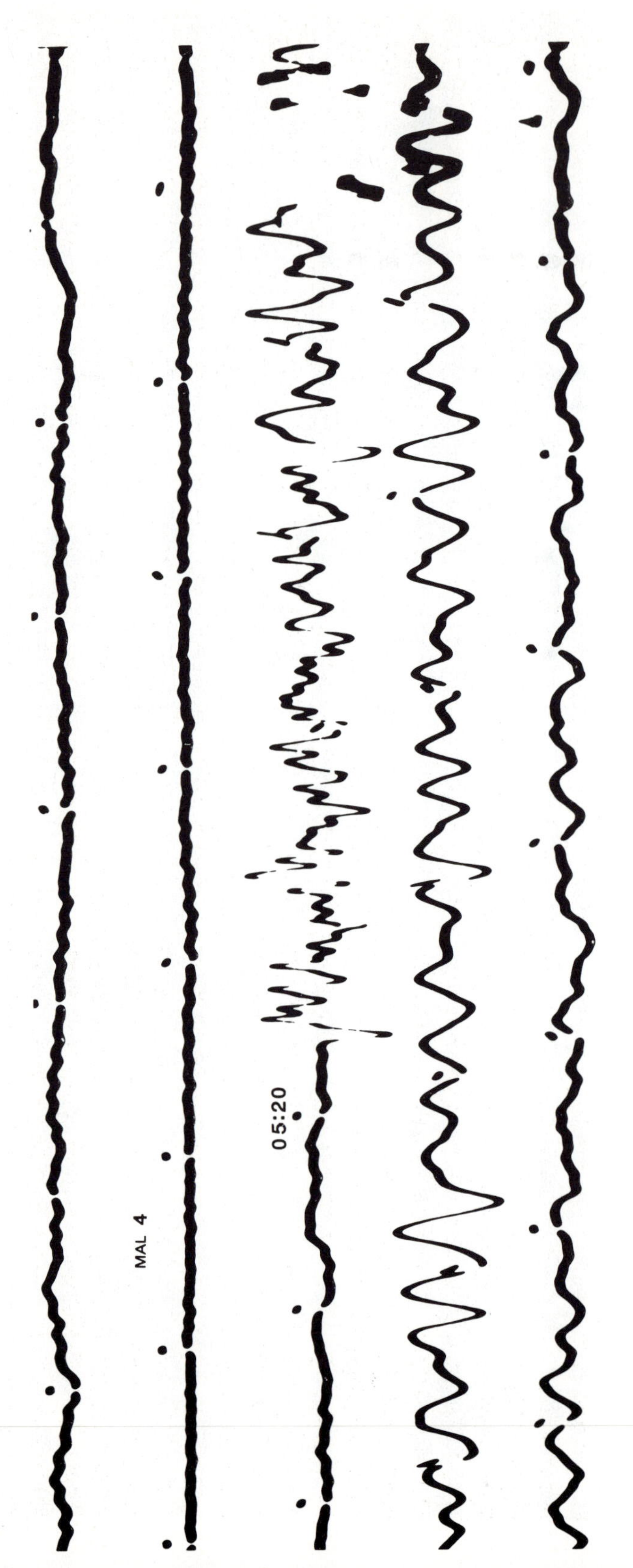

Figure 6.25 (*continued*)

Figure 6.26: Seismogram for Practical 4. MAT(1,4)

Figure 6.26 (*continued*)

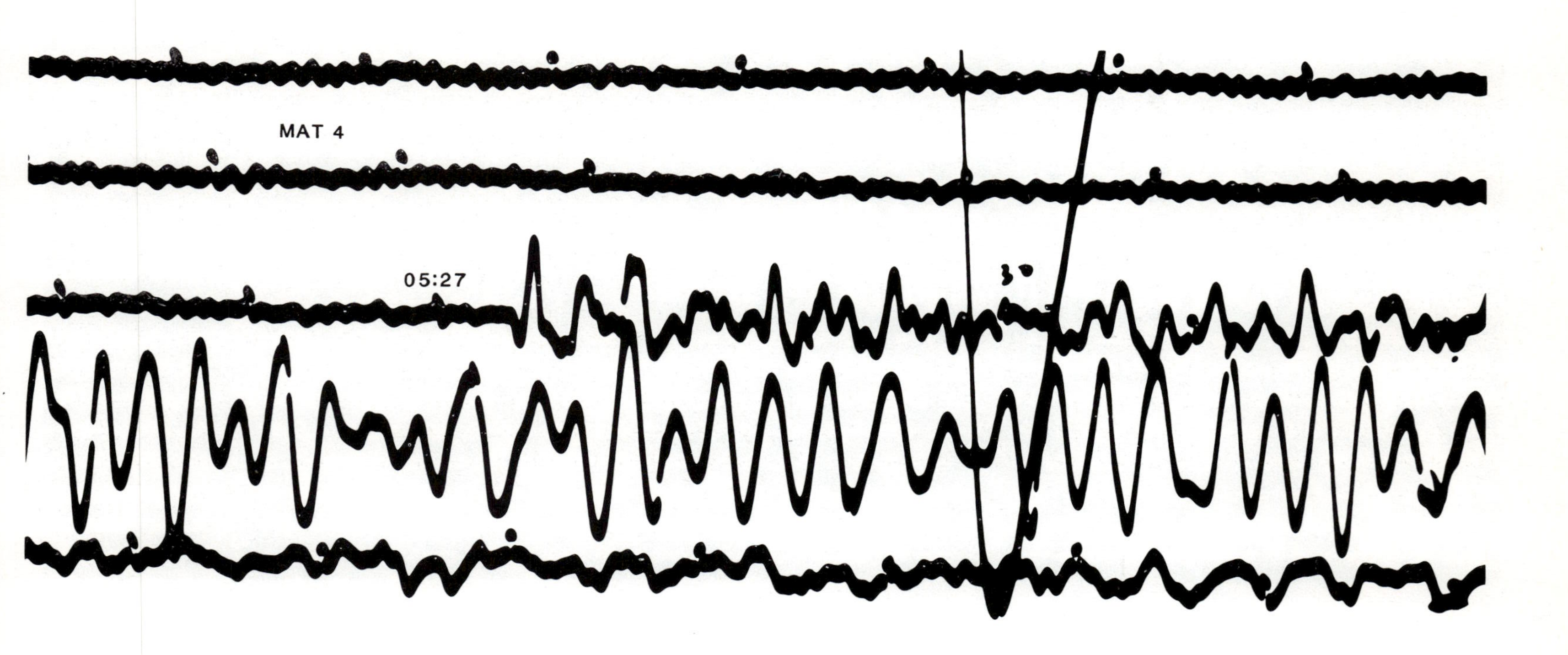

Figure 6.27: Seismogram for Practical 4. MSH(1,4)

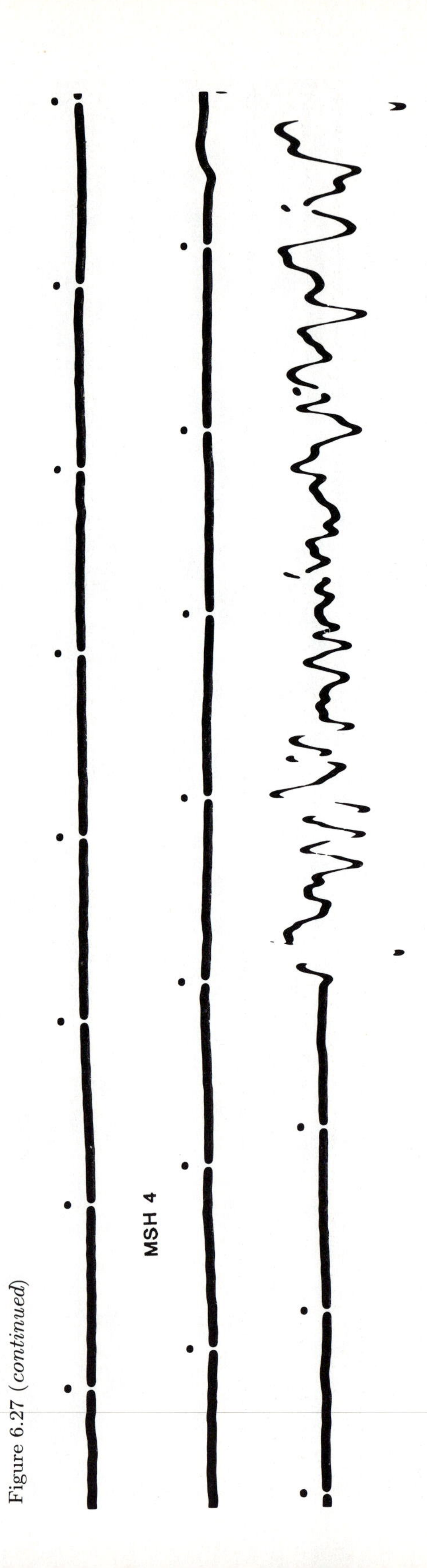

Figure 6.27 (*continued*)

Figure 6.28: Seismogram for Practical 4. NAI(1,4)

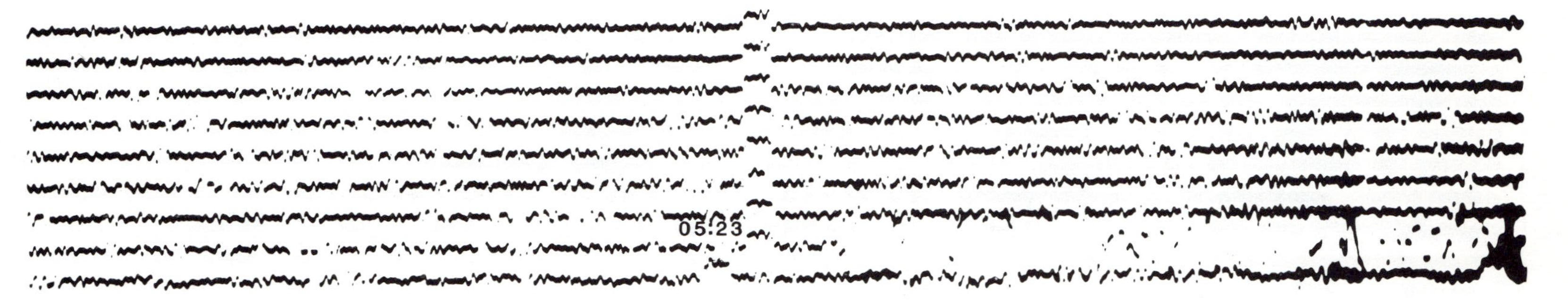

NAI 1

NAI 4

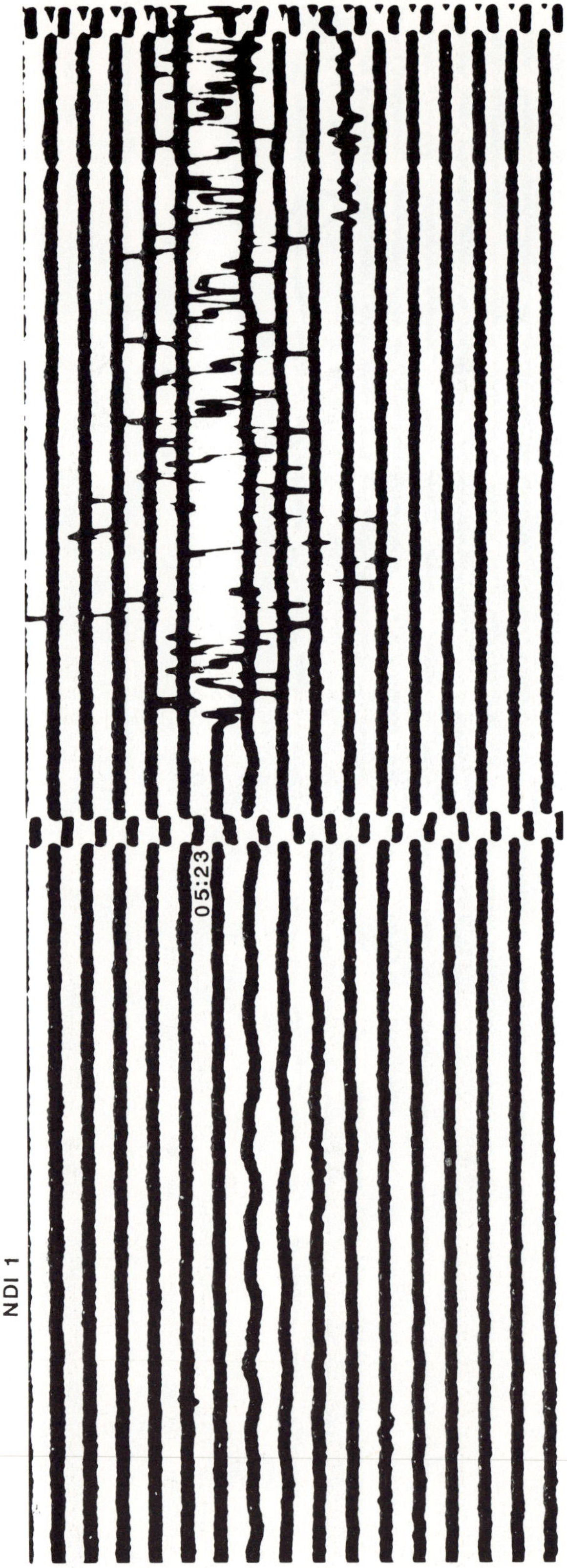

Figure 6.29: Seismogram for Practical 4. NDI(1,4)

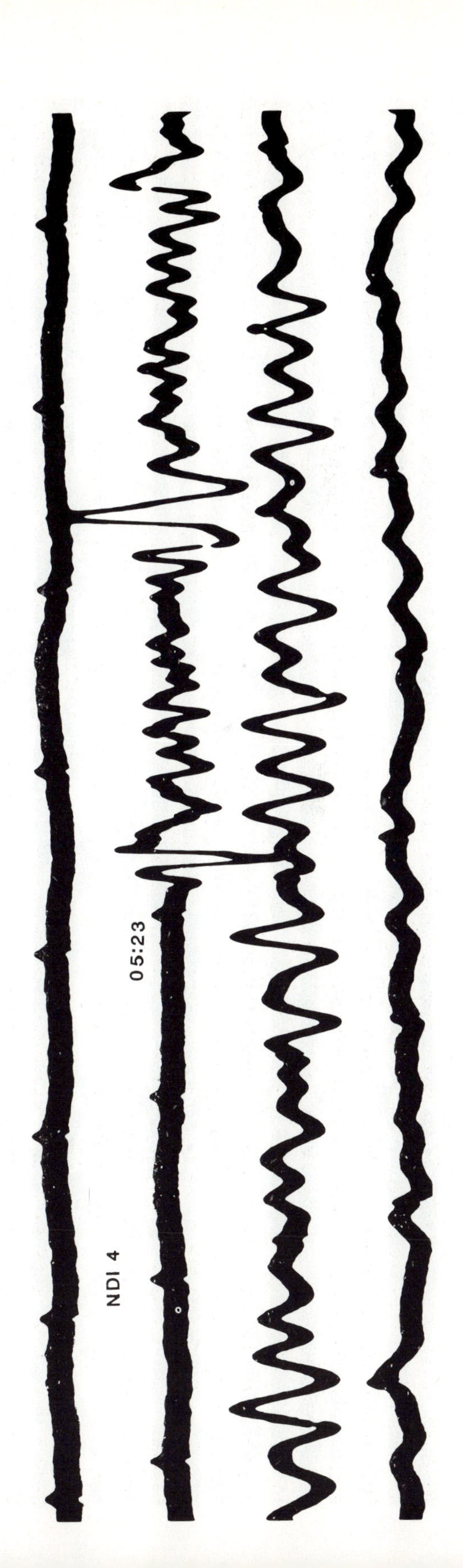
NDI 4
05:23

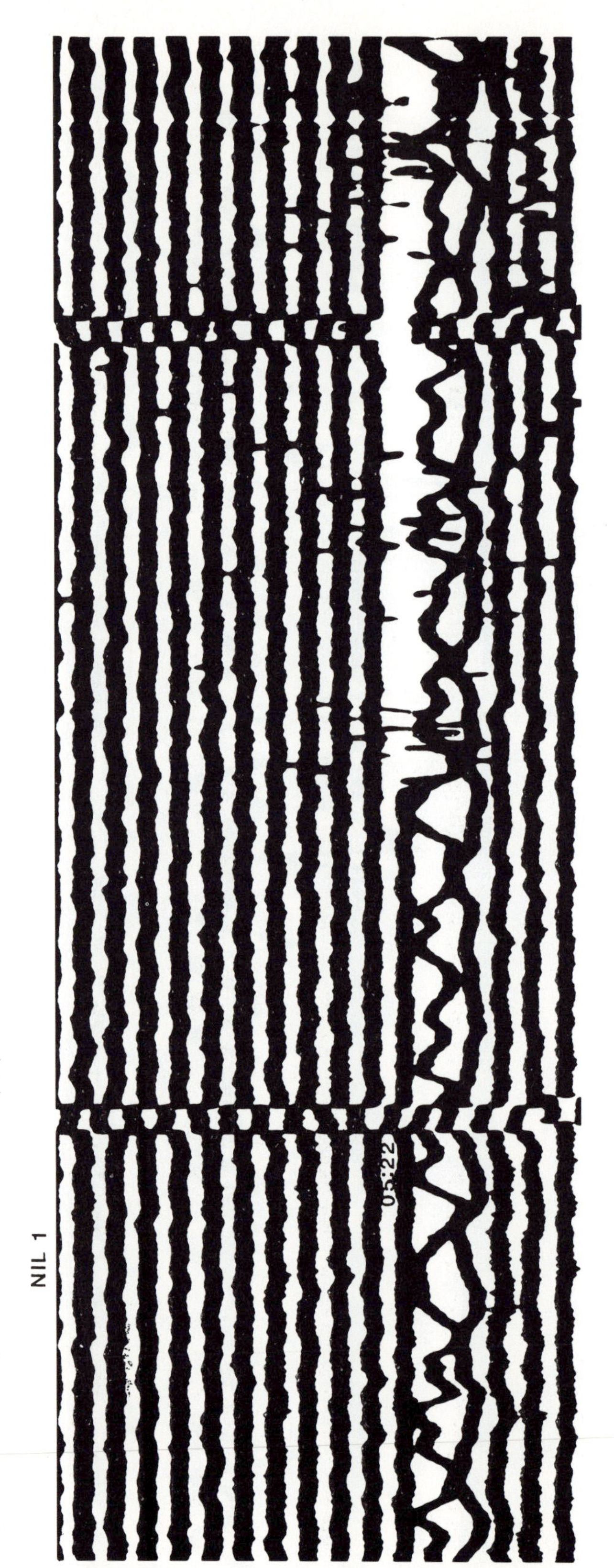

Figure 6.30: Seismogram for Practical 4. NIL(1,4)

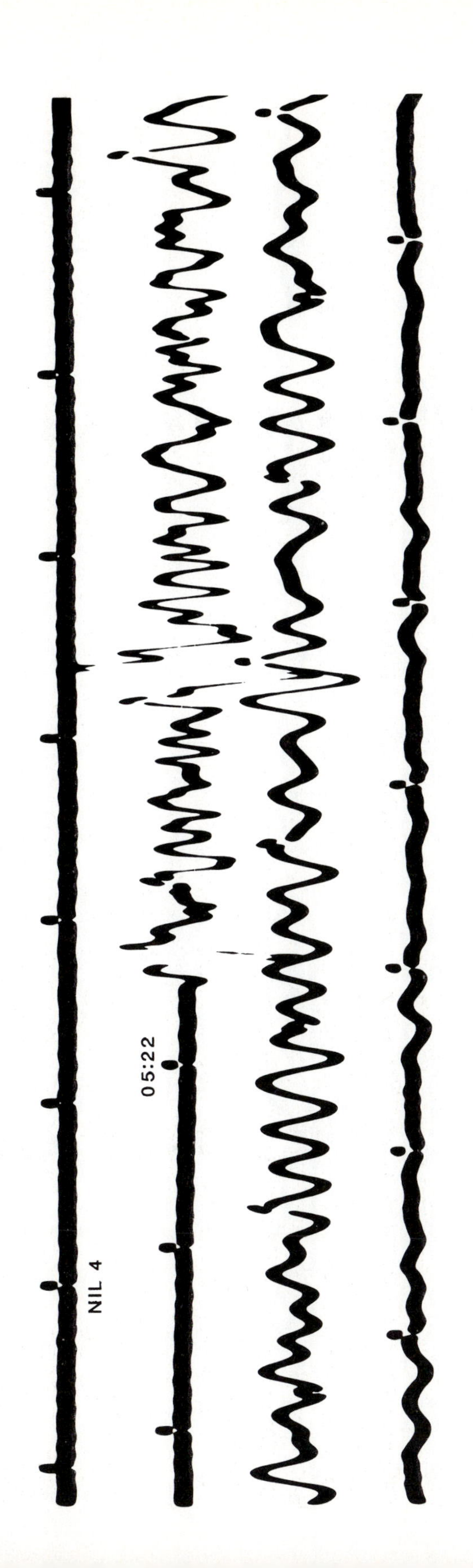
NIL 4
05:22

Figure 6.31: Seismogram for Practical 4. NUR(1,4)

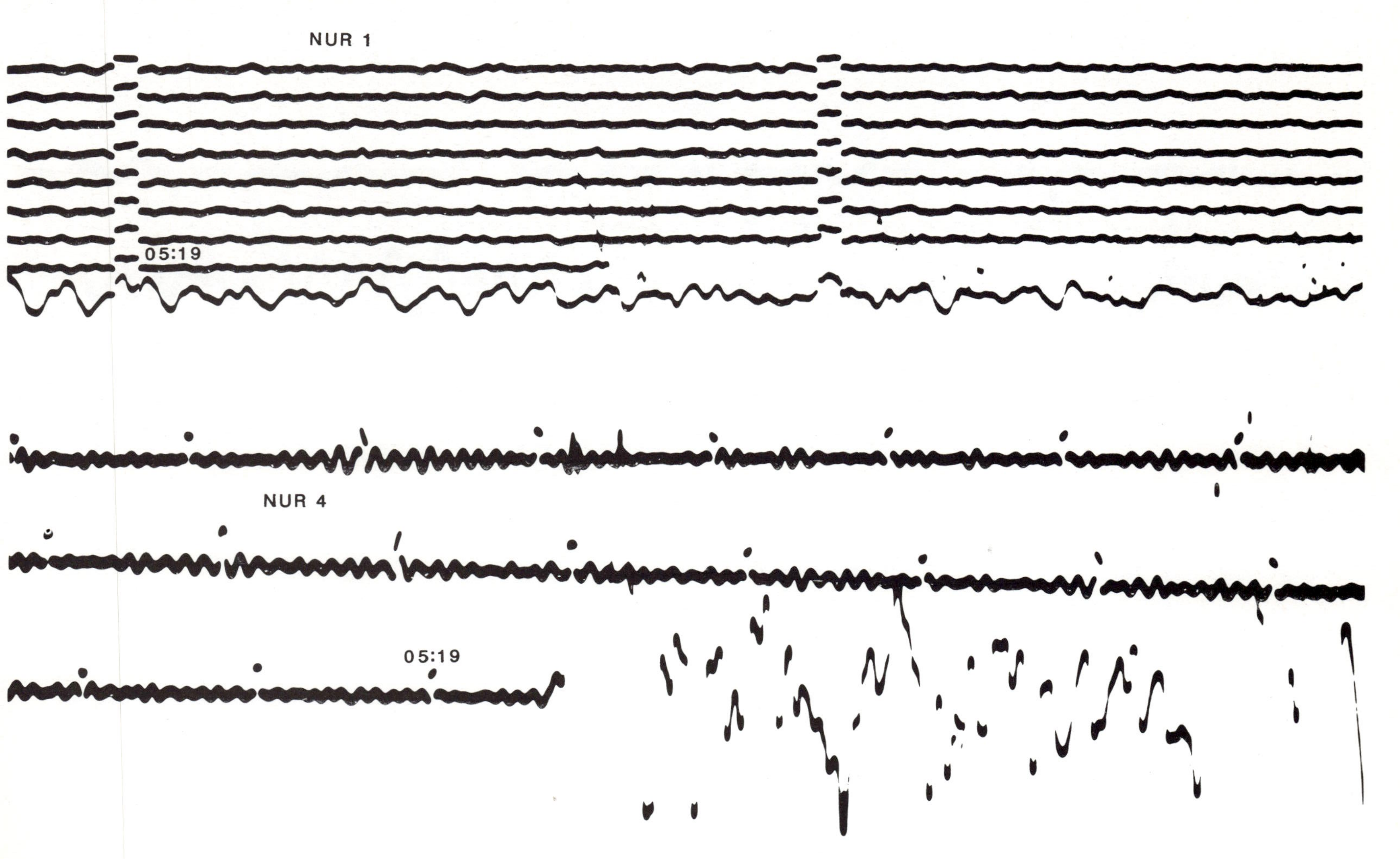

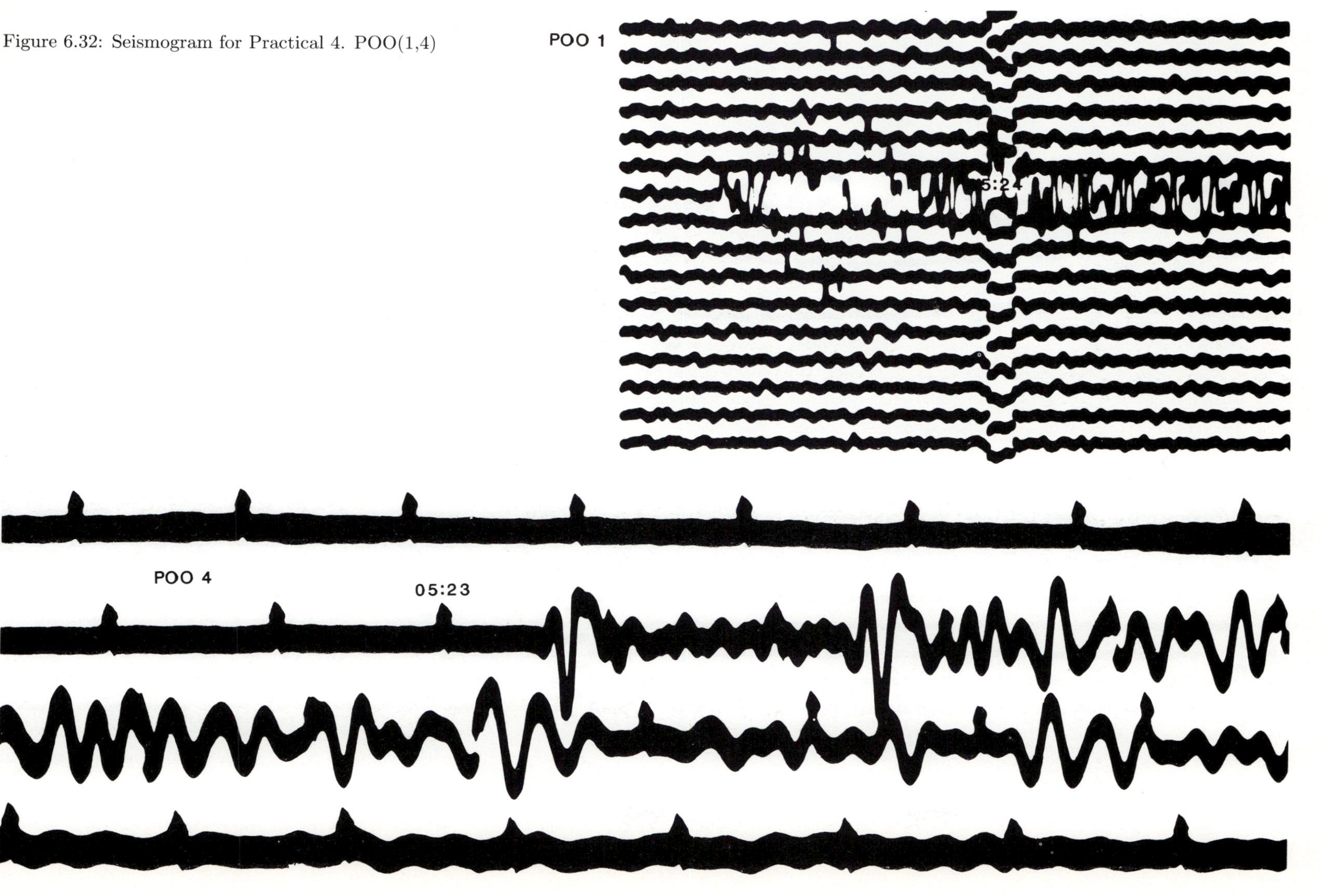

Figure 6.32: Seismogram for Practical 4. POO(1,4)

Figure 6.33: Seismogram for Practical 4. PRE(1,4)

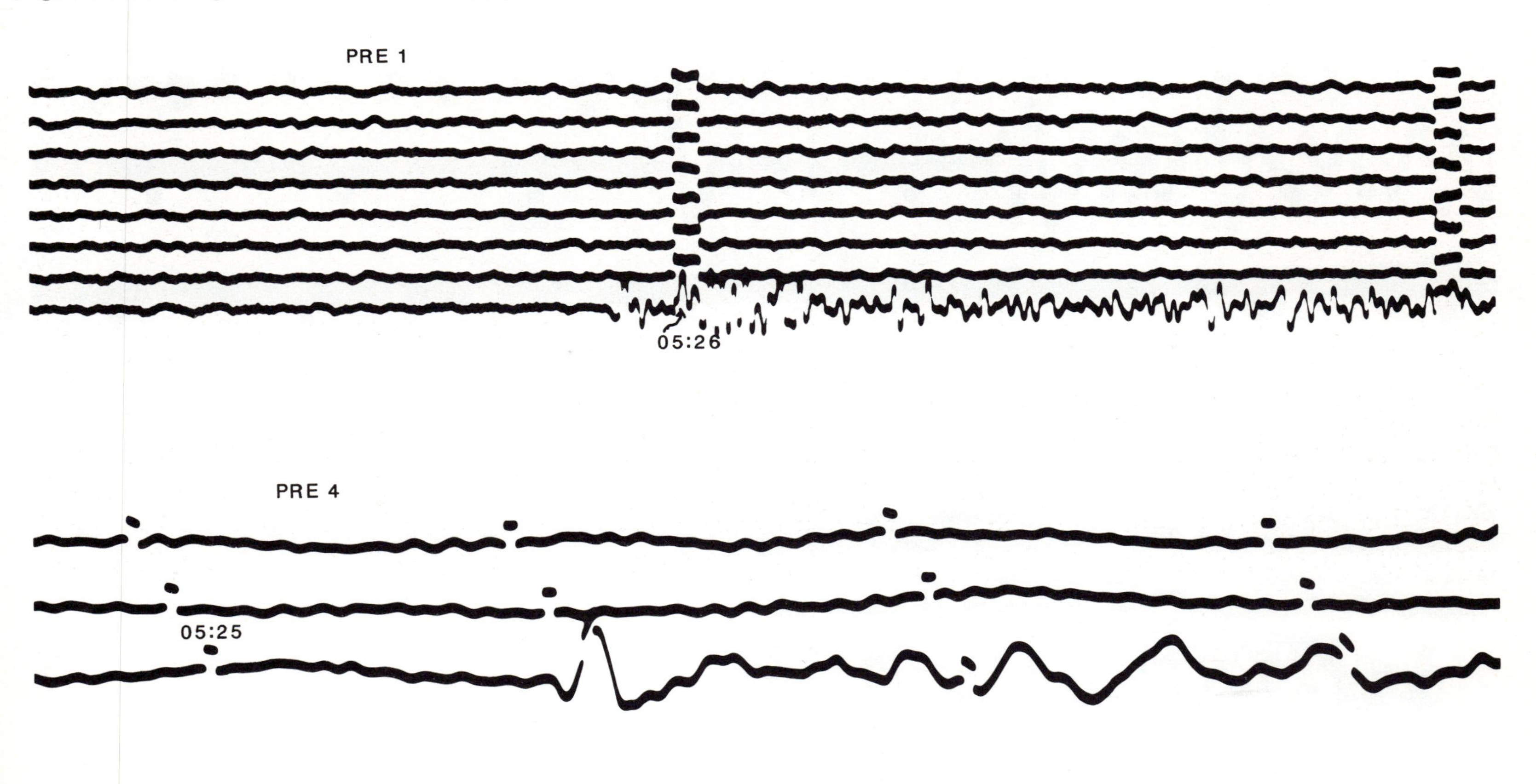

Figure 6.34: Seismogram for Practical 4. QUE(1,4)

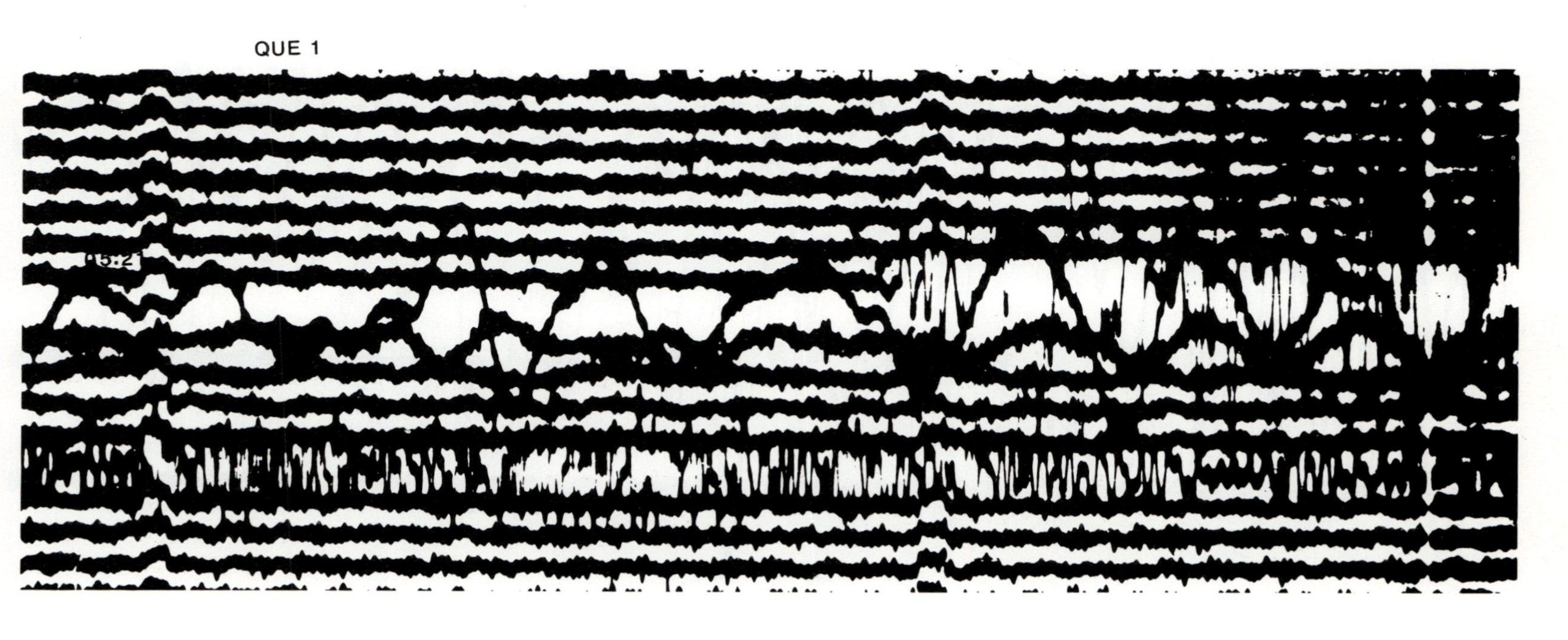

Figure 6.34 (*continued*)

Figure 6.35: Seismogram for Practical 4. SHI(1,4)

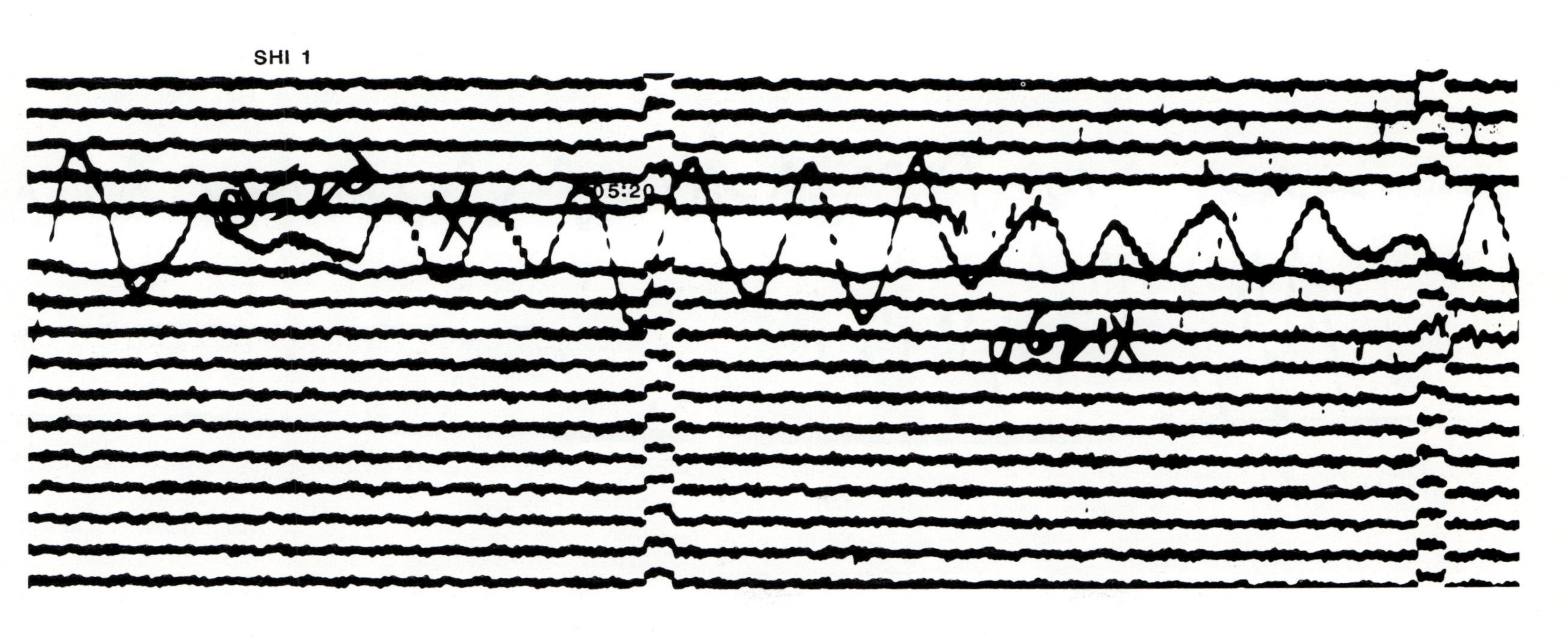

Figure 6.35 (*continued*)

Figure 6.36: Seismogram for Practical 4. SHL(1,4)

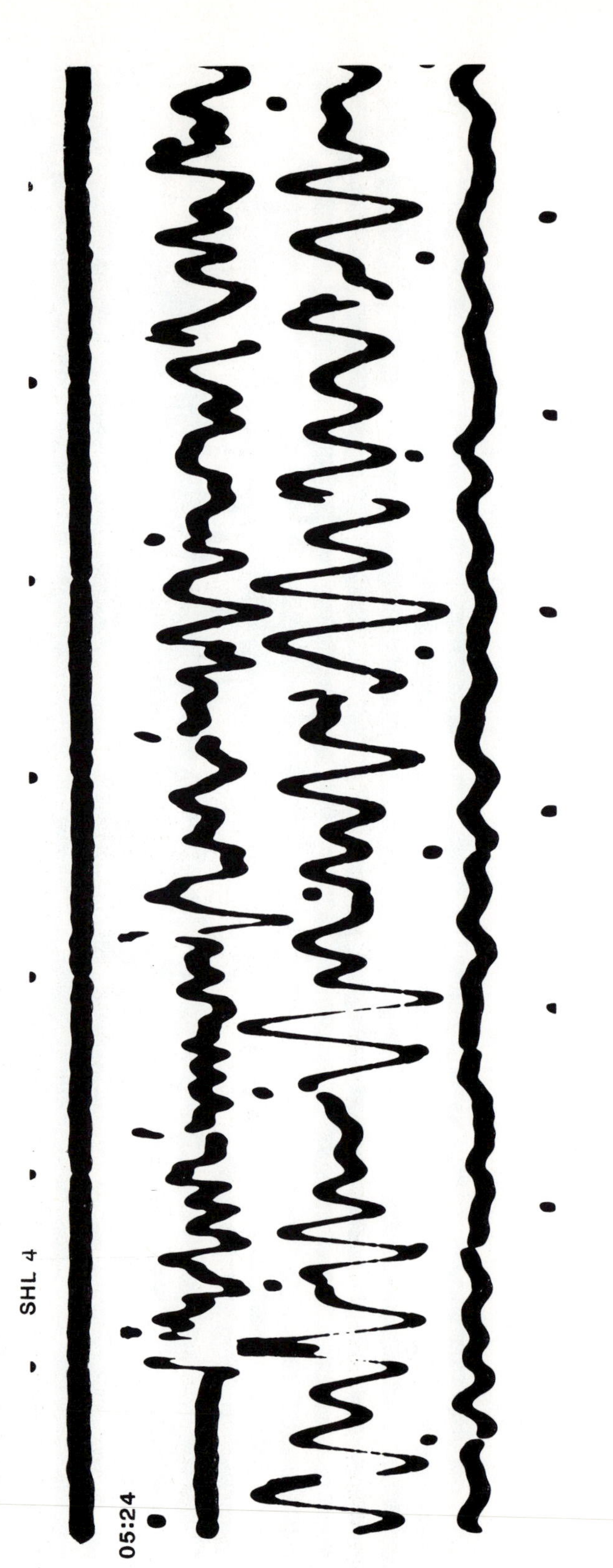

Figure 6.36 (*continued*)

Figure 6.37: Seismogram for Practical 4. SNG(4)

Figure 6.38: Seismogram for Practical 4. STU(1,4)

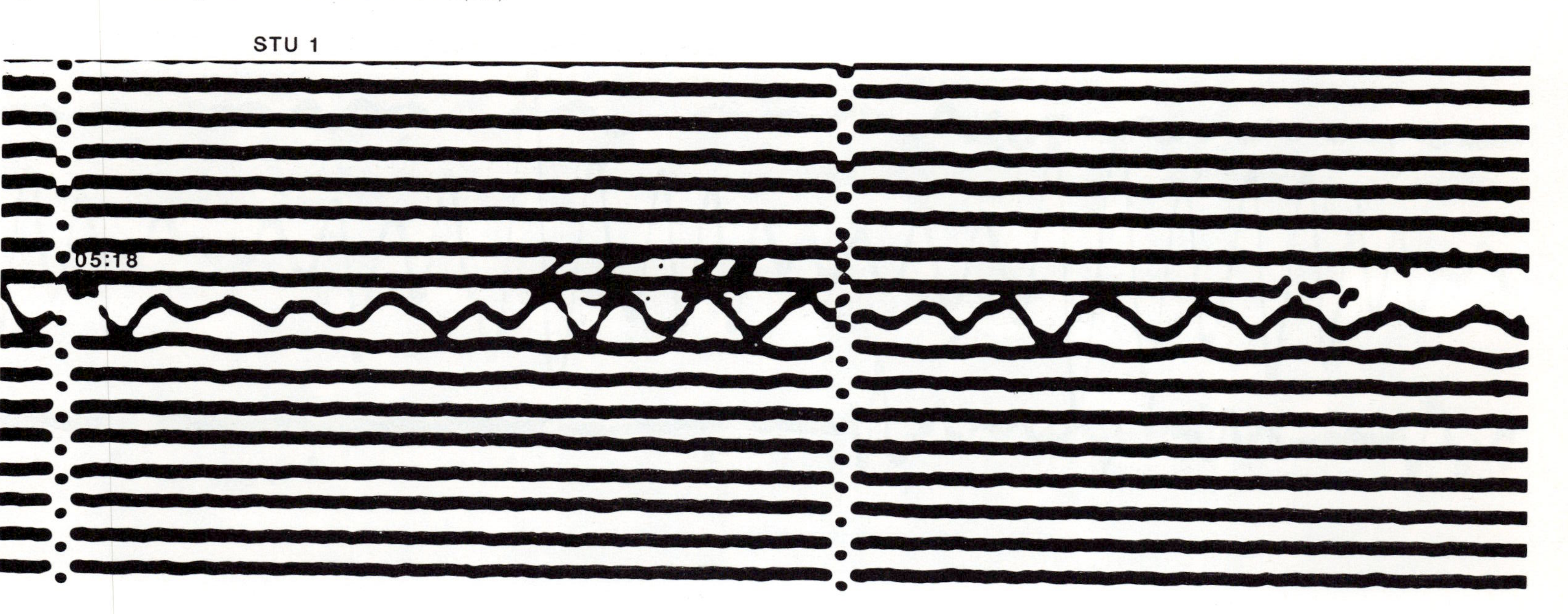

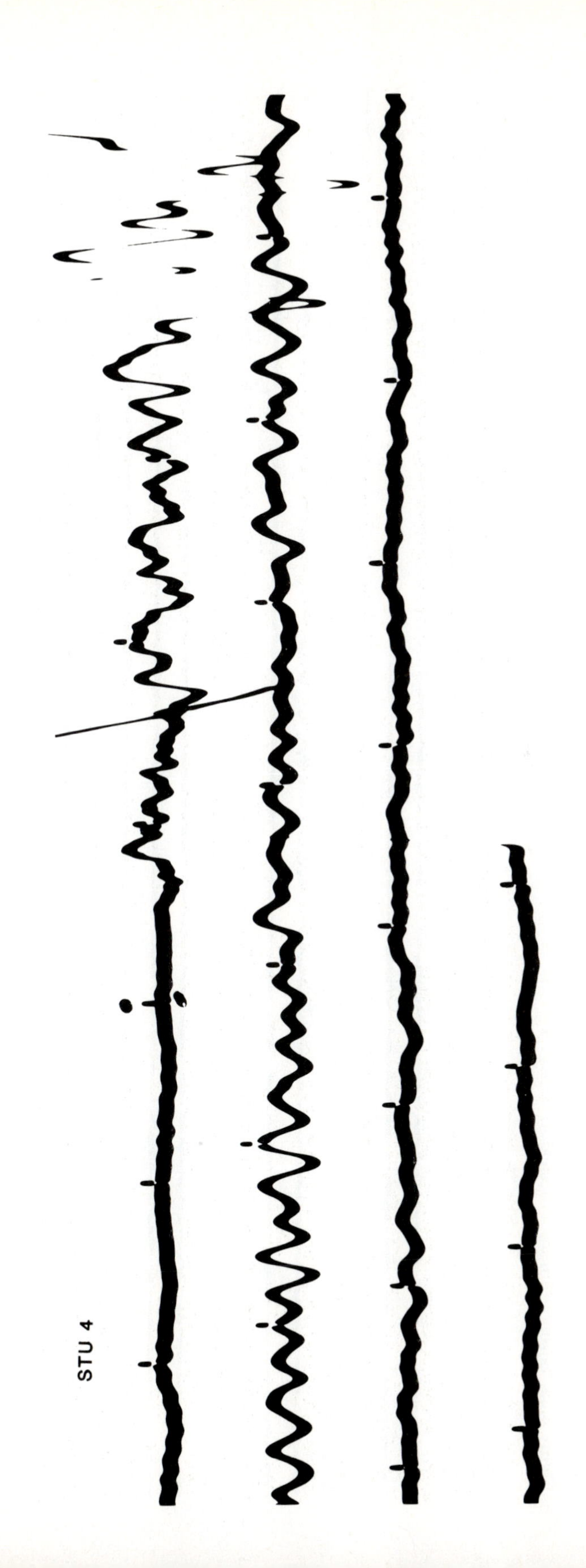
STU 4

Figure 6.39: Seismogram for Practical 4. SHK(1,4)

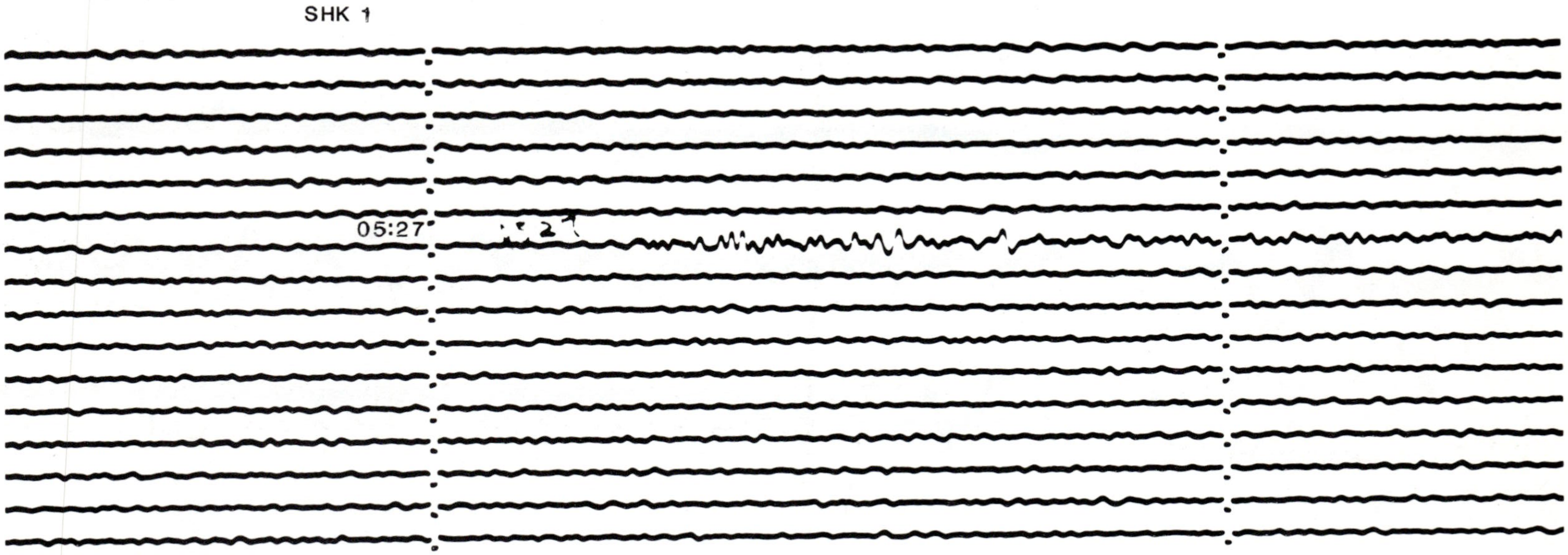

SHK 1

05:27

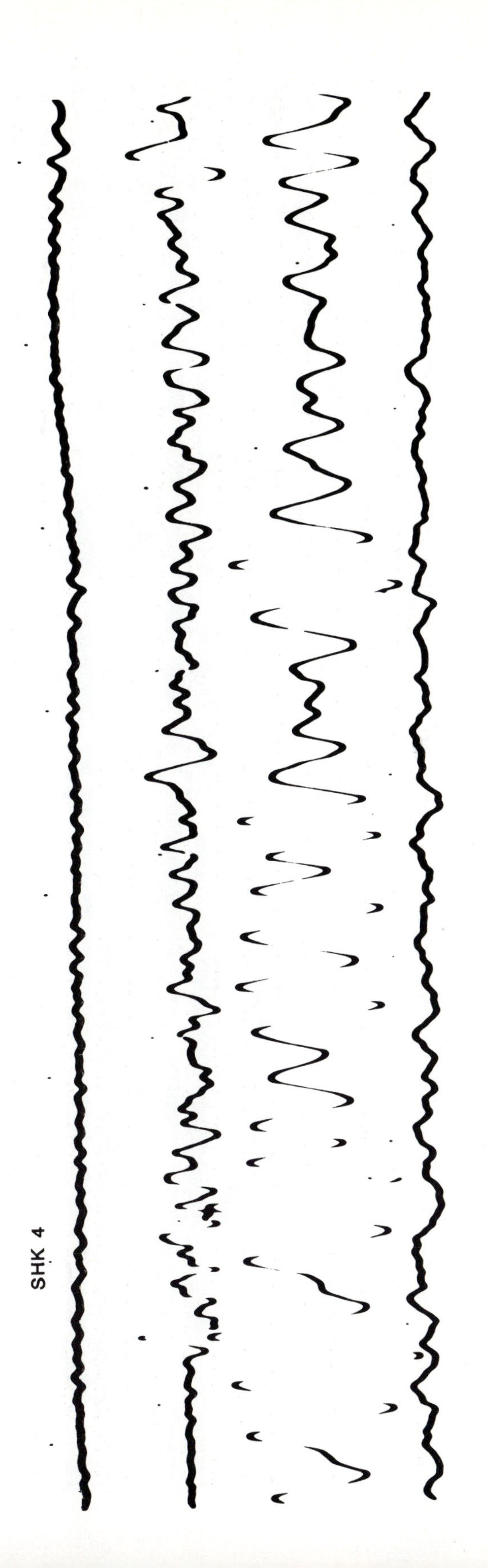
SHK 4

Figure 6.40: Seismogram for Practical 4. TAB(1,4)

TAB 4
05:18

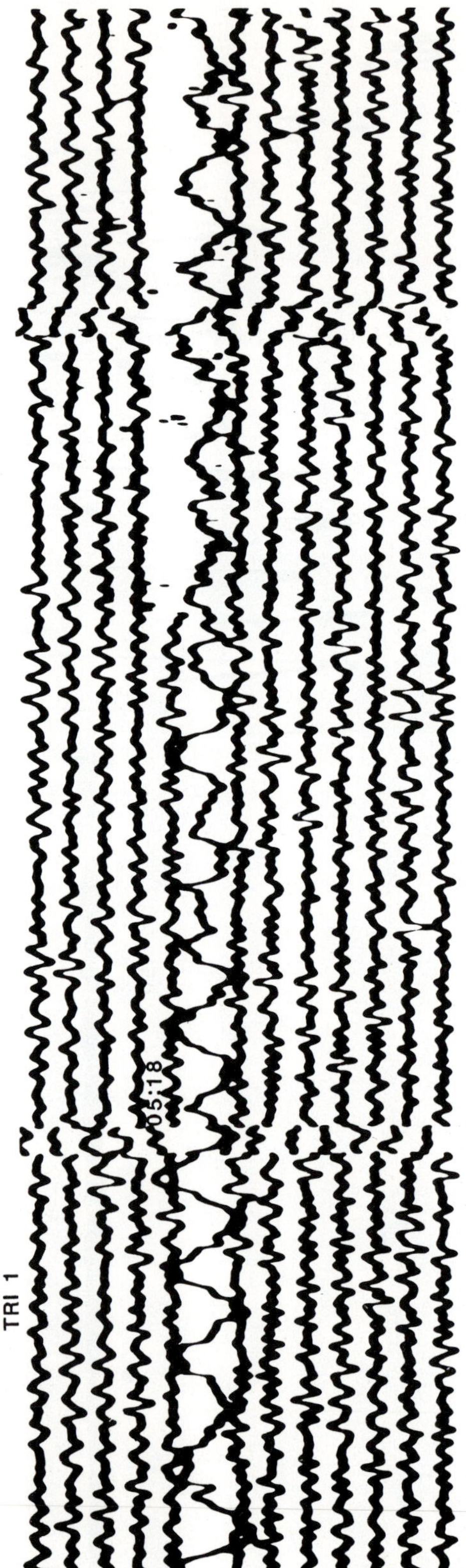

TRI 1

05:18

Figure 6.41: Seismogram for Practical 4. TRI(1,4)

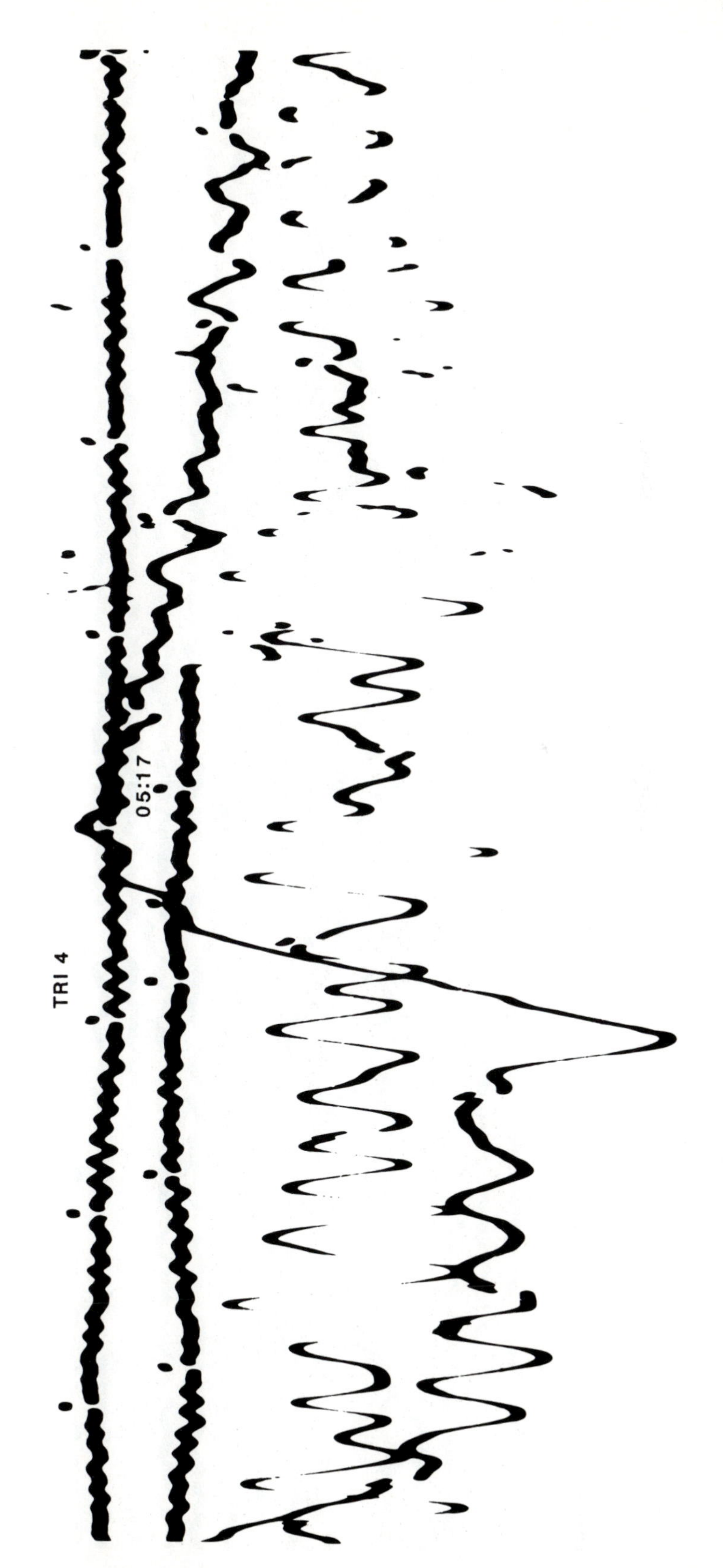
TRI 4
05:17

Figure 6.42: Seismogram for Practical 4. UME(1,4)

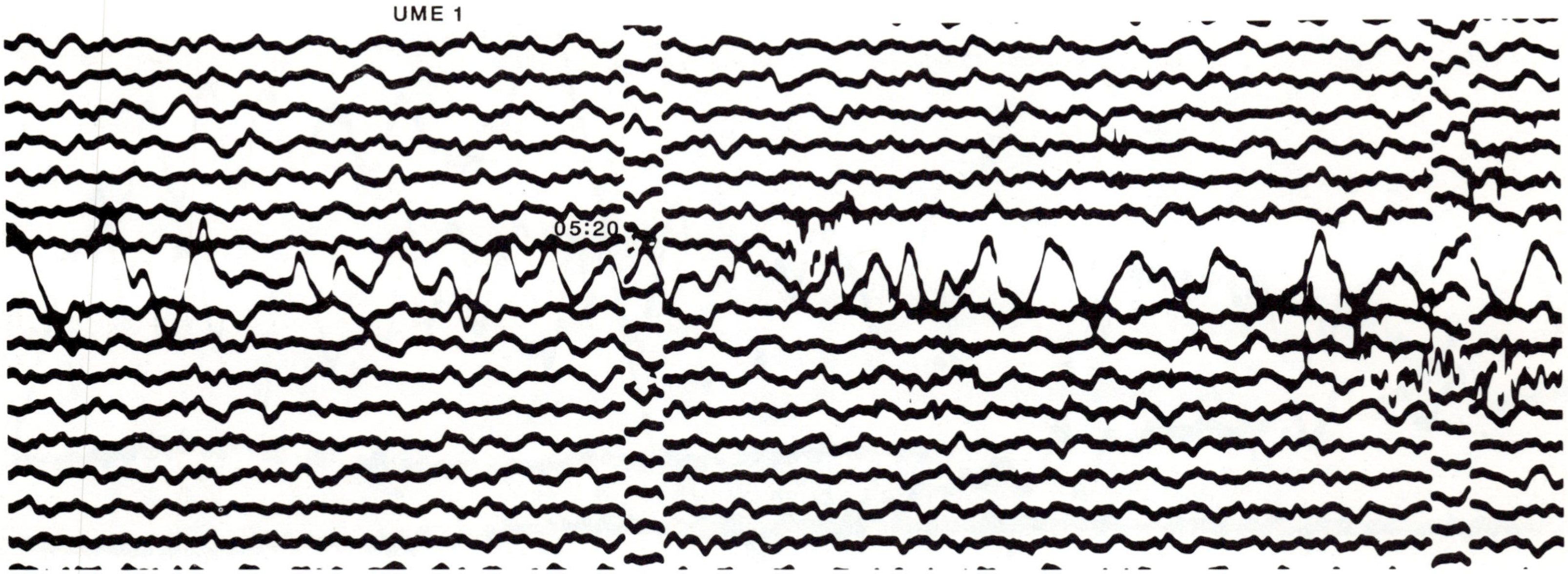

UME 4
05:20

Figure 6.43: Seismogram for Practical 4. VAL(1,4)

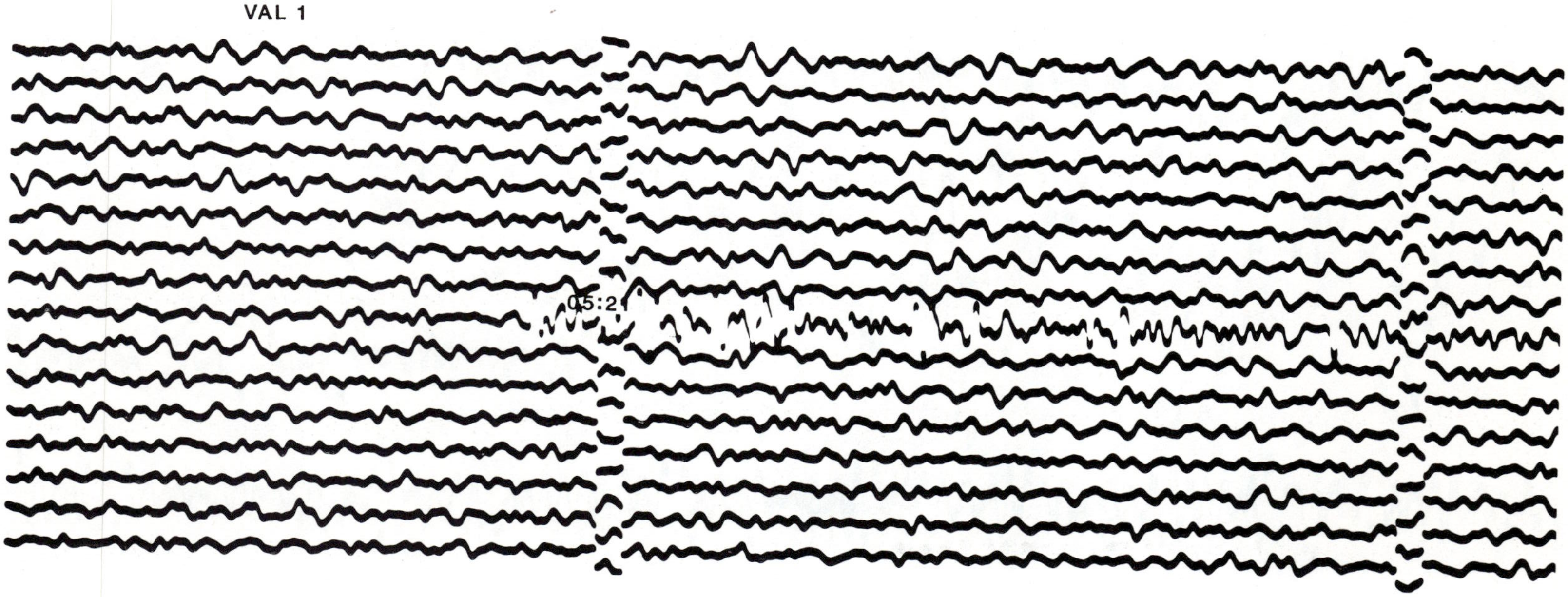

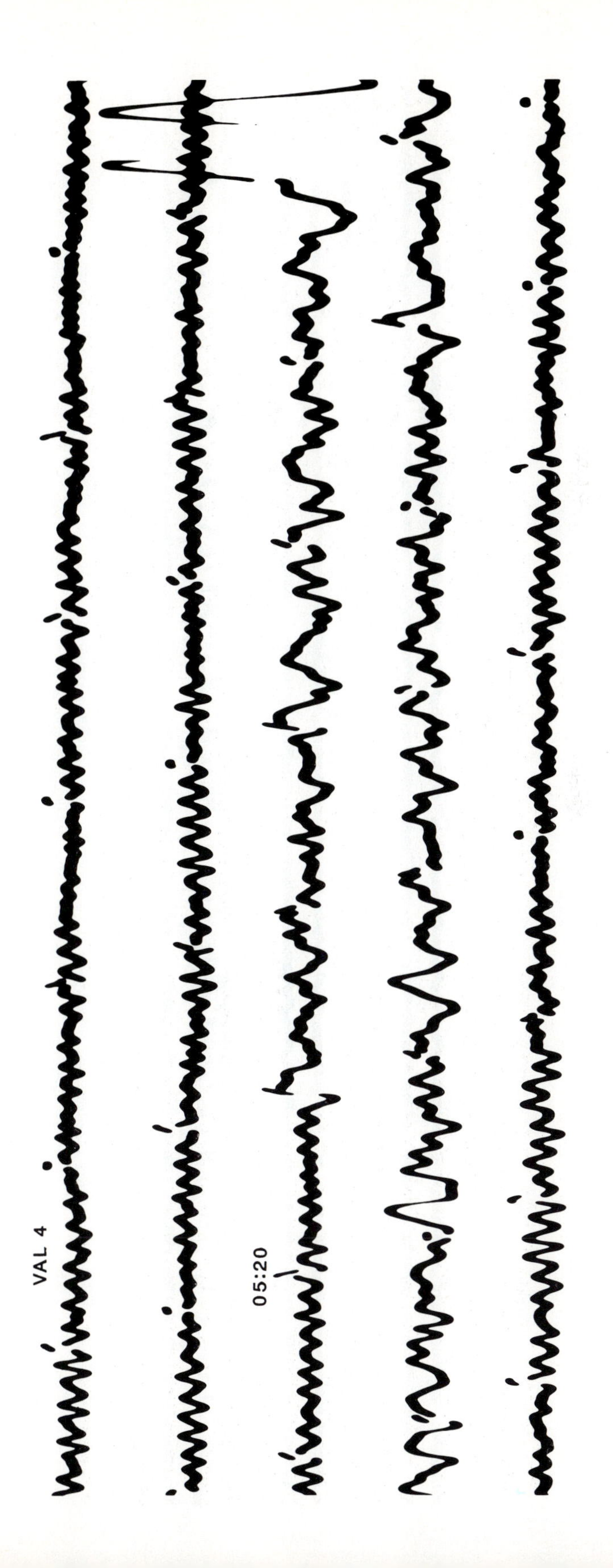
VAL 4
05:20

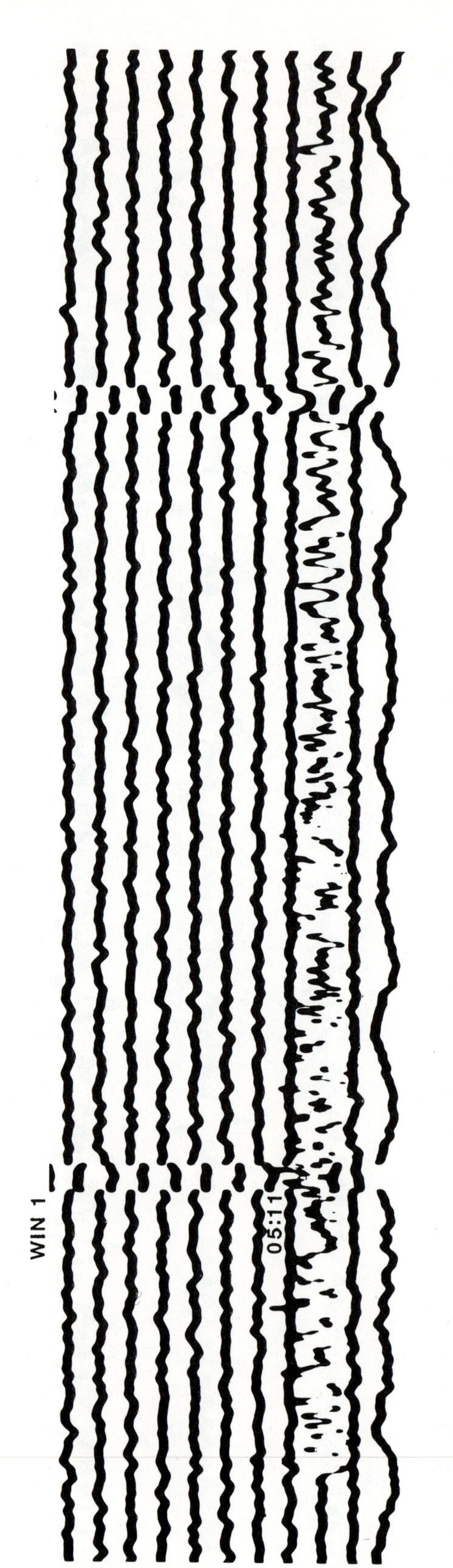

Figure 6.44: Seismogram for Practical 4. WIN(1)

Exercises

1. Use the reciprocal theorem to argue that seismic rays are reversible.
2. The moment tensor for shear displacement is given by (6.36). Explain why it is impossible to determine $\boldsymbol{\nu}$ and $[\mathbf{u}]$ separately from teleseismic observations.
3. An earthquake involving shear displacements has moment tensor

$$\mathrm{M} = \begin{pmatrix} 0 & 1 & 0 \\ 1 & 0 & 1 \\ 0 & 1 & 0 \end{pmatrix}$$

Find the possible orientations of the fault plane and the auxiliary plane.

4. Determine the equivalent force system for the moment tensor in the previous question.
5. Show that the displacement for slip $([u_1], 0, 0)\delta(t)$ on a fault with normal $(0, 0, 1)$ is

$$u_n(\mathbf{x}, t) = \int_{\xi_3=0} \mu[u_1] \left(\frac{\partial G_{n1}}{\partial \xi_3} + \frac{\partial G_{n3}}{\partial \xi_1} \right) d\xi_1 d\xi_2$$

and that

$$u_n(\mathbf{x}, t) = \int_{\xi_3=0} \left\{ \mu[u_1] \frac{\partial G_{n1}}{\partial \xi_3} - \frac{\partial}{\partial \xi_1} \left(\mu[u_1] \right) G_{n3} \right\} d\xi_1 d\xi_2$$

and determine the two different equivalent body forces. This shows the equivalent force system is not unique.

6. Show for an isotropic solid

$$M_{kl} = \int_{\Sigma} \left[\mu \left([u_k]\nu_l + [u_l]\nu_k \right) + \lambda [\mathbf{u} \cdot \boldsymbol{\nu} \delta_{kl}] \, dS \right.$$

7. Determine the trace of the moment tensor in the previous question and relate it to the change in volume.
8. Construct a moment tensor for an event with slip vector parallel to $(1, 1, 1)$ and normal to the fault plane along $(1, 0, -1)$, with seismic moment M. Verify the moment tensor has zero trace.
9. Show that a general symmetric moment tensor is equivalent to two double couples and a change in volume.
10. Sketch the fault plane solutions for (a) an explosion (b) a rapid phase change involving volume reduction (c) a tear about a vertical plane.

Summary

Source theory is based on two fundamental theorems: UNIQUENESS and RECIPROCITY. The first of these is self- explanatory; the displacement can be calculated provided we have either displacement or tractions, or a linear combination of both, on the surface. The second asserts that forces and their consequent elastic displacements can be exchanged, provided we reverse their times of application to conform with requirements of causality. The theorem depends on conservation of angular momentum. One consequence is that the elastic response to any applied force can be written as the superposition of displacements caused by point forces — the REPRESENTATION THEOREM. Earthquake sources are not body forces but movement across breaks in the elastic medium. We showed they are equivalent to body forces — the EQUIVALENT FORCE SYSTEM. These fictitious forces serve merely as an alternative mathematical representation of the source. Conservation of momentum and angular momentum by an intrinsic source requires the equivalent force system to have no net component or couple; the simplest configuration is two opposed couples, the DOUBLE COUPLE SOURCE.

In general a source is represented as the MOMENT DENSITY TENSOR, defined across the fault surface as a function of time for the duration of the event. For small, sudden events only the integrated MOMENT TENSOR need be considered, it has six independent elements. The simplest type of faulting is SHEAR FAULTING in which the displacement across a fault lies in the plane of the fault. This has a double couple equivalent force system and radiates energy with upward vertical FIRST MOTION in two quadrants, and downward vertical first motion in the other two quadrants. By observing the first motions at teleseismic distances we can map the quadrants and, taking account of the refraction of waves in the earth, determine the FAULT PLANE SOLUTION, the orientation of the fault plane and SLIP VECTOR.

The response of the earth to a point force is given by GREEN'S TENSOR. A tensor is needed because it relates a vector (force) to another vector (displacement). Calculation of the Green's tensor for realistic earth models is the main challenge facing theoretical seismology. Three main methods are used: ray theory (geometrical and physical optics); REFLECTIVITY, in which reverberations between layers are summed; and normal mode summation. Each method has advantages for certain applications.

Further reading

AKI & RICHARDS give detailed theory for the uniqueness, reciprocal and representation theorems. KASAHARA gives a good elementary description of seismic source theory. KENNETT gives a formalism for the reflectivity method; HUDSON describes some aspects of ray theory.

7

Plate tectonic theory

We now know how to use seismology to determine elastic properties within the earth, to locate earthquakes and hence map active faults, and to determine qualitatively the direction of relative movement across faults using fault plane solutions. The striking concentration of earthquakes into narrow belts, shown in Figure 1.5, is itself strong evidence for restricting deformation to linear boundaries between large aseismic blocks, as plate tectonic theory does, but the excellent locations achieved by the current network of seismometers was not available in 1960, and mislocation made the seismic belts very diffuse. Magnetic anomalies gave the first quantitative measure of plate motions; seismology was used subsequently in a qualitative confirmation of the plate motions. Today seismology is used in understanding active tectonic processes, which are far more complex on a small scale than the following discussion of global tectonics would suggest. Much of this chapter is concerned with the fundamental notions of plate tectonic theory, which must be developed before understanding how it relates to seismology.

The similarity between the coastlines of Africa and South America has prompted speculation about drifting continents since the time of Wegener, who wrote on the subject at the beginning of this century. However, the idea did not receive widespread approval until about 1960, despite support from such distinguished geologists as Arthur Holmes and the continual efforts of paleomagnetists who were obtaining paleomagnetic directions which could only be explained by large lateral movements. What is now known as the Plate Tectonic, or New Global Tectonic revolution, began in about 1960. It took only a decade for the subject to progress from the proposal of sea floor spreading by H. Hess to the development of the whole theory of plate tectonics and verification of its predictions using magnetic anomalies, seismology, and the deep sea drilling program.

Why did this development have to wait so long? In the author's view, as a graduate student starting research in 1968 in the Department of

Geodesy and Geophysics in Cambridge where many of the discoveries had recently been made, there were three reasons. The first is that a new breed of scientist — some marine geologists and some geophysicists with a physics background — had begun to work in the subject: they were prepared to take the possibility of continental drift seriously and were prepared to push simple concepts to their limit.

A good example of this approach is the "Bullard fit" of the American continents to Africa and Europe (Figure 7.1). There had been many previous attempts to fit these and other continents back together (indeed Wegener is said to have hung maps upside down in his search for new reconstructions) but they were all flawed in some way, and none of them were quantitative. For example, some early reconstructions involved translating coastlines on a Mercator map, which does not reproduce the true coastline on the earth's spherical surface (the Mercator projection preserves relative local angle but not length). Bullard, Everitt & Smith (1965) introduced several important new ideas into the reconstructions:

Figure 7.1: A statistical fit of the American continent to Africa and Europe. The fit used the continental slope rather than the coastline, rotated Spain to close the Bay of Biscay, and was carried out properly on a sphere. From Bullard *et al.* (1965).

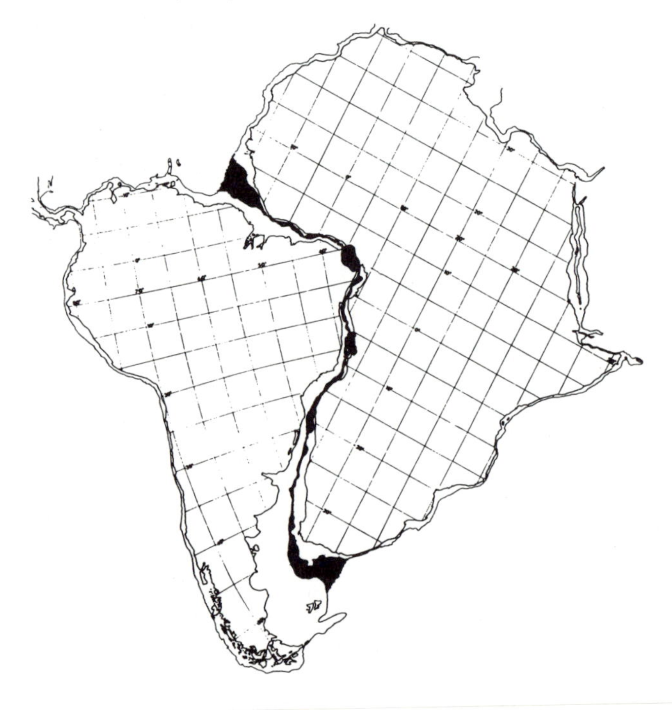

they used the edge of the continental slope, rather than the coastline, as the relevant geological boundary (this was very important in the north Atlantic), they found the best fit using the correct spherical geometry, and they regarded Iceland as oceanic crust, closed the Bay of Biscay by rotating the Iberian peninsula, and accounted for the gap near the Niger delta by erosion, which could only occur after formation of the Atlantic Ocean. Thus continental drift was no longer a curiosity: in the right hands it had become a theory capable of predicting tectonic events. Indeed subsequent work led to the discovery of magnetic stripes in the Bay of Biscay and clear proof of its opening subsequent to the opening of the Atlantic Ocean, and evidence that, geologically, Iceland is oceanic rather than continental.

The second and most important reason for the development of plate tectonic theory at that particular time was the rapid development of marine geology after the second world war. Marine seismology was essentially unknown before the war, and on the rare occasions when the geology of the deep ocean floor had been contemplated by geologists it was in the way one might have studied the other planets before the space age — without any hope of ever sampling it. Marine research changed all that, and the relative freedom with which a ship can sail international waters, compared with the political restrictions placed on land-bound seismic experiments, meant the oceans could, in some respects, be mapped more completely than the land. The geology of the ocean floors is very much simpler than that of the continents: the crust is thin, less than a fifth the thickness of continental crust; it is young, representing less than 5% of the earth's history; and the mid-ocean ridge system is a most impressive topographic feature. Sediment thickness increases with distance away from the ridge, suggesting the crustal age increases away from the ridge. This prompted Hess' hypothesis of sea floor spreading.

The third reason is the deployment of the *WWSSN* system of seismometers which were described in Chapter 4. The good global distribution of these seismometers and their prompt and reliable reporting allowed earthquakes to be located far more accurately than hitherto. Prior to the *WWSSN* the global seismicity map showed earthquakes occurring in very diffuse seismic zones. *WWSSN* data provided seismicity maps like that in Figure 1.5; the seismic zones are much narrower and the errors in the hypocentre locations are probably smaller than the size of one of the dots used to mark them on the map. These zones tell us the outline of the plates; deformation is restricted to a few very narrow zones.

7.1 Plates

What is a plate? And why use the term plate tectonics rather than the earlier term of continental drift? Long ago H. Jeffreys raised serious objections to the idea of continents ploughing their way through the ocean floor because it is too strong to allow deformation on the scale required. These valid objections are circumvented in modern plate tectonics, in which deformation is taken up by slip at the edges of rigid blocks. The change in terminology is useful in distancing us from the idea of continents ploughing along like ships in an olivine sea — the mantle. A second reason for the change in terminology is that the continents are to a large extent irrelevant to plate tectonics, as we shall see. Plate tectonic theory was devised originally to explain the observations of marine geologists, and despite its unprecedented success in explaining global tectonics it can only be applied to continental regions with great caution.

The fundamental hypothesis of plate tectonics is that the major part of the earth's surface moves rigidly and deformation only occurs at the edges of these rigid blocks in the thin zones delineated by the seismicity as in Figure 1.5. Earthquakes are our major diagnostic for deformation, but they only occur in brittle materials that are capable of fracturing: ductile material will flow under stress and there is no opportunity for large stresses to build up and cause brittle failure.

All earthquakes occur within the uppermost 700 km of the earth, in the upper mantle, but those events which are deeper than about 100 km are clearly associated with a few thin, anomalous zones. It is therefore reasonable to suppose that only the top 100 km or so is sufficiently brittle to sustain earthquakes. This is because only this top layer is sufficiently cold to be brittle: it is called the *lithosphere*. Deep earthquakes occur within cold lithospheric material that has sunk deep into the mantle. It is this lithospheric layer of some 80–100 km thickness that constitutes the *plate* that moves rigidly without fracturing or deformation, at least in the elementary form of plate tectonic theory which we shall study.

Beneath the lithosphere the mantle is hotter and flows more easily; it can deform to accommodate the motion of the overlying lithosphere without fracturing. It is called the *asthenosphere*. During the last ice age much of northern Canada and Fennoscandia were glaciated. As this ice load melts, the land in these regions rises. Measurements of the uplift suggest part of the deep earth, about 100 km down, must flow more readily than the lithosphere. This is the main evidence for the asthenosphere. "Mantle convection" is an exciting area of current research but we shall not study it here; we restrict ourselves to the study of those deformations which we can detect — those that cause earthquakes. Plate tectonics is a purely kinematic theory: it seeks to describe the motions of the plates at the earth's surface but it does not address the cause of the motions or the underlying forces driving them. The dynamics of plate motions, the forces that drive mantle convection, is beyond the scope of this book.

The lithosphere is therefore a thermo-mechanical boundary layer at the top of the mantle. Its thickness is determined by the depth of the isotherm at which it loses its strength and ceases to be able to sustain earthquakes. This isotherm varies in depth, being shallow near mid-ocean ridges and deeper in older oceanic regions. The lithosphere consists of both crust and upper mantle; the oceanic crust is a mere 5 km thick skin of basaltic material at the top of the lithosphere, whereas the upper mantle extends to 600 km and includes the asthenosphere (Figure 7.2).

The major plates are outlined and named in Figure 1.7. The boundaries are delineated by the seismicity map. Some smaller plates have been proposed in certain areas, but these will not concern us here.

Plate boundaries in the oceans are observed to be of three distinct types, each with its own tectonic style. These are *spreading rifts* or *ridges*, so-named because of their association with the mid-ocean ridge system, which generally open perpendicular to their strike; *transform faults* with motion parallel to their strike; and *trenches*, so-named because of the deep ocean trenches associated with them, where the motion is convergent but not necessarily normal to the strike. These three types of boundary are drawn as shown in Figure 7.3.

Figure 7.2: A "plate" is about 80 km thick; its bottom is a thermo-mechanical boundary where the high temperature removes the essential brittle quality of the lithosphere. The underlying asthenosphere is ductile. Note the lithosphere is distinct from, and very much thicker than, the crust, which in the oceans is only 5 km thick.

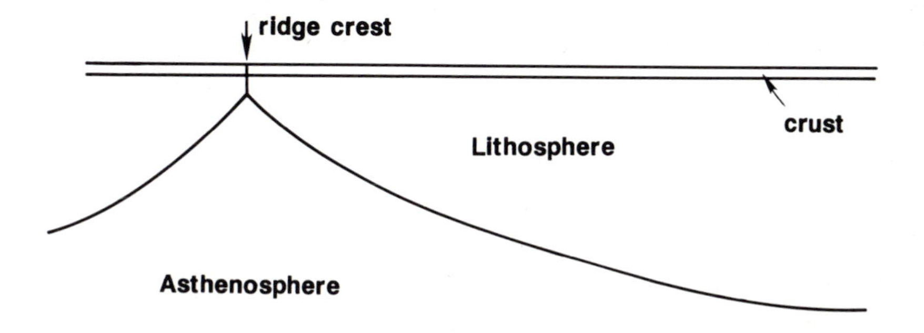

We discuss each type of boundary in turn, but we point out here the unusual nature of the spreading motion at ridges, which generally produces plate on either side at the same rate and does not have any oblique motion; likewise that transform faults never exhibit anything other than pure shear motion. The fundamental ideas of plate tectonics may be summarised as four simple postulates.

1. The earth's surface is divided into a small number of plates, which move rigidly and independently without internal deformation.
2. Plate boundaries are of three types: ridges, transforms, and trenches; with spreading, shear, and convergent motions respectively.
3. Ridges spread normal to their strike and the spreading rates are symmetrical.
4. Transform faults have only shear motion with no convergence or spreading.

As with all sets of rules there are exceptions: for example, there are cases of "leaky" transform faults that allow some spreading; but overall these rules are remarkable for their consistency with observed behaviour rather than their exceptions.

7.2 Poles of relative rotation

The plate tectonic hypothesis means that we can describe all movement on the earth's surface simply by specifying the relative motion of each of the plates. The spherical shape of the earth's surface is

Figure 7.3: The standard representation of the three types of plate boundaries. Arrows show relative motion. By convention the arrow is drawn on the underthrusting (subducted) plate at a convergent boundary (c).

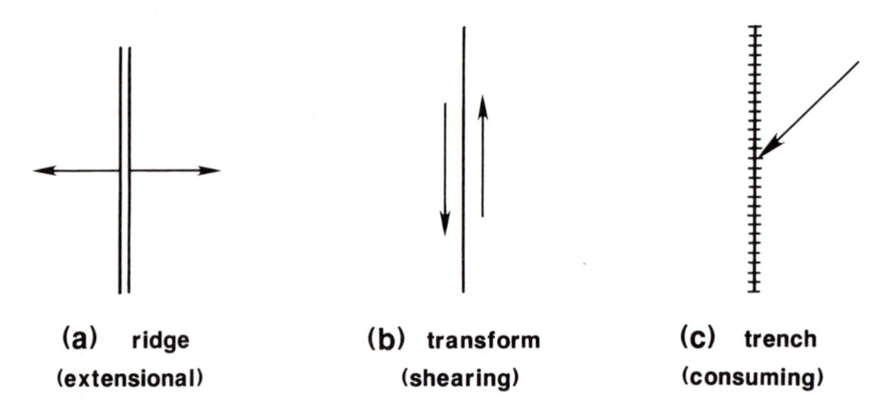

important, and we must bear in mind that each plate is a spherical cap. We make use of an old result of geometry:

> **Euler's theorem**: *movement on a sphere can be represented by rotation about an axis.*
> It is easy to see that any point on a sphere can be rotated into any other point; for example we can draw the great circle path between the two points and take the rotation axis to be perpendicular to the diametric plane. The surprising aspect of Euler's theorem is that the same applies to a patch of the surface of the sphere. Consider a patch containing two fixed points A and B; these points move to A', B' (Figure 7.4(a)). Let R_1 be the pole of the rotation which takes A into A'. In general this rotation will not take B into B'; assume it takes it to B''. The patch is rigid so the lengths AB, $A'B'$, and $A'B''$ are equal. Therefore there is a rotation R_2 with pole at A' which takes B'' into B'. B was an arbitrary point, so the operation R_2R_1 gives the required movement. It will only be a rotation if it leaves two points on the sphere unmoved; those points will be where the rotation axis meets the surface of the sphere. They can be constructed as shown in Figure 7.4(b). All the lines in Figure 7.4(b) are great circle paths. Rotation R_1 about angle θ takes all points on R_1P_1 into the line R_1P_2. The spherical triangles $R_1R_2P_1$ and $R_1R_2P_2$ are similar (2 sides and included angle equal) and therefore $R_2P_1 = R_1P_2$. Consequently rotation R_2 takes all points on R_2P_2 into R_2P_1. The stationary point is therefore the intersection point P_1. It is easy to show there is another intersection point on the opposite side of the sphere. P_2 is the pole for the rotation R_1R_2, which is different from R_2R_1 in general because rotations do not commute.

Let us fix one plate arbitrarily. The movement of a second plate relative to this fixed plate can then be specified by an axis and an angle of rotation. The axis cuts the sphere in two points; we adopt the convention shown in Figure 7.5 and define the pole of relative motion to be that point where the axis cuts the sphere and the relative motion is in the sense required by the right hand rule. At this stage our definition appears to require that we fix one plate arbitrarily, but this is not so: the axis of rotation is a line that remains fixed relative to both plates regardless of our frame of reference. The two points where the axis cuts the sphere are the only two points that remain stationary relative to both plates. We can therefore make a more general and succinct definition. The *pole of relative rotation* between two plates is one of two points on the sphere that remain stationary with respect to both plates; the point is selected by convention from the right hand rule as shown in Figure 7.5. Finite rotation is not a vector, although it can be represented by magnitude (the angle of rotation) and direction (the axis of rotation); such a "vector" would not satisfy the parallelogram law of addition (or more strictly the commutative property of vector addition). Angular velocity or infinitesimal rotation can be represented as a vector, however, and in plate tectonics when we wish to treat instantaneous motions we shall find the vector representation of instantaneous angular velocity most useful. We define the *pole of instantaneous angular velocity* in the same way as the pole of relative rotation. Finite rotations relate to relative positions; instantaneous rotations to relative velocity. If we wish to use instantaneous rotation poles to calculate positions in the geological past then we implicitly assume the poles have remained fixed. There is no reason this should be so, and examples are known when poles of instantaneous angular velocity have moved. It is therefore a dangerous assumption.

Note here that any finite rotation could be represented as a single axis even though the complete time evolution might not be capable of representation by a single angular velocity vector. A simple example of a pole of rotation is given in Figure 7.6. The pole of instantaneous motion lies on the extension of the trench segment shown. Relative motion is

Figure 7.4: Movement of a finite rigid patch on a spherical surface can always be represented as a rotation. P_1 is shown in (b) to be an invariant point which is therefore the axis of the rotation.

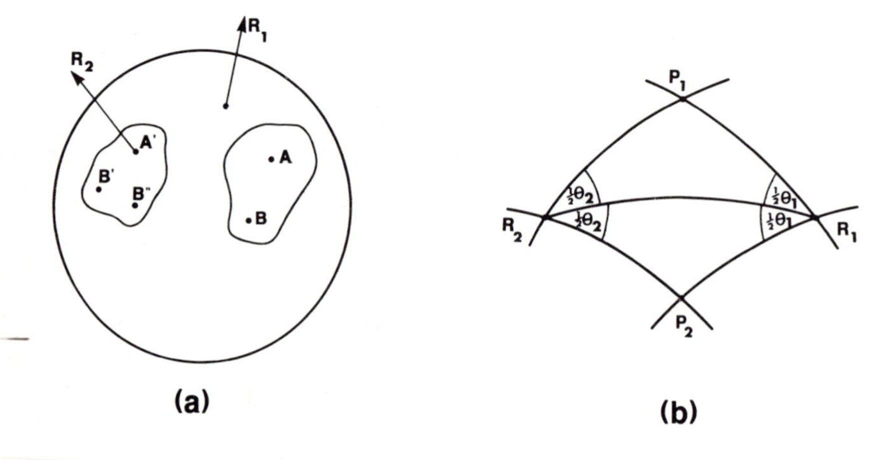

Figure 7.5: ω is the angular velocity vector of plate B relative to plate A. If plate A is regarded as fixed then plate B moves as shown by the arrow. The right hand convention required the vector to be drawn in the direction shown. The rotation pole is the point P where $\boldsymbol{\omega}$ cuts the sphere.

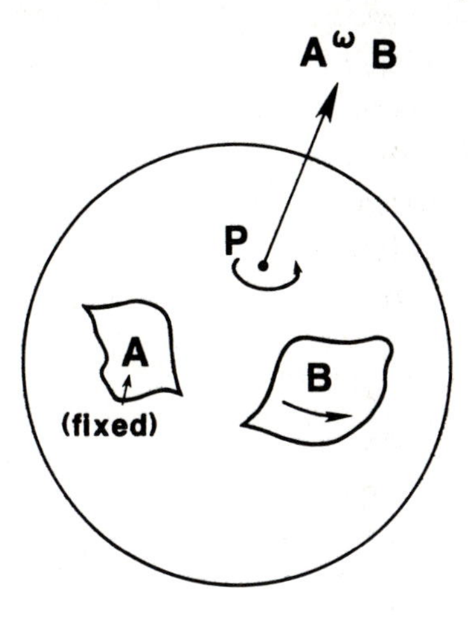

Figure 7.6: Convergence rates at a consuming boundary depend on the distance from the rotation pole P. In this simple case P lies on the extension of the trench separating plates A and B and the convergence is normal to the strike of the trench.

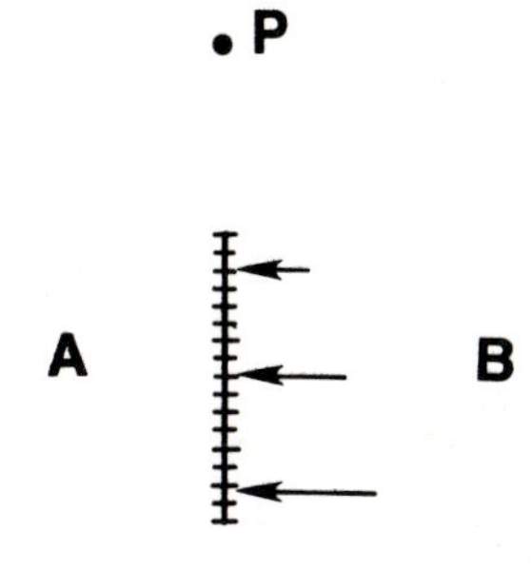

normal to the strike of the trench, and the convergence rate increases with distance from the trench.

Angular velocities can be combined by vector addition. Take three plates A, B and C as in Figure 7.7. Write the angular velocity of B relative to A as ${}_A\omega_B$; that of A relative to C as ${}_C\omega_A$; and that of C relative to B as ${}_B\omega_C$. Clearly the motion of B relative to C is the combined effect of rotating B relative to A and A relative to C, so combining by vector addition we have

$$ {}_A\omega_B + {}_B\omega_C + {}_C\omega_A = 0 \quad (7.1) $$

This equation is used in determining poles of instantaneous rotation when two angular velocity vectors are known but the third is not, perhaps because of inadequate knowledge about the plate boundary (there are few methods for accurate measurements of relative motion across trenches, for example). Relative motion between Pacific and Indian plates in the south Pacific is a good case in point.

A word of caution about relative and absolute plate motions is in order. The techniques we shall discuss apply only to relative motion across plate boundaries. Any plate may be fixed arbitrarily but to establish absolute motions we need an absolute frame of reference. The earth's rotation axis is an absolute reference point (although it precesses every 26500 yr) and paleolatitudes may be determined from paleomagnetism (assuming the earth's magnetic field has always been an axial dipole, as it is today)

Figure 7.7: Angular velocities sum as vectors. Thus for any three plates the angular velocity of C relative to A is the sum of C relative to B and that of B relative to A.

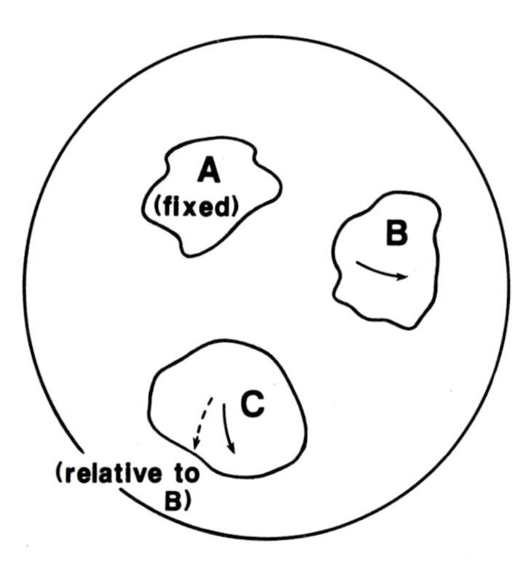

or paleoclimate; but there is still a problem with the longitude. Some absolute geological frame is needed but all attempts to find one have been fraught with difficulties. We do not consider them further in this book: the interested student is referred to the reading list at the end of the chapter.

Another confusing term that has been used is *polar wander.* A paleomagnetic determination of latitude involves measuring the local magnetic field direction and determining the distance from the magnetic pole assuming the magnetic field to have a dipolar configuration. The result can be interpreted variously as plate movement, movement of the magnetic pole, or movement of the rotation pole (or perhaps a non-dipole magnetic field). All the evidence suggests that changes in the local magnetic field direction are caused by plate movement. The term polar wander does not imply that the pole moves, but that the apparent position of the pole moves. We shall not use the term polar wander.

7.3 Ridges

The mid-ocean ridge system stands several thousand metres above the deep ocean floor (Figure 7.8). It has only been surveyed in detail comparatively recently because of the difficulty in making depth measurements in the deep ocean before the advent of echo sounding. The mid-ocean ridges are young; the overlying sediment cover is very thin or non-existent, and rocks dredged from near the ridge crest are mainly fresh basalts. Deep sea photographs show pillow lavas and have occasionally recorded lava eruptions in progress; eruptions have also been heard on seismic recording instruments towed by ships.

Large earthquakes occur on ridges (up to surface wave magnitude 7) although the overall seismicity is not as high as in trench boundaries.

Figure 7.8: Schematic diagram of the bathymetry of a mid-ocean ridge. The central median valley is characteristic of slowly spreading ridges such as the mid-Atlantic; it is typically 5 km across.

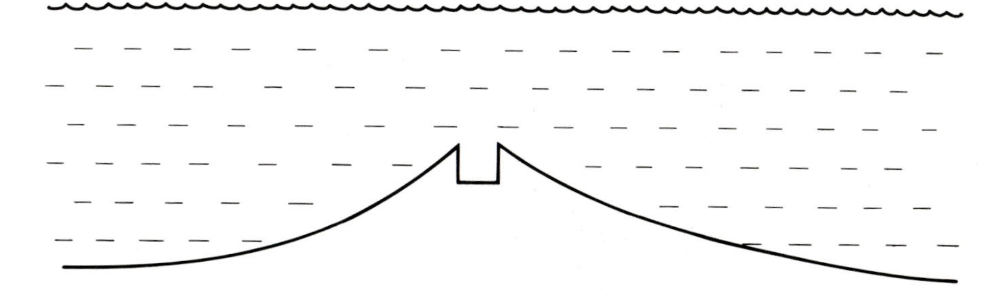

Ridge earthquakes are difficult to detect and locate accurately because of their great distance from recording stations, which are mainly on continents. None of the earthquakes are deep, and most show normal faulting focal mechanisms. These mechanisms are evidence of spreading motion at the ridge.

With such a high topography at the ridge crest we might expect to see a gravity anomaly because of all the excess mass, but the free-air gravity anomaly[1] is virtually zero. The excess mass above the mean depth of the oceans must be compensated by a mass deficit at depth, beneath the mid-ocean ridge mountains. Such compensation is known from other mountain ranges and is called *isostatic compensation.* Stresses are needed to support a topographic feature that sits on a surface. As the stresses relax the load sinks into the surface, reducing the topographic feature. If the load is made of less dense material than the underlying rock then it will "float" after all shear stresses have relaxed: the pressure at depth will be hydrostatic. In this case the free-air gravity anomaly will be zero, and we must deduce that this is the situation at the ocean ridges.

The young crust and lack of sediment at the ridges shows that crust is being formed there from mantle material that wells up to fill the gap left by the spreading plates. The lithosphere-asthenosphere boundary is close to the surface near ridges; it is not surprising, therefore, that the mantle cannot support the load of the topographic feature of the mid-ocean ridge mountains and there is no free-air gravity anomaly.

7.4 Magnetic anomalies

All this would be weak evidence for plate movement were it not for the magnetic recording left on the ocean floor: the *magnetic stripes* that parallel the ridges are a complete history of the plate motion and are our most valuable data in the reconstruction of former plate configurations. Magnetic surveying at sea has gone on since the time of Halley, at the end of the seventeenth century, but routine surveying of total magnetic intensity was made much easier by the invention of the proton magnetometer. A magnetometer can be towed behind a ship on a routine survey to give a trace of intensity as a function of position along a ship's track. The actual measurement is the intensity of $\mathbf{H}_t$, the main core field plus the anomaly field: $|\mathbf{H}_t| = |\mathbf{H}_c + \mathbf{H}_a|$. An *anomaly* is calculated from the measurement by subtracting the component of $\mathbf{H}_c$ along $\mathbf{H}_t$. These fluctuations are attributed to variations in the magnetisation of the crustal

[1]The value of g is corrected to a reference height, assuming no mass between observation and reference levels.

rocks (sediments are less strongly magnetised than basalts) and must be converted to an estimate of the magnetisation of the rock.

Originally it was presumed that the basaltic layer (layer 2) was magnetised to a depth of a few hundred metres, and that this provided the major part of the observed anomaly. Subsequent drilling suggests a more complicated picture: the magnetisation of the rocks recovered was weak, requiring magnetisation to greater depth to explain the observed anomaly strength, and vertical sections contain reversal boundaries, suggesting that the system of dykes and intrusions at the mid-ocean ridge is complicated, or that the edges of the magnetised blocks are tilted.

If the ship steams perpendicular to the strike of the mid-ocean ridge a sequence of peaks and troughs appears on the magnetometer trace; by correlating adjacent peaks one can map a pattern of stripes of alternatively strong and weak total intensity (For example Figure 7.9). The pattern of magnetic stripes corresponds to the time scale of reversals of the earth's magnetic field which is based on results from lava flows on continents (Figure 7.10). The magnetic stripes were explained in 1963 in what is now known as the Vine-Matthews hypothesis of sea-floor spreading. The sea floor spreads at the ridges, allowing deeper mantle material to rise to fill the gap. This material cools through the Curie temperature and becomes magnetised; the magnetisation it acquires depends on the direction of the magnetic field at the time. As the earth's magnetic

Figure 7.9: Magnetic anomaly traces as measured by a proton magnetometer towed behind a ship. These traces have been processed by subtracting a mathematical model of the core field. The profile across the Juan de Fuca ridge is a composite of several tracks. From Vine (1966).

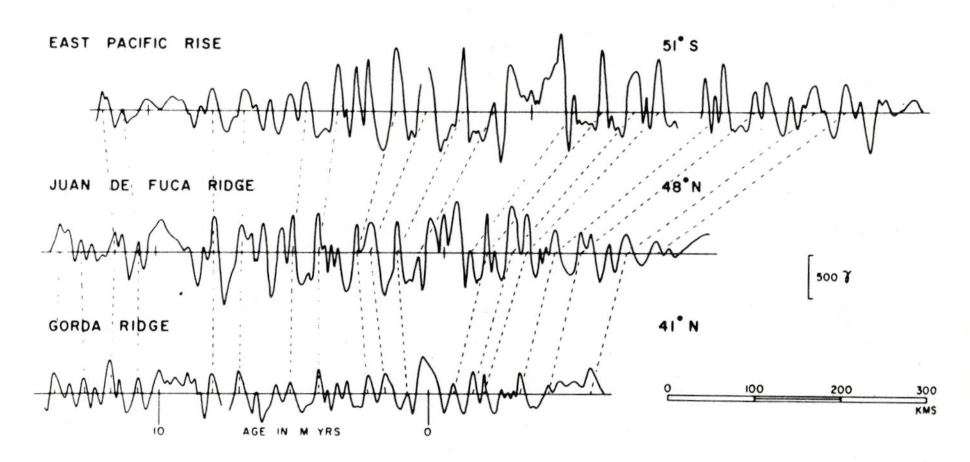

field reversed irregularly the new oceanic crust became magnetised in alternating directions, leading to the observed pattern of stripes.

The ridge acts as a large tape recorder; the oceanic crust is the tape, the magma production system at the ridge crest the magnetising head, with the earth's varying magnetic field providing the signal. We can take the analogy further and calculate the resolution of our ridge-tape recorder. An ordinary domestic tape recorder has a head size of about 0.1 mm and a tape speed of 50 mm s^{-1}. The highest frequency the recorder can respond to is typically 10 kHz; it is obviously proportional to tape speed and inversely proportional to head size. Scaling up to the proportions for the earth we take a ridge "recording head" to be 5 km wide, a tape speed of 5cm/yr (typical for the mid-Atlantic, a slow spreading ridge) and obtain a frequency response of 2×10^{-4} cycles per year: one reversal every 5000 yr can be recorded.

The appropriate "head size" depends on the width of region near the ridge crest where near-surface rocks are still above the Curie temperatures. A median valley some 5 km across is often seen at the very middle of the ridges, particularly when they are slow-spreading. Dykes and faulting occur within this median valley, particularly near its edges. Our figure of 5 km comes from this width of the median valley. Slowly spreading ridges give poorly resolved anomalies; the Pacific-Antarctic ridge is currently the fastest spreading boundary anywhere: it is also the highest fidelity magnetic recorder. The Atlantic gives rather poor anomalies.

We are indeed fortunate that the frequency of reversals is right for the recording mechanism at the ridges. More frequent reversals and the recording mechanism would saturate and not be able to record a pattern of stripes; less frequent and we would not have enough anomalies to trace the plate movement. Indeed the "quiet interval" in the Cretaceous, when there were no reversals for nearly 40 Myr, has left a large area of the western Pacific and some of the Atlantic Ocean without any magnetic

Figure 7.10: The geomagnetic reversal time scale. Periods during which the field was normal (same direction as today) are coloured black. Redrawn from Heirztler, Dickson, Herron, Pitman & LePichon (1968).

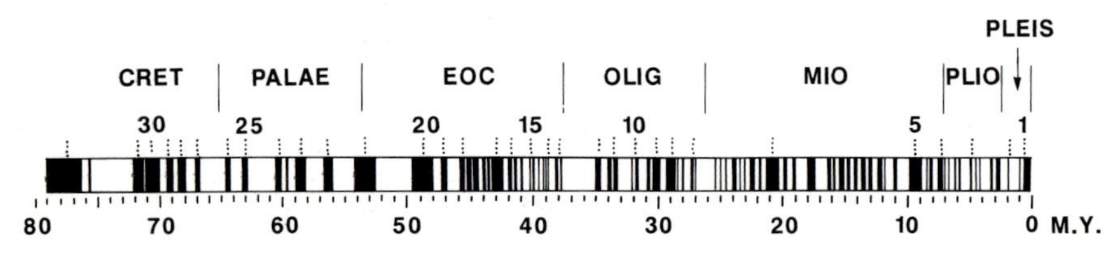

anomalies, which has obstructed attempts to reconstruct the plate motions for these early times. The end of the Cretaceous quiet zone can be seen at the edge of the anomalies in the Pacific in Figure 7.26. Periodic reversals, such as occur on the sun, would not help in determining plate motions either — the stripes would be indistinguishable from each other. Magnetic anomalies allow us to measure the spreading rate at ridges simply by measuring the distances between stripes and dating them by the reversal time scale based on records from dated lavas on land. The time scale is far from perfectly known, particularly for older anomalies, but a reliable scale has now been built up.

This gives our first and most reliable method for determining the positions of poles of rotation. Ridges spread normal to their strike, and therefore the pole must lie on an extension of the ridge (Figure 7.11). The spreading rate will increase with distance from the pole and by measuring the separation of the anomalies we can extrapolate back to obtain the pole position. In practice things may not be quite so simple and the information furnished by the magnetic anomalies sometimes gives only a partial guide to the pole position.

7.5 Cooling of oceanic plates

The topography of the mid-ocean ridges is caused by the high temperature of the lithosphere there. Plates are created at the ridges at about the temperature of the top of the asthenosphere, 1200 K say,

Figure 7.11: Determining the pole position from magnetic anomalies. The magnetic stripes lie along great circle paths passing through the pole; the pole can therefore be found by tracing them back along great circles to their intersection point.

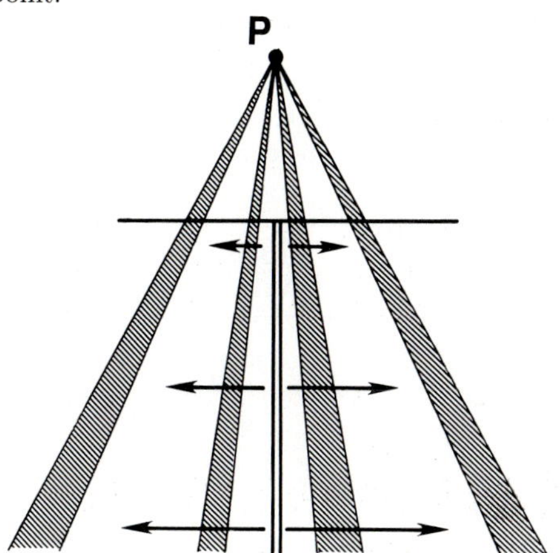

and they cool as they move away from the ridge. Thermal contraction can then explain the increase in the ocean depth with distance from the ridge. This mechanism only works because the thermal time constant for the lithospheric plate is comparable with the time it takes the ridge to spread an appreciable distance. Applying the simple calculations of conduction times given in Chapter 2 (page 42) we find the time constant for lithosphere 100 km thick to be $\tau_k = 260$ Myr (in fact a lower value of 65 Myr is widely accepted). A spreading rate of 5 cm yr^{-1} will take a plate 13,000 km in that time — the width of the Pacific Ocean. The lithosphere in the western Pacific has the oldest oceanic crust in the world; it has probably cooled to thermal equilibrium by now, whereas younger lithosphere is still cooling.

The simplest model of plate creation at a ridge is one that produces plate at the same uniform temperature, T_1, equal to that at the top of the mantle, on both sides of the ridge. Consider the model shown in Figure 7.12, in which two uniform slabs form at a ridge at $x = 0$ at temperature T_1 and spread symmetrically at velocity v. The lithosphere-asthenosphere boundary does not correspond to the depth $z = 0$; this depth simply denotes that of the bottom of the cooling slab under consideration. As the slab cools the isotherm defining the lower boundary of the lithosphere will migrate downwards and the lithosphere will thicken with distance from the ridge.

As the plates move away they lose heat to the sea floor and cool. The temperature of the sea floor remains constant at $T = 0$ and the temperature at the bottom remains a constant T_1. This is a standard problem in heat flow, similar to that studied in the problems to Chapter 2

Figure 7.12: Cooling of oceanic plates. The two plates pull apart at the ridge and allow material at temperature T_1 to well up and fill the gap, accreting onto the plate. The plate cools by conduction and moves away from the ridge.

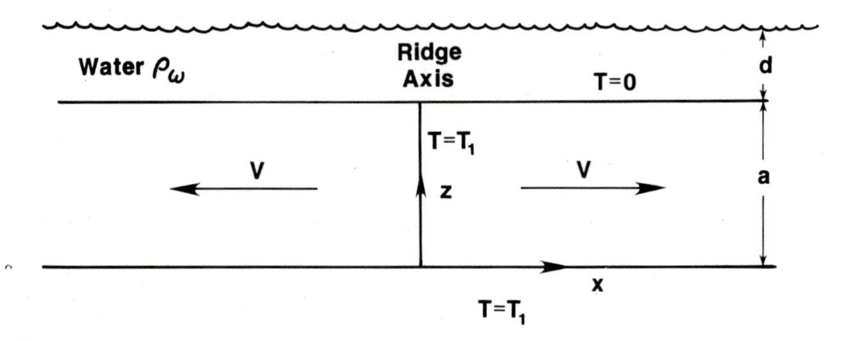

except that here we have a moving medium. The equation governing heat loss is

$$\frac{DT}{Dt} = \frac{\partial T}{\partial t} + \mathbf{v} \cdot \nabla T = \kappa \left(\frac{\partial^2}{\partial x^2} + \frac{\partial^2}{\partial z^2} \right) T \tag{7.2}$$

We seek a steady solution to this Eulerian equation: the ridge is supposed to have spread at a uniform rate and in the same direction for all time, and there is no overall cooling or heating of the system. The heat equation becomes

$$v \frac{\partial T}{\partial x} = \kappa \left(\frac{\partial^2}{\partial x^2} + \frac{\partial^2}{\partial z^2} \right) T \tag{7.3}$$

It is now convenient to put the equations into dimensionless form by setting $x = ax'$, $z = az'$, $T = T_1 T'$, where a is the thickness of the cooling slab, to give

$$\frac{va}{\kappa} \frac{\partial T'}{\partial x'} = \left(\frac{\partial^2}{\partial x'^2} + \frac{\partial^2}{\partial t} z'^2 \right) T' \tag{7.4}$$

Hence the solution depends only on the dimensionless combination

$$P_e = va/\kappa \tag{7.5}$$

P_e is called the *Peclet number*. When the plate has cooled completely the temperature profile will be a linear decrease from T_1 at the bottom to zero at the top of the slab; we subtract this from the temperature profile and solve for the transient by assuming a separable solution of the form $X(x')Z(z')$. This gives the simple harmonic motion equation for both parts $X(x')$ and $Z(z')$. We anticipate a temperature which decays with distance from the ridge and choose the separation constant to give exponential solutions for $X(x')$. The solutions for $Z(z')$ are sines and cosines and we set the cosine terms to zero to satisfy the boundary condition at $z' = 0$. Superimposing solutions for all allowed values of the separation constants gives

$$T' = 1 - z' + \sum_{n=1}^{\infty} A_n \exp \alpha_n x' \sin k_n z' \tag{7.6}$$

where

$$k_n = n\pi \tag{7.7}$$

to satisfy the boundary condition on $z' = 1$, and

$$\alpha_n = \frac{P_e}{2} - \left(\frac{P_e^2}{4} + n^2 \pi^2 \right)^{\frac{1}{2}} \tag{7.8}$$

to satisfy the differential equation. The exponentially growing solutions, corresponding to positive α_n, have been excluded, since we require the transient to tend to zero as $x' \to \infty$.

The boundary condition on $x' = 0$ is $T'(0, z') = 1$, the temperature of newly-formed lithosphere. Applying this to (7.6) and solving for the Fourier coefficients $\{A_n\}$ gives

$$A_n = \frac{2\,(-1)^{n+1}}{n\pi} \tag{7.9}$$

Substituting (7.9) back into (7.6) gives

$$T' = 1 - z' - 2 \sum_{n=1}^{\infty} \frac{(-1)^n}{n\pi} \exp\left\{ \left[\frac{P_e}{2} - \left(\frac{P_e^2}{4} + n^2\pi^2 \right)^{\frac{1}{2}} \right] x' \right\} \sin n\pi z' \tag{7.10}$$

The two observations we have are the heat flux through the sea floor and the ocean depth. The heat flux is obtained from (7.10) by differentiation. Reverting to dimensional variables gives

$$-k \left(\frac{\partial T}{\partial t} z \right)_{z=a} = \frac{kT_1}{a} \left\{ 1 + 2 \sum_{n=1}^{\infty} \exp\left[\frac{P_e}{2} - \left(\frac{P_e^2}{4} + n^2\pi^2 \right)^{\frac{1}{2}} \right] \frac{x}{a} \right\} \tag{7.11}$$

To calculate the bathymetry we need the thermal expansion coefficient but we must also take account of isostacy. The absence of a free-air gravity anomaly proves the topographic features are "floating" in the asthenosphere, and therefore we should take the pressure at the base of the lithosphere to be constant. Let the upper surface be

$$\ell\,(x) = a + e(x), \; e \ll a \tag{7.12}$$

Constant pressure at the base of the lithosphere gives

$$(d - e)\rho_w + \int_0^{\ell(x)} \rho\,(x, z)\,dz = \text{constant} \tag{7.13}$$

where α is the thermal expansion coefficient, $\ell(x)$ is the upper surface of the slab, d its depth, and ρ_w is the density of sea water. The density is given by

$$\rho = \rho_0 \left[1 - \alpha T(x, z)\right] \tag{7.14}$$

Note here that we are making an approximation: we calculated the temperature assuming a uniform thickness of the layer and yet we are now allowing for a variation in its thickness. This is a good approximation however, because the thermal contraction turns out to be very small compared with the overall thickness a.

In particular as $x \to \infty$ we have

$$d\rho_w + a\rho_0 \left(1 - \frac{\alpha T_1}{2}\right) = \text{constant} \tag{7.15}$$

The solution for $e(x)$, the bathymetry, becomes

$$\begin{aligned} e(x) &= \frac{4\alpha\rho_0 T_1}{\pi^2 (\rho_0 - \rho_w)} \sum_{k-0}^{\infty} \frac{1}{(2k+1)^2} \\ &\quad \times \exp\left\{\left[\frac{P_e}{2} - \left(\frac{P_e^2}{4} + (2k+1)^2 \pi^2\right)^{\frac{1}{2}}\right] \frac{x}{a}\right\} \end{aligned} \tag{7.16}$$

When $P_e \gg 1$

$$\left\{\frac{P_e}{2} - \left[\frac{P_e^2}{4} + (2k+1)^2 \pi^2\right]^{\frac{1}{2}}\right\} \frac{x}{a} \approx -(2k+1)^2 \frac{t}{\tau} \tag{7.17}$$

where $\tau = a^2/\pi^2\kappa$ is a thermal time constant and $t = x/v$ is the age of the lithosphere. Figure 7.13 shows heat flow data for the north Pacific Ocean plotted against the theoretical curve based on (7.11) for the parameters given in the caption and a thickness $a = 75$ km. The agreement is remarkably good considering the crude nature of the model used. Figure 7.14 shows topography for the north Pacific against the theoretical prediction of equation (7.16). Apart from anomalies caused by seamounts and oceanic island chains such as the Hawaiian chain, the agreement is again very good. This is perhaps the most remarkable prediction of plate tectonic theory — that the age of the ocean floor can be related to its depth with such reliability is a very surprising result.

Figure 7.13: Variation of heat flow with distance from the ridge, in km, in the north Pacific. The line gives the prediction from the simple conductive cooling model. The model does not give a good prediction near the ridge, where also heat flow measurements are difficult to make because of the thin sediment cover. Redrawn from Sclater & Francheteau (1970).

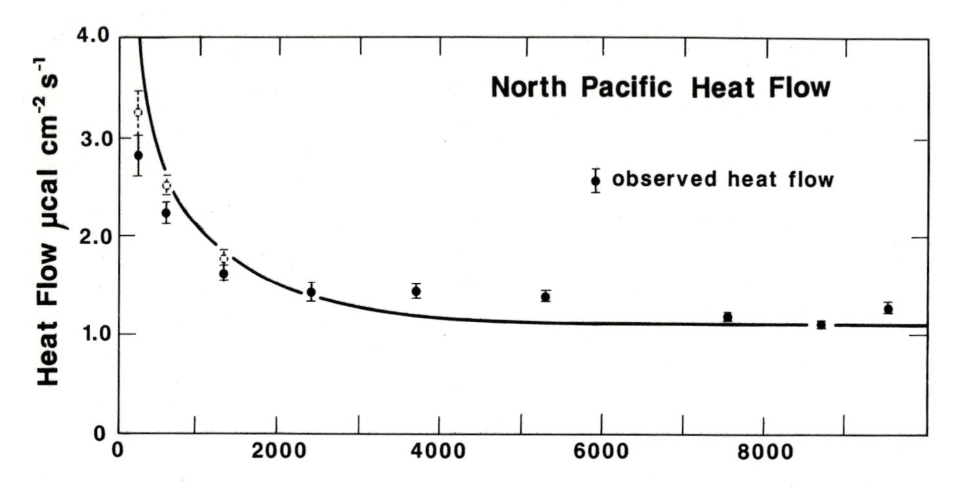

Figure 7.14: Ocean depth with distance from the ridge in the north Pacific. The dashed lines give the theoretical prediction for two different choices of the numerical values of the parameters. From Sclater & Francheteau (1970).

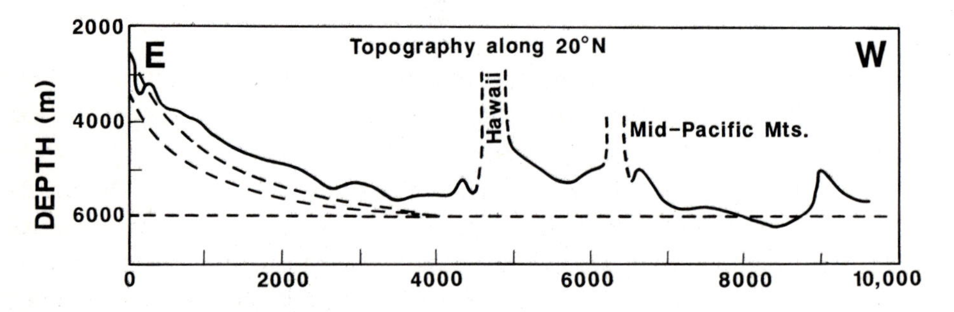

7.6 Transform faults

Transform faults appear as a scarp on the sea floor; the magnetic anomalies are offset across the fault (Figure 7.15 for example). The offset of the magnetic anomalies tells us there is a difference in age across the fault, and this in turn suggests we should expect a difference in bathymetry because the lithosphere on the older side will have contracted more. The height of the scarp depends on the age difference and may be predicted from the magnetic anomalies using the depth-age relation. The bathymetry is shown symbolically in Figure 7.15.

The seismicity is shallow on transform faults, as on ridges, but the focal mechanisms show strike-slip motion rather than normal faulting. Transform faults are most commonly seen offsetting ridge sections; the section of the fault between the ridges is seismically active but there is also a scarp usually seen as an extension of the active part of the transform fault. The whole system of active fault plus scarp extension is called a *fracture zone*; the inactive part of the fracture zones can be traced on the sea floor for many thousands of kilometres, and appear to run into continents and disappear without any sign of deformation at their ends.

Fracture zones were very puzzling before the transform fault was devised by J.T.Wilson in 1965. The simplest explanation of an offset ridge is a *transcurrent fault*, one across which the relative motion is strike slip, that has offset an initially straight ridge segment. In this case the sense

Figure 7.15: The lithosphere on either side of a fracture zone is of a different age. The difference in age is associated with a scarp on the sea floor because of the depth-age relationship. From "The Deep Ocean Floor" by H.W.Menard.

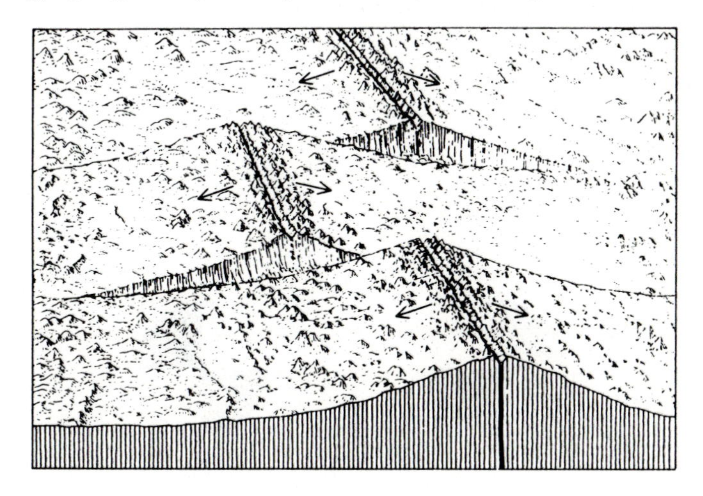

of motion is that shown in Figure 7.16(a). The sense of motion across a strike-slip fault of this nature is described uniquely as follows: imagine yourself standing on one side of the fault watching the movement of the ground on the other side; if it moves to your left the fault is *left-lateral*, if it moves to the right it is *right-lateral*. It is easy to verify that both types of fault obey this rule regardless of which side of the fault you stand on. The problem with the transcurrent fault configuration depicted in Figure 7.16(a) is that it requires deformation at the ends of the fracture zones, or at the end of the active part of the fault if we suppose the inactive extensions are relics of previous faulting. This deformation violates the first postulate of plate tectonics, that the plates move as rigid blocks. The concept of the transform fault overcomes this problem. Both sections of ridge are spreading; if we stick to our postulates of plate tectonics and allow no deformation within the plate whatever, the relative motion across the fault must be as shown in Figure 7.16(b), the opposite sense to that of the transcurrent fault. The ridges are not a static feature of the ocean floor because new ocean floor is being created at them continuously; therefore the distance between ridges may change with time depending only on the spreading rates.

The focal mechanisms of earthquakes on a transform fault, as in Figure 7.16, are opposite to those of a transcurrent fault. Focal mecha-

Figure 7.16: Relative motion across (a) transcurrent fault and (b) transform fault. The transcurrent fault produces an offset of any topographic feature but deformation must occur around the ends of the fault to accommodate the relative motion. The transform fault accommodates relative motion of the two plates without deformation at the ends.

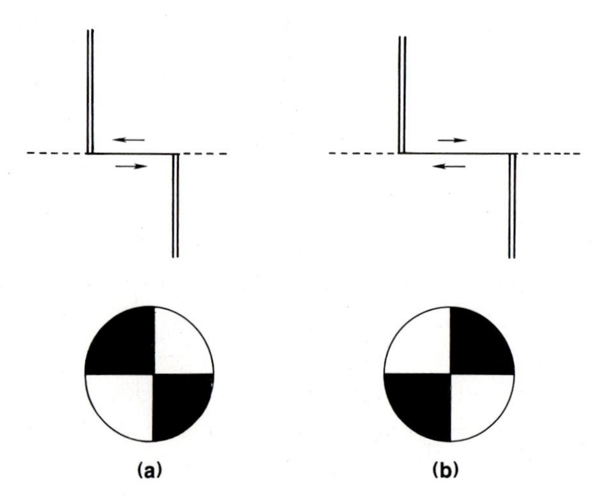

nisms of earthquakes were used to discriminate between these two ideas in favour of the transform fault idea. The scarps in the inactive parts of the fracture zones must be a consequence of the difference in the age of the ocean floor across them. There must be some relative vertical movement across these scarps in order to accommodate the changing height of the scarp; we must presume that this movement is too slow to produce any detectable seismicity.

If the relative motion across a transform fault plate boundary is horizontal strike slip the boundary must describe a small circle centred on the pole of relative motion between the two plates. This gives us a second means of determining rotation poles: if we construct a series of normals to the transform fault they should intersect at the pole. Alternatively we can plot the plate boundaries on a Mercator projection with the relevant rotation pole at the pole of the projection. The transform faults then appear as straight lines because latitude lines are straight lines on a Mercator projection.

7.7 Trenches

By trenches we mean those plate boundaries where two plates are converging. They are not always associated with a deep topographic feature on the ocean floor; at slowly convergent margins there is frequently a very thick sediment cover on the crustal rocks at the boundary, which can fill the trench and conceal the topographic feature. At rapidly converging boundaries there is a deep ocean trench with an arcuate structure of volcanic islands behind; examples are the Aleutian Island chain in the north Pacific and the south Sandwich Islands in the south Atlantic Ocean.

At these boundaries oceanic lithosphere on one side descends into the mantle. One side of the boundary, the one carrying the volcanic islands, is permanent, while the other side is progressively lost as the plate disappears. This process of consuming oceanic lithosphere is called *subduction.* As oceanic lithosphere is consumed we lose half the record of the history of the ocean floor: magnetic anomalies, sea mounts and island chains all disappear into the mantle.

Old oceanic lithosphere is cold and therefore dense; it might therefore sink into the mantle under its own negative buoyancy without any additional force being applied. Continental crust is much lighter than oceanic crust and it cannot sink into the mantle. Trenches, unlike ridges, have a substantial free-air gravity anomaly, showing that strong forces are at play supporting the topographic features such as the trench and volcanic arcs.

Convergent boundaries are essentially the only sites of very deep earthquakes. If we keep to our postulates of plate tectonics then we must suppose that the deep earthquakes occur in the rigid lithosphere, and therefore the locations of these earthquakes define the position of the subducted plate. Location of deep earthquakes is difficult because of the trade-off between depth and origin time, particularly for trench areas where there is unlikely to be good coverage in the predominantly oceanic region immediately above the event. The master event technique described in Chapter 4 can be used to relocate the events relative to a single large earthquake with greater accuracy that an absolute location can afford. Relative locations show the seismic zone is very thin indeed and probably restricted to the upper surface of the down-going plate (Section 4.12). The problem is complicated still further by local anomalies in the seismic velocities associated with the cold plate.

Focal mechanisms of thrust earthquakes on convergent boundaries provided the first vindication of the theory of plate tectonics by confirming a major prediction. Plate motions can be found at ridges, from the magnetic anomalies, and at transform faults, by their orientation, but no such determination is possible at trenches. The relative motions of the plates can be predicted at trenches from the first two methods and combined using equation (7.1). In 1968, Isacks, Oliver & Sykes published an extensive and definitive set of fault plane solutions showing the directions of the slip vectors on all the major plate boundaries. The results are shown in Figure 7.17.

The slip vectors on the convergent boundaries around the Pacific are in complete agreement with the predictions of the plate movements: the theory was finally confirmed. Since then it has occupied the central position in modern geology. We should pause to realise what a remarkable result this really is: using fortuitous observations of magnetic stripes and ocean bathymetry (and a few assumptions) we have been able to predict tectonic motions on faults tens of thousands of kilometres away; the plate really does remain rigid over distances a hundred times its thickness.

The deepest events lie at at depth of about 700 km. What controls this depth and why are there no events any deeper? There is a phase change in the mantle at about this depth, the one that separates the upper and lower mantle, and we may expect it to have some influence on the down-going slab. Again we can do a simple thermal calculation for the temperature of the subducted lithosphere, which we expect to be gradually warmed by thermal conduction. This calculation resembles that for the lithosphere spreading from an oceanic ridge and we do not repeat it in detail here, but there is an essential difference. The lithosphere descends deep into the upper mantle and it would be a poor approximation to assume a constant mantle temperature with depth. However, the temperature variation in the upper mantle is likely to be controlled by the pressure, to be close to the adiabatic value. It can be accounted for in the calculation by using the *potential temperature*, which is defined as the temperature of the material after it has been reduced adiabatically to standard pressure. We shall only be concerned with temperature differences between the subducted slab and the surrounding mantle at the same depth, so we need only refer to potential temperature. The full calculation depends on the assumed angle of subduction. One result is plotted in Figure 7.18. The isotherms are of potential temperature, not absolute temperature, scaled as for the problem in Figure 7.12.

Earthquakes can only occur in the lithosphere, which is assumed to lose its ability to sustain brittle fracture when it warms. The level of the deepest earthquakes may therefore be determined by the temperature of the subducted lithosphere — perhaps when it reaches a scaled potential temperature of 0.9 in Figure 7.18 for example. The depth of

Figure 7.17: Slip vectors around the Pacific determined from fault plane solutions. They confirm the relative motions at convergent boundaries around the Pacific that is predicted by the poles of rotation, and that the plates behave rigidly across the whole of the Pacific sea floor. From Isacks *et al.* (1968).

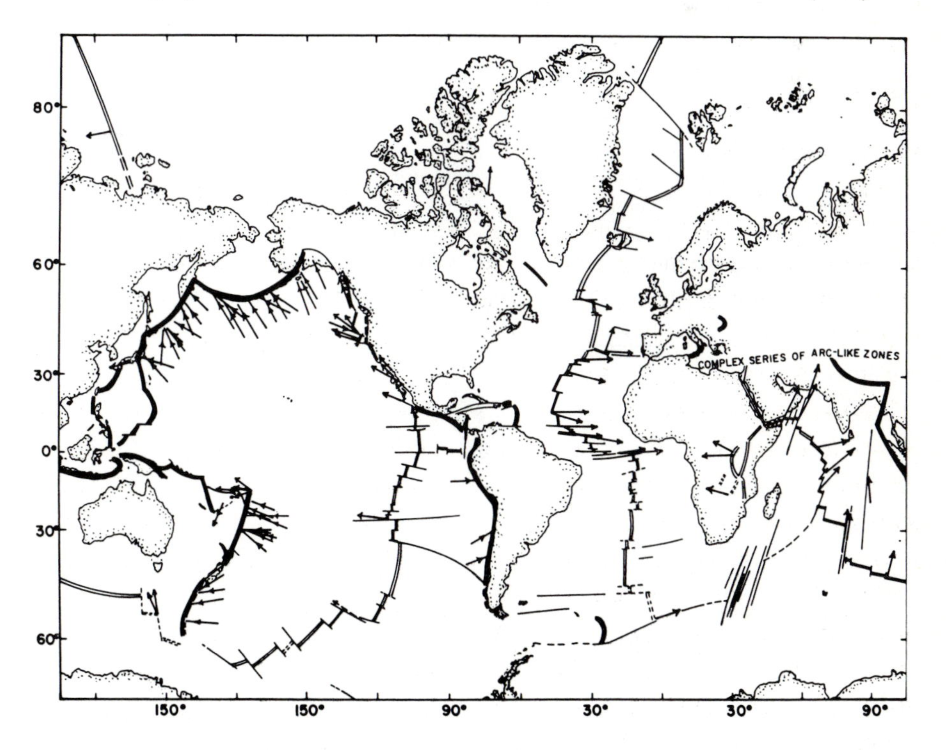

the deepest events should therefore obey a simple formula involving the convergence rate at the boundary and the thermal time constant. This has been applied to the Tonga-Kermadec trench in the south Pacific (Figure 7.19, inset). The pole of rotation lies to the south of the trench and the convergence rate increases as one moves northwards. The events occur at progressively deeper depths to about 20°, where the strike of the boundary changes. The deepest events lie quite close to the predicted 0.9 isotherm.

The focal mechanisms of earthquakes within the subducted lithosphere tell yet another story. Either the compression or the tension axis is found to be aligned with the dip of the slab, but the nature of the stress changes with depth and varies between different subduction zones. The results are shown in Figure 7.20. Very long slabs like "Tonga" appear to be in compression throughout their length, while somewhat shorter slabs like "Kermadec" are in tension at intermediate depths and compression deeper down; short slabs appear to be in tension.

This behaviour is explained in terms of a cold, dense plate experiencing a negative buoyancy force and tending to sink into the mantle, but meeting resistance to material of increasing strength (Figure 7.21). The phase change at the boundary between upper and lower mantle may represent a change to very strong material which resists further penetration of the slab.

The depth to which slabs penetrate is a topic of some controversy still.

Figure 7.18: Potential temperature contours in a subducted slab. The slab heats adiabatically as it sinks because of the increase in pressure, but remains colder than its surroundings because the thermal conduction time is slow compared with the time it takes to consume the slab. For a rapidly convergent boundary (10 cm yr^{-1}) the slab reaches about 700 km before warming up.

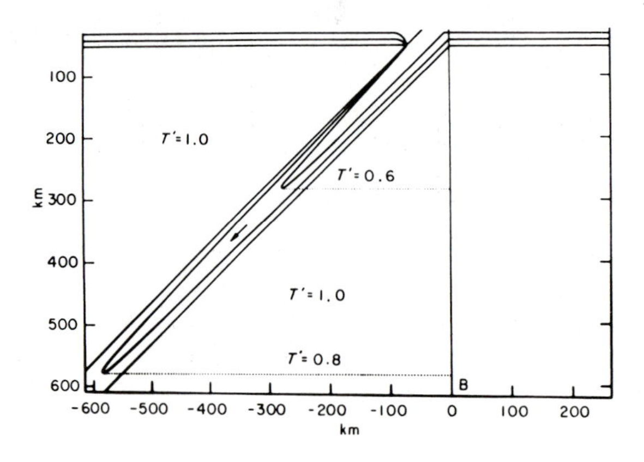

Figure 7.19: The deepest earthquakes appear to be restricted to a single temperature boundary, in this model a potential temperature 0.9 that of the mantle. For the Tonga-Kermadec trench the convergence rate is greater in the north, where the deepest earthquakes occur.

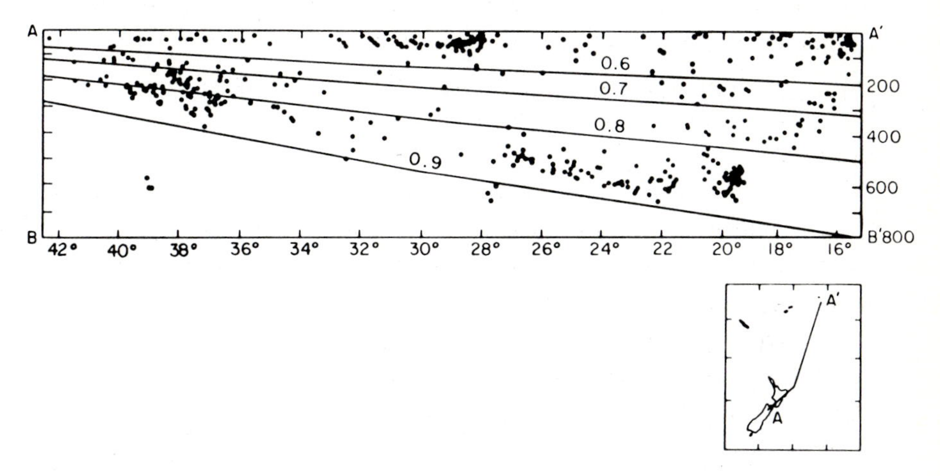

Figure 7.20: Fault plane solutions in 14 subducted slabs. The pressure (open circles) or tension (solid circles) axes tend to be aligned down the slab. From Isacks & Molnar (1969), copyright ©1969 Macmillan Magazines Ltd. Reprinted with permission.

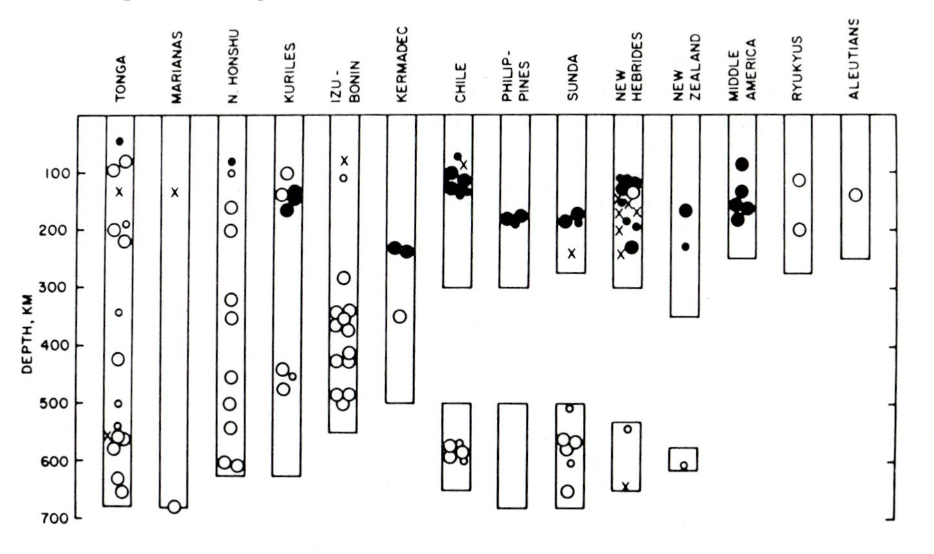

The two views are (a) that the lower mantle is too dense or viscous to allow penetration, or (b) that the plate does penetrate but warms or undergoes a phase change so that it no longer supports earthquakes. The debate centres around the detection of seismic velocity anomalies below the deepest earthquakes, which would suggest a cold lithosphere there. An example of recent work is Creager & Jordan (1984).

7.8 Triple junctions

Many configurations of plate motion must be transitory to be consistent with our postulates, and therefore they are unlikely to be seen in nature: they would require a freak occurrence to be present at this particular time. As an example consider a trench with relative plate motion as shown in Figure 7.22. We adopt the usual notation of drawing the arrow depicting relative motion on the underthrusting plate. Above the point a we have convergence with the plate on the right underthrusting that on the left; the permanent boundary is on the left. Below the point a the plate on the left is underthrusting that on the right; the permanent boundary lies on the right here. This configuration is consistent with our postulate of no deformation away from the plate boundary, but only for an instant in time. The plate boundary moves with the plate on the right above a and with the plate on the left below a and will therefore shear, opening a transform fault to accommodate the relative motion between the two plates.

Figure 7.21: An explanation of the fault plane solutions in subducted slabs. The slab initially sinks under its own weight, but later encounters resistance of a stronger medium. Some slabs may break off, giving a gap in the seismicity and deep events under compression that appear separate from the shallower events that exhibit tension. From Isacks & Molnar (1969), copyright ©1969 Macmillan Magazines Ltd. Reprinted with permission.

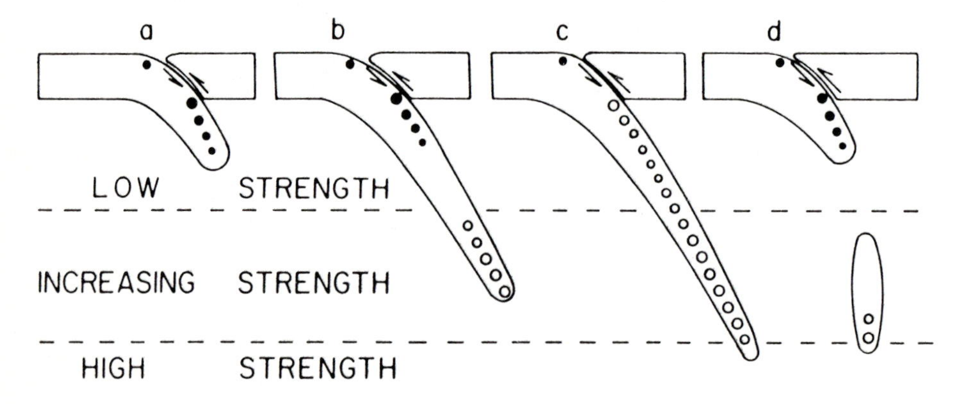

This is the situation in New Zealand at the moment, with subduction of the Pacific plate beneath the Indian plate to the north, with associated volcanism in the North Island of New Zealand and the Kermadec; strike-slip motion across the Southern Alpine fault in the South Island of New Zealand; and subduction of the Indian plate beneath the Pacific plate to the south of New Zealand, with associated deep seismicity dipping from west to east beneath the Fjordland district. Actually the Southern Alpine fault is a poor example of a transform fault: New Zealand is a piece of old continent and the rules of plate tectonics do not apply well; a great deal of compression has been taken up in a crush zone around this fault.

Points where three plates meet are called *triple junctions*. A triple junction where three trenches meet is shown in Figure 7.23. We shall use the notation TTT for such a triple junction, with R denoting ridges and F denoting transform faults in other types of triple junctions. The relative motions of the three plates A, B and C are shown by arrows. The permanent side of the AB boundary is part of the A plate, and so is the permanent side of the AC boundary. The permanent side of the BC boundary is on the C plate and therefore moves with the C plate. The relative motion of the BC and AB boundaries is therefore given by the

Figure 7.22: The plate boundary in (a) is unstable and will evolve immediately into the trench- transform-trench configuration (b). Unstable configurations occur only for an instant in time and are therefore only seen in nature as a fluke. The situation (b) is seen in New Zealand today (c). From McKenzie & Morgan (1969), copyright © 1969 Macmillan Magazines Ltd. Reprinted with permission.

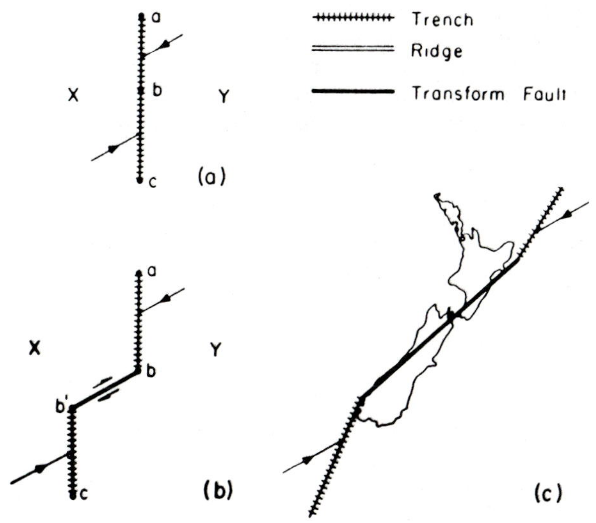

motion of the C plate relative to A: the boundary will migrate upwards as shown in the figure.

The geometry of the triple junction shown in Figure 7.23(a) is transient; it immediately evolves into a triple junction in which two of the boundaries are parallel, as in Figure 7.23(b). The transient type of triple junction is *unstable* and we would not expect to encounter it in nature, except perhaps by a fluke. The configuration of the triple junction shown in Figure 7.23(b) is *stable* however, and therefore we might hope to find an example in nature. A good example is at the northern end of the Marianas trench.

We need a general approach that would enable us to investigate the stability of all possible triple junctions. As a first example consider the triple junction formed by three spreading ridges, as in Figure 7.24. It is not at all obvious by inspection whether this configuration is stable or unstable. Rather surprisingly, it turns out to be stable for all geometries. The vectors of relative velocities between the plates form a closed *velocity triangle* as shown in Figure 7.24. The velocities must form a triangle because the three relative velocities sum to zero: the velocity of A seen from B plus that of B seen from C plus that of C seen from A is zero: it is just the velocity of A seen from A. Thus

$$_A\mathbf{v}_B + _B\mathbf{v}_C + _C\mathbf{v}_A = 0 \tag{7.18}$$

Figure 7.23: Evolution of an unstable triple junction. The BC boundary is on the C plate; the B plate is consumed here. The C plate has a northward component of motion relative to A. The BC boundary therefore moves north to form the stable configuration shown in (b).

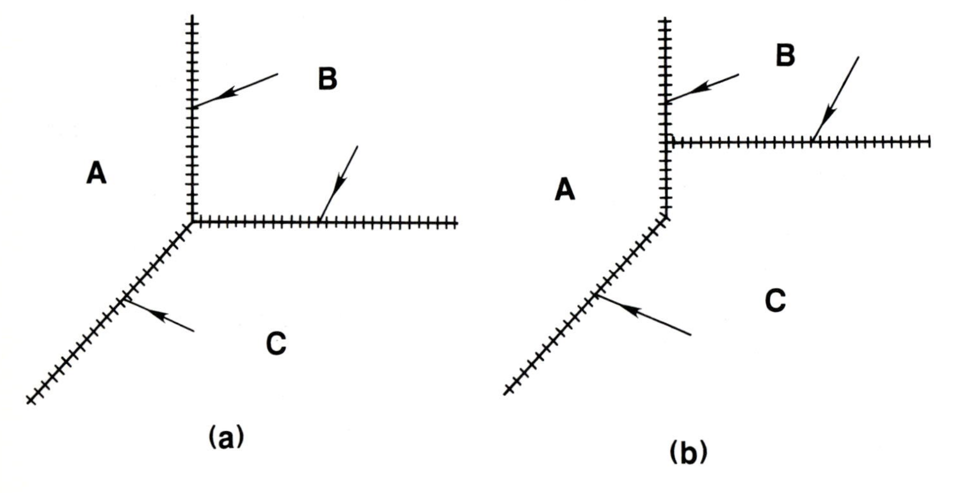

The velocities add vectorially and therefore form a triangle.

We define a two-dimensional velocity space in which we plot the velocity of any point. Thus in Figure 7.24 the point A may represent the velocity of plate A. We have not specified an origin and therefore we have not specified the absolute velocity of plate A. B represents the velocity of plate B, and the line AB represents the relative velocity ${}_B\mathbf{v}_A$ of B with respect to A. Similarly C represents the velocity of the plate C.

Now consider where we must plot the triple junction itself. This would define its velocity relative to each of the three plates. Suppose it plots to the point J. The AB ridge spreads symmetrically normal to its strike and therefore the relative velocity of the triple junction with respect to either plate must be equal: the point J must lie on the perpendicular bisector of AB in velocity space. The same result applies to the other two ridges and J must also lie on the perpendicular bisectors of both AC and CB.

J can lie on two lines at once if it is at their intersection point; but it can lie on three lines at once only if those lines meet in a single point. It just so happens that this condition is met by the velocity triangle for RRR triple junctions because of the familiar result from Euclidean geometry, that the perpendicular bisectors of a triangle meet in a point. The triple junction is therefore stable, and lies at the intersection of the perpendicular bisectors of the velocity triangle. Moreover, we can find its velocity relative to any of the plates by constructing the velocity triangle and measuring off the

Figure 7.24: Stability of a RRR triple junction. The lines AB, BC, and AC represent the relative plate velocities ${}_A\mathbf{v}_B$, ${}_B\mathbf{v}_C$, ${}_A\mathbf{v}_C$, respectively. *ab*, *bc*, and *ca* are parallel to the ridges. The junction is stable because the perpendicular bisectors meet in a point, T, which defines the motion of the triple junction.

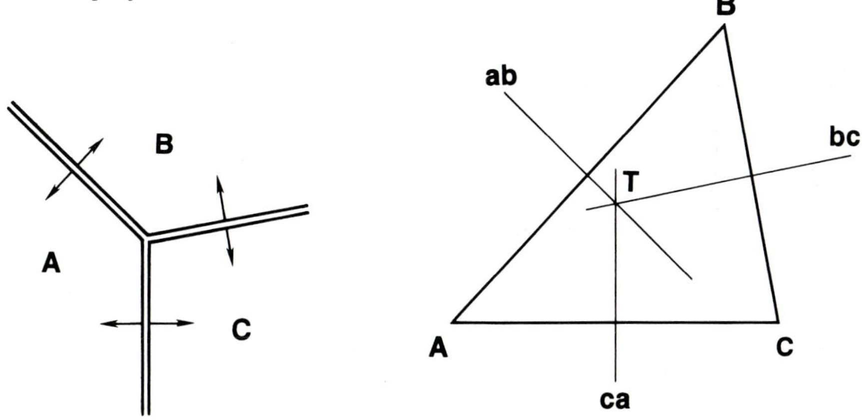

relevant distances. We conclude that all *RRR* triple junctions are stable, provided the ridges spread symmetrically perpendicular to their strike. There is a good example of one in the central Indian Ocean.

Another example will illustrate the procedure well. Consider the general *TTT* triple junction shown in Figure 7.23. The corresponding velocity triangle is shown in Figure 7.25. We have already seen that this type of triple junction can be unstable, but the velocity triangle allows us to investigate all possible configurations and their stabilities. The velocity triangle is constructed by drawing *AC* parallel to the arrow denoting relative velocity between *A* and *C* and of length proportional to its speed; the sides *AB* and *BC* are constructed in the same way. The triple junction is attached to the permanent side of the *AB* boundary and therefore plots on a line through *A* parallel to the *AB* boundary — it can move along the boundary but not away from it. This is the line *ab* in Figure 7.25. Similarly it must lie on line *ac* which runs parallel to the *AC* boundary and through *A*. Finally, the triple junction is also attached to the *C* plate at the *BC* boundary and must plot in velocity space on the line *bc*, through *C* and parallel to the *BC* boundary. In general the three lines *ab*, *bc*, *ac* will not meet in a point and the triple junction must therefore be unstable. There are just two exceptional cases when the triple junction is stable. These are:

1 the case already met, when *ab* and *bc* are coincident and the three lines meet in *A*: the *AB* and *BC* boundaries are parallel and the triple junction runs along the straight trench as in Figure 7.23;

Figure 7.25: Velocity triangle for a general *TTT* triple junction. The triple junction must lie on all three lines *ab*, *bc*, and *ca*, which meet in a point only in two special cases. Except for these special cases the triple junction is unstable.

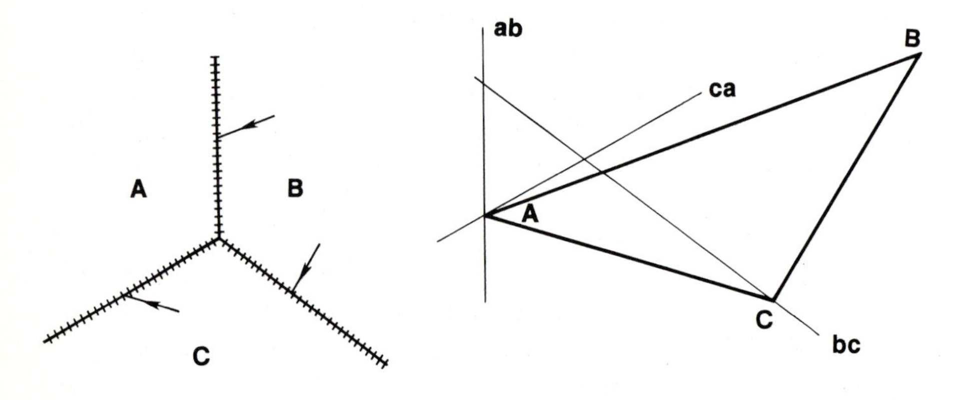

2 when bc lies along AC; the relative motion of C with respect to A is parallel to the BC boundary, and the junction remains stationary.

McKenzie & Morgan (1969) give a summary of all possible types of triple junction geometry and their velocity triangles.

7.9 The tectonic evolution of the north-east Pacific

We are now in a position to put plate tectonic theory to work in understanding the tectonic evolution of an area. The classic example of this is the north-east Pacific as reported by Attwater (1970; in COX). The north-east Pacific Ocean is probably the best-surveyed piece of ocean floor anywhere. It was formed at a fast spreading ridge and consequently the magnetic anomalies are well formed. They are shown in Figure 7.26. Note the clean offsets by the fracture zones, which we expect from the theory of transform faults.

There are two peculiar features of these anomalies which appear at first sight to violate the postulates of plate tectonics: only half of them are present, and those in the upper left hand corner of the map are bent through almost a right angle (this feature is called the *Great Magnetic Bight*). Magnetic anomalies are generated symmetrically by ridges, and both sets ought to be present. Anomalies are generated parallel to a ridge and are therefore straight; they must remain straight because the plates do not deform. How can these anomalies be reconciled with plate tectonics? Consider first the Great Magnetic Bight. Magnetic anomalies always lie on great circle paths, and two sets of great circles suggest two poles, and therefore three plates. There must have been a third plate present when these anomalies were formed. In fact this pattern of stripes is formed by an *RRR* triple junction (Figure 7.27).

There is no longer any sign of a triple junction: it must therefore have been subducted somewhere. Subduction would also explain the missing half of the magnetic anomalies. Subduction occurs along the Aleutian Islands trench today and the relative motion suggests this convergent boundary as a candidate for the consumption of the missing triple junction. We need to postulate a now-vanished plate; it is called the Kula plate.

Figure 7.28(a) shows a reconstruction of the Pacific ocean floor 40 Myr ago. The Pacific plate has the same motion relative to the Aleutians boundary, which is fixed in this figure, as it does today, but the Kula plate is subducted at a different direction and with a different velocity. The magnetic anomalies that remain today are marked in this figure. There are three triple junctions marked in this figure; their motion determines

Figure 7.26: Magnetic anomaly map of the north-east Pacific. Note the absence of anomalies to the west, due to the Cretaceous quiet zone, the bent stripes of the Great Magnetic Bight, and the absence of half the anomalies. Reprinted, with permission, from Attwater & Severinghaus, Tectonic Map of the North Pacific, in Winterer, E.L., Hussong, D.M. & Decker, R.W., eds., DNAG : The Eastern Pacific Ocean and Hawaii, *Geol. Soc. Am. Publ.* in press 1989.

the evolution of the plates, which can be calculated using the similar techniques as those described in the practical for planar geometry.

After 20 Myr the plates reach the configuration shown in Figure 7.28(b). As the ridges are subducted beneath the Aleutians trench the relative motion across the boundary changes to that determined by the relative motion of the Pacific plate; there is a change in tectonic style from convergence to strike-slip motion at the western end near the triple junction marked 2 in the figure: this region is characterised by extinct volcanoes which lends some support to this reconstruction. Likewise the relative motion across the boundary with the American plate changes as the triple junction 3 sweeps along, changing it to shear motion as is observed today.

The present configuration is shown in Figure 7.28(c). The boundary with the American plate is mainly strike-slip. Subduction still occurs everywhere along the Aleutian Island arc except near its western end. This still does not explain the loss of the magnetic anomalies that should appear in the south: a second subduction zone is needed as well as the boundary with the American plate in the south.

A possible configuration is shown in Figure 7.29. The Farallon plate, of which a relic still exists today, is presumed to have been subducting beneath the American plate at 5 cm yr^{-1} in the direction shown. That half of the magnetic anomalies that resided on the Farallon plate will be

Figure 7.27: Magnetic anomalies produced at an *RRR* triple junction. Only the anomalies on the plate *A* are illustrated. The spreading rate is slower on the *AC* boundary and the anomalies more closely spaced on that branch. The bend in the stripes indicates the presence of three plates rather than two.

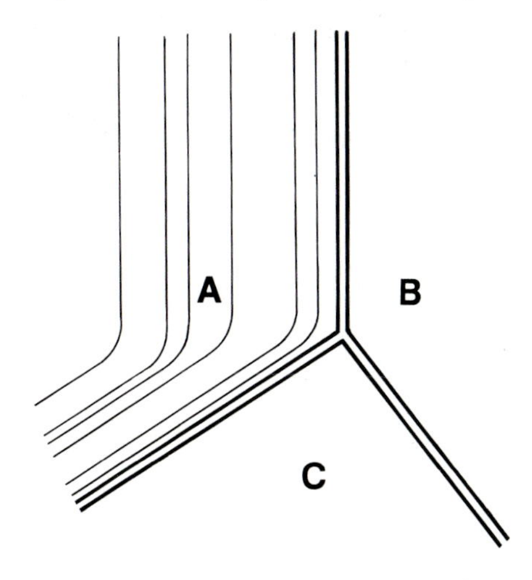

Figure 7.28: Schematic diagram of possible evolution of the Aleutian Island arc (a) 40 Myr ago, (b) 20 Myr ago, (c) present.

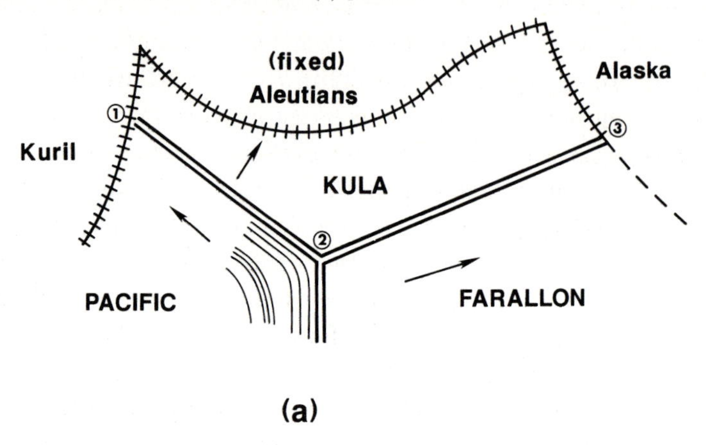

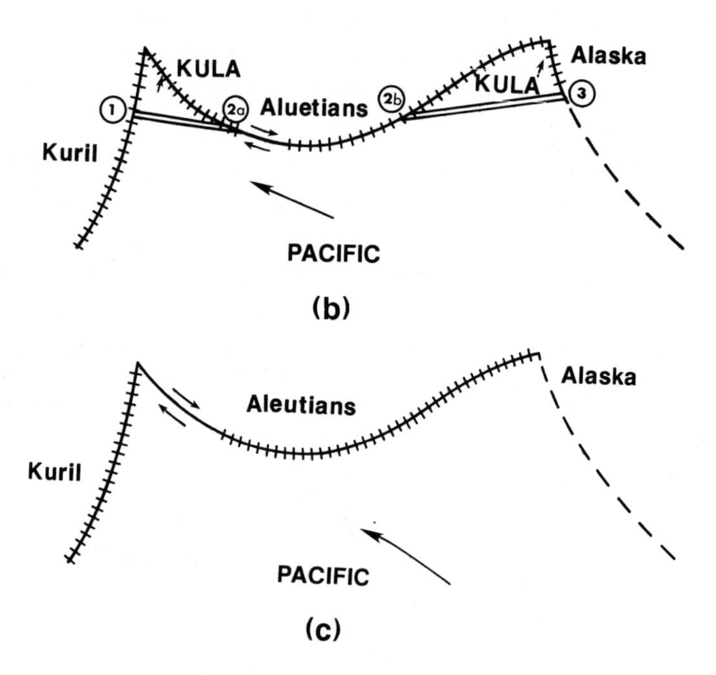

lost by the subduction process, as will the ridge, but half the magnetic anomalies and the fracture zones have remained to tell the story.

The evolution of a plate boundary is represented in a *velocity-age diagram* as shown in Figure 7.30 for the simple case of subduction of a ridge-transform junction. The velocity-age diagram shown in (b) has distance in the horizontal direction and the points marked A and B correspond with the two points A and B in (a). If we plot the evolutions of the tectonic style of the plate boundary with time increasing downwards then each type of motion occupies an area on the diagram (b) as shown.

The relative motion across the boundary changed to strike-slip, as is observed across the San Andreas fault section of the boundary today, when the ridge was subducted. Of this ridge system only the Juan de Fuca ridge, off the coast of Oregon, remains. Inshore from this there is active volcanism associated with the weak subduction that is still going on. The boundary separating strike-slip motion from subduction is the locus of the two triple junctions that are formed when the ridge is first subducted; again the crucial factor in following the evolution is the motion of the triple junctions.

The picture is not quite this simple: the oldest anomalies offshore now are some 30 Myr old (Figure 7.26); subduction did not stop immediately but anomalies continued to be overrun by continental material. Strike-slip motion across the San Andreas should therefore have begun 30 Myr

Figure 7.29: Proposed subduction of the Farallon plate beneath the American plate. This convergent boundary has now largely disappeared, but we know that some oceanic plate must have been subducted because some magnetic anomalies are missing. From Attwater (1970).

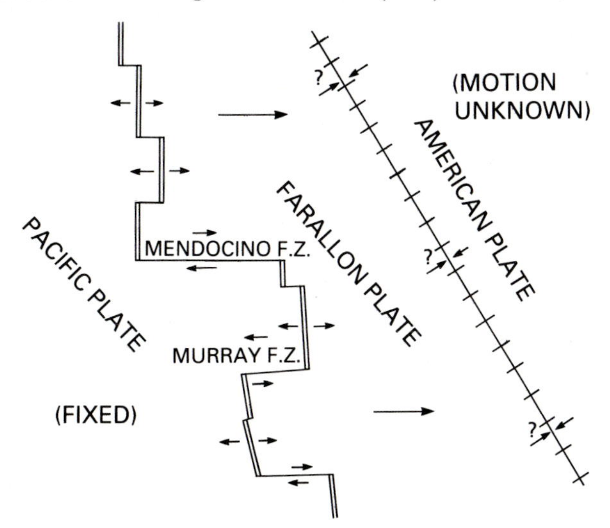

Figure 7.30: Evolution of a tectonic boundary described in a velocity-age boundary. An offset ridge (a) is subducted giving two new triple junctions and a change in tectonic style from subduction to a transform fault (b). Age increases downwards in the velocity-age diagram. The lines separating different tectonic styles are the loci of the triple junctions.

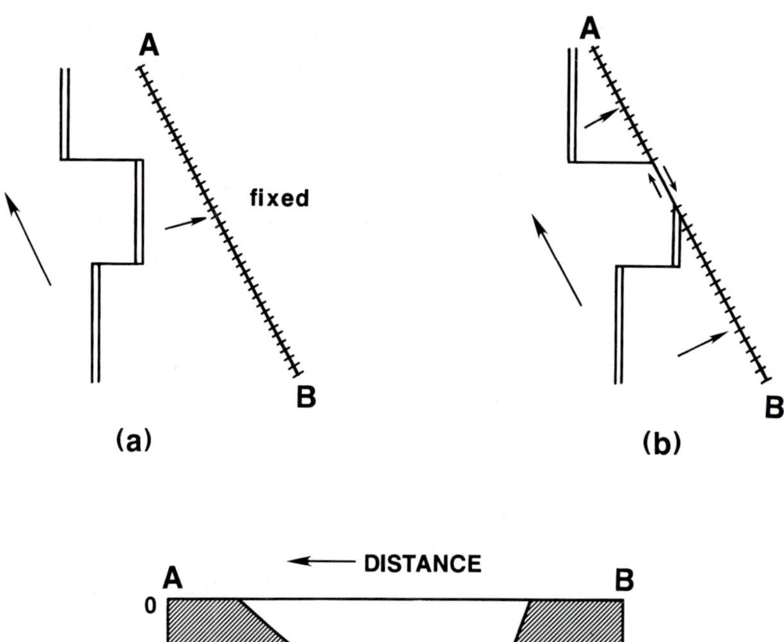

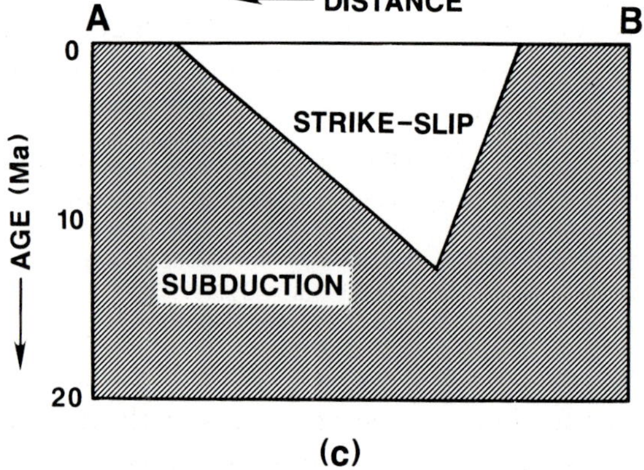

ago. This presents a serious problem because geological offsets across the San Andreas fault are some 350–500 km, which at the estimated rate of relative movement of the Pacific and American plates, of 6 cm yr^{-1}, would take only 5 Myr, not 30 Myr.

There are two ways out of this problem; to assume the slip is taken up on faults other than the San Andreas (making the geological offsets irrelevant), or that the San Andreas is a recent feature which only started 5 Myr ago, before which the Pacific and American plates were welded together. For the second hypothesis we must assume the San Andreas is an older re-activated feature, in order to explain the larger, older offsets. The first, constant-motion, hypothesis seems the more satisfying. The San Andreas is on a continent and we know that plate tectonics does not apply there. Continental lithosphere is prone to large scale deformation and the width of the "boundary" may be very wide indeed, incorporating many faults which may have taken up the slip in varying degrees in the geological past.

This is a good place to emphasise the limitations of the plate tectonic theory. It is very successful in explaining the geology of the deep oceans but the postulates do not seem to hold on continents. Where plate boundaries appear to occur on land, such as at the San Andreas system, the Southern Alpine fault in New Zealand, and the Alpide belt, the rules simply do not apply. Deformation occurs and there is clear evidence of massive shortening of the crust in some places. Deformation makes the unravelling of past tectonics very much more difficult.

7.10 The Deep Sea Drilling Program

The *D*eep *S*ea *D*rilling *P*rogram (*DSDP*) was begun in 1968. It grew out of the unsuccessful project *MOHOLE* which was an attempt to drill through the thin oceanic crust to sample the mantle. The new project was to drill a large number of shallow holes into the sediment and then a few into basement, a much less ambitious project in terms of its drilling aims than *MOHOLE*, but scientifically far more successful. The project continues as *IPOD* — the *I*nternational *P*hase of *O*cean *D*rilling. The drilling barge was the *Glomar Challenger*; it was equipped with satellite navigation for accurate positioning and set transponders on the sea bed in order to maintain accurate positioning while drilling. After a hole was drilled it was fitted with a re-entry cone that enables drilling to be re-started at a later date. Each hole could be sampled by continuous coring.

The first aim and major acheivement of the *DSDP* was the verification of the sea floor spreading hypothesis and the postulates of plate tectonics. At many of these holes the sediment was drilled to basement; the age of the crust predicted by the magnetic anomalies was confirmed and the reversals time scale found to be remarkably accurate. The drilling has helped with the interpretation of magnetic anomalies in many areas, particularly the older ones. If the magnetic anomalies gave the first confirmation of plate tectonic theory and the fault plane solutions of seismology gave the second, the drilling programme gave the third.

Before the drilling programme, the sediments on the ocean floor could only be investigated by acoustic profiling. In many cases drilling showed that reflectors, which had been interpreted as a new layer of sediments, were in fact the same sediment interrupted by a thin opaque layer. A simple model of the formation of deep sea sediment is shown in Figure 7.31. Sediments are deposited at a ridge at a fairly rapid rate — several cm per thousand years. As the ridge spreads the crust is carried down the flank of the ridge and becomes deeper. At one point it reaches what is called the *carbonate compensation depth* (*CCD*); below this depth calcareous remains do not reach the bottom but dissolve before reaching it. The only deposit at this depth is red abyssal clay, which forms at the much slower rate of about 1 mm per thousand years. If we drill we expect to penetrate the red clay and find pelagic sediment, followed by basaltic basement. This is the case in the north-east Pacific.

In the north-west Pacific four layers of sediment are found. At the equator (and in the arctic) the production of sediment is high because the plankton is abundant. The effective *CCD* is lower because so much more sediment falls in these regions. Now the crust that is presently in the

Figure 7.31: Pelagic sediments (light shading) form at a shallow ridge crest. As the ocean depth increases the sea floor passes below the carbonate compensation depth (*CCD*), pelagic sedimentation stops, and abyssal clay deposits form, giving a second layer of sediments in the deeper oceans.

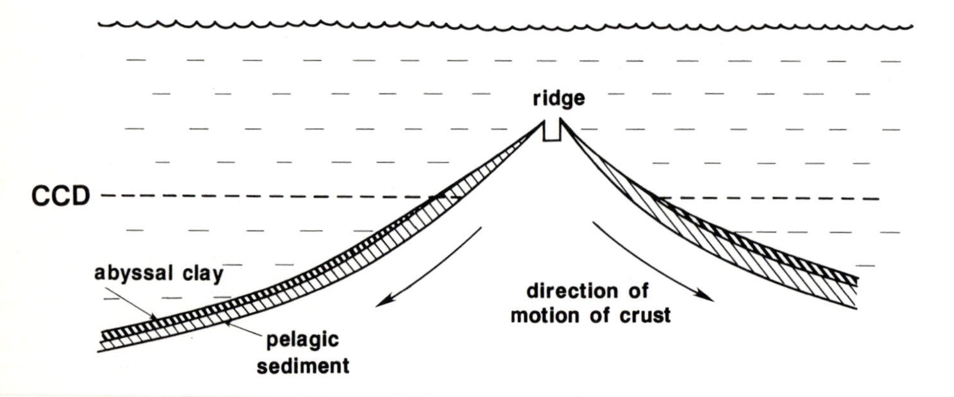

north-west Pacific was formed in the southern hemisphere, collected two layers of sediment but then passed into equatorial waters and collected a second layer of pelagic sediments, then another layer of red clay as it passed north of the equatorial region. So the sediment record could perhaps tell us whether the plate has passed the equator, and how fast it was going at the time.

Practical 5. Triple junctions

This practical is designed to give some idea of how to calculate relative motion from plate tectonic theory and use the concepts of the evolution of triple junctions to predict the future of a system of plates. In Figure 7.32 the complication of spherical geometry has been removed by treating the earth as flat, and you should not take the correspondence with the real earth too seriously. The Erratic plate is particularly well named.

The scale is everywhere the same and angles are true. The plates are subject to displacements but not rotations. The practical can be done graphically with the aid of a ruler and sheets of tracing paper. A protractor is not required since angles may be transferred from the map in Figure 7.32.

You are required to:

1. Determine the relative motion vector of the Beautific-Jokers-Nasty triple junction (*BJN*-junction) to the Allende plate.
2. Determine the relative motion vector of the Beautific-Nasty-Erratic triple junction (*BNE*-junction) to the Allende plate.
3. Using these vectors draw the locus of the future position of these two junctions relative to the Allende plate. Mark points at 10 Myr intervals.
4. At 10 Myr intervals redraw the positions of the plate boundaries until both the Jokers plate and the Nasty plate have been consumed at the trench bounding the Allende plate. This is rather easy as the plates do not rotate.

Exercises

1. A section of the boundary between two plates consists of two rifts offset by a transform fault. The junctions between rifts and the transform fault are at A (0°N,60°E) and C (45°N,0°E). The transform fault also passes through the point B (30°N,30°E). If P is the pole of relative rotation between the two plates and O is the centre of the earth show that

$$\vec{OA} \cdot \vec{OP} = \vec{OB} \cdot \vec{OP} = \vec{OC} \cdot \vec{OP}$$

Hence find P.

Figure 7.32: Plates on a flat earth, for Practical 5.

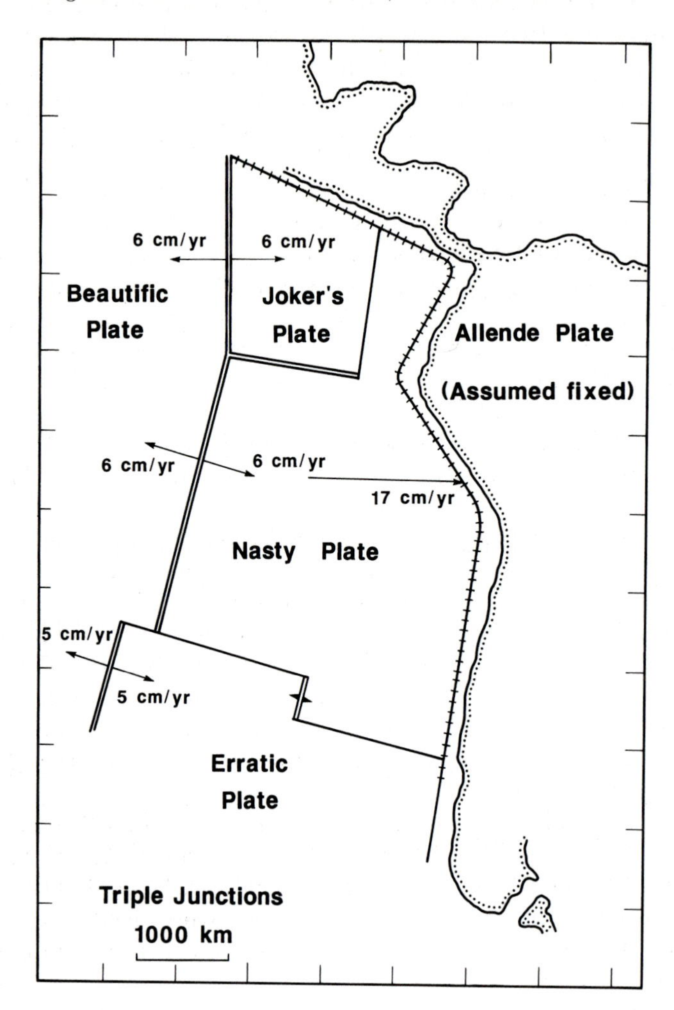

2 Magnetic anomalies show that the spreading rate at the ridges of the previous question is 10 cm yr^{-1}. Find the vector of relative instantaneous rotation in radians per year. (Earth's radius is 6371 km.)

3 The three plates India, Africa and Antarctica meet at a triple junction of three ridges at (-20°N,70°E). The angular velocity vectors between plates are:

India-Antarctica	0°N 45°E	at	6.10^{-7}°yr^{-1}
India-Africa	15°N 45°E	at	6.10^{-7}°yr^{-1}

where in both cases the angular velocity is positive if the second plate is fixed. Find the angular velocity of Africa relative to Antarctica.

4 The pole of rotation for plate A relative to plate B is at 45°N, 0°E, with present velocity 2.10^{-8}rad yr^{-1}. A convergent boundary runs north-south through the point C (0°N,45°E). Calculate the relative motion between the plates normal to the boundary.

5 Estimate the depth of the deepest earthquakes for the convergent boundary in exercise (4) assuming the plate descends at an angle of 45° to the vertical and takes 10 Myr to warm up.

6 Plot the velocity triangle for the India-Antarctica-Africa *RRR* triple junction. Use the angular velocity vectors from exercise (3). Use vector methods: note that a unit vector pointing in an easterly direction at the point (λ°N,ϕ°E) is $(-\sin\phi, \cos\phi, 0)$ and in a northerly direction $(-\sin\lambda\cos\phi, -\sin\lambda\sin\phi, \cos\lambda)$, relative to a fixed cartesian set of axes with z along the rotation axis and x in the Greenwich meridian.

7 Is the India-Africa-Antarctica triple junction stable? Determine its motion relative to each plate. Approximately how fast is the length of each ridge changing? Sketch the resulting magnetic anomalies.

8 Determine the conditions which must be satisfied if the following triple junctions are to be stable: *RRF, FFR, RTF* (two cases). Sketch the fault plane solution you would expect on (a) a ridge segment (b) a "leaky" transform fault (c) the "tear" in the Pacific plate, where part of the plate is underthrust beneath the Tonga trench and part moves past the northern boundary of the trench.

9 Explain why fault plane solutions on a particular plate boundary will have similar slip vectors but dissimilar normals to the fault planes, but for aftershocks the normal to the fault plane is consistent.

Summary

Plate tectonics is a vast subject and I cannot to do justice to its implications for geology: nor would such a treatment be appropriate. I have stripped plate tectonic theory to the bare bones and presented a simplified view of global tectonics, but the theory has been so successful that perhaps this is not a bad thing.

Plate tectonics rests on four simple postulates:

1 There are a small number of rigid PLATES which move independently without internal deformation.

2 Plate boundaries are either RIDGES, TRENCHES or TRANSFORM FAULTS,depending on the relative motion across their boundaries.

3 Transform faults have only shear motion.

4 Ridges spread symmetrically and normal to their strike.

EULER'S THEOREM, that any motion on a sphere can be represented as a rotation, is central to the mathematical description of plate motions. The POLE OF ROTATION between two plates is stationary relative to both plates. It is determined from spreading rates at ridges, which increase with distance from the pole, and the orientation of transform faults, which lie on small circles centred on the pole of rotation.

By far the most important quantitative evidence for plate motion comes from the magnetic stripes that form on the ocean floor parallel to ridges. The ridge acts as a giant tape recorder, with the newly-formed crust acting as the tape, registering the reversals of the earth's magnetic field. Fortunately the reversal frequency, about one per million years, is often enough to give a good collection of stripes but is not too frequent to saturate the tape recorder.

The remarkable successes of plate tectonics are its predictions for global tectonics. As plates move away from the ridge they cool, become more dense, and sink. A simple conduction model gives a good fit to the heat flow observed in the deep ocean, which decreases as the plate cools, and the ocean depth, which increases as the plate sinks. This also explains the scarp associated with FRACTURE ZONES, even the inactive part at the ends of a transform fault that offsets two ridges, by the age difference across them.

At trenches one plate is SUBDUCTED and part of the evidence is lost. There is less information about the relative motion across convergent boundaries than the other two. Fault plane solutions of the earthquakes in trenches show slip vectors that are in line with plate motions calculated from magnetic anomalies and transform fault orientations. Seismology gave plate tectonics its second confirmation, after magnetic anomalies.

Reconstructing plate motions in the distant past is much more difficult than finding current instantaneous motions, because we do not have seismological evidence for the location or nature of boundaries in the past. Subduction removes much of the evidence. Plate motions are followed, surprisingly, by calculating the motion of TRIPLE JUNCTIONS — points where three plates meet. Some triple junctions are UNSTABLE — they evolve immediately into another configuration. We would only see such triple junctions by a fluke; if we happened to have one at the present time that was in the process of evolving. STABLE triple junctions are therefore the most important in nature, and we gave a complete list of the stable configurations. The evolution of the north-east Pacific was unravelled using these techniques.

The final confirmation of plate tectonics has come from the DEEP SEA DRILLING PROGRAM. A programme of drilling has run since 1970; drilling to basement in many places throughout the world has shown the age of the oceanic crust is as predicted by plate tectonic theory.

Further reading

The physicist must take the trouble to read the geological background to plate tectonics: PRESS & SIEVER and HOLMES are good starting points. HOLMES is a classic but now very out-of- date in some respects. The theory of plate tectonics as described here is set out in LE PICHON *et al.*, although this book does not delve very deeply into the seismological details. The collection of original papers in COX is imperative reading; several of the papers are referred to in the text.

Appendix A

Spherical geometry and projections

Formulae from spherical geometry are needed in calculating distances and directions on a sphere, and latitudes and longitudes of points on a sphere. They are needed in using travel time tables and in calculating poles of rotation in plate tectonics. They are best derived using vectors. For example, a useful formula gives the angular distance between two known points on a sphere. Let the coordinates of the points A and B be (θ_A, ϕ_A) and (θ_B, ϕ_B). Their position vectors in terms of cartesian axes with z axis aligned with the polar axis are, for a unit sphere,

$$(\sin\theta_A \cos\phi_A,\ \sin\theta_A \sin\phi_A,\ \cos\theta_A)$$

and

$$(\sin\theta_B \cos\phi_B,\ \sin\theta_B \sin\phi_B,\ \cos\theta_B)$$

respectively. These are unit vectors and the angle between them is the required distance, Δ. Forming the scalar product and simplifying using the formula for the difference of cosines gives

$$\cos\Delta = \cos\theta_A \cos\theta_B + \sin\theta_A \sin\theta_B \cos(\phi_A - \phi_B) \tag{A.1}$$

Distances on a sphere are always less than 180°, and are therefore uniquely defined by the cosine function (but not by the sine function, which is always positive for angles in this range).

The other angle relating two points on a sphere is the azimuth, Z. If you stand on point A and look towards point B along a great circle path, the angle between the northward direction and this great circle, measured positive clockwise from north, is the azimuth. It varies from 0° to 360°. In this case sine or cosine functions are ambiguous to within a quadrant. We can find the azimuth uniquely by specifying both sine and cosine functions. The *FORTRAN* function *ATAN2(S,C)* is most useful in these calculations. The sine of the angle is proportional to S and the cosine proportional to C; the angle is determined in the correct quadrant in one step provided the S and C have the same signs as those of sine and cosine respectively.

Let $\mathbf{n}$ be the unit vector from the centre of the sphere, O, to the north pole, N. Then $\mathbf{n} = (0,0,1)$. Let A, B be two points on the surface of the sphere and $\mathbf{a}$,$\mathbf{b}$ the unit vectors along the lines OA, OB. The azimuth is then the angle between the planes ONA and ONB, and the angle between the normals to these two planes is $\pi - Z$. The lengths of the cross products $\mathbf{n} \times \mathbf{a}$ and $\mathbf{a} \times \mathbf{b}$ are $\sin\theta_A$ and $\sin\Delta$ respectively, so the dot product gives

$$\begin{aligned} -\sin\theta_A \sin\Delta \cos Z &= -\sin^2\theta_A \cos\theta_B \\ &\quad + \sin\theta_A \sin\theta_B \cos\theta_A \cos(\phi_B - \phi_A) \end{aligned}$$

which on rearrangement gives

$$\cos Z = \frac{\sin\theta_A \cos\theta_B - \sin\theta_B \cos\theta_A \cos(\phi_B - \phi_A)}{\sin\Delta} \tag{A.2}$$

The modulus of $\sin Z$ is obtained from the length of $(\mathbf{n} \times \mathbf{a}) \times (\mathbf{a} \times \mathbf{b})$. The vector is

$$\begin{aligned} &[\sin^2\theta_A \sin\theta_B \cos\phi_A \sin(\phi_B - \phi_A), \\ &\sin^2\theta_A \sin\theta_B \cos\phi_B \sin(\phi_B - \phi_A), \\ &\sin\theta_A \sin\theta_B \cos\theta_A \sin(\phi_B - \phi_A)] \end{aligned}$$

The length of this vector gives

$$\sin\theta_A \sin\Delta \sin Z = \pm \sin\theta_A \sin\theta_B \sin(\phi_B - \phi_A)$$

All terms in this expression are positive except for $\sin Z$ and $\sin(\phi_B - \phi_A)$, and by inspection it is clear that, if Z increases clockwise from north and ϕ increases east, then $Z > 0$ when $\phi_B > \phi_A$, and $+$ is the appropriate sign. This gives

$$\sin Z = \frac{\sin\theta_B \sin(\phi_B - \phi_A)}{\sin\Delta} \tag{A.3}$$

Equations (A.1)–(A.3) can be used to determine the coordinates of point B given the distance and azimuth from a given point A. (A.1) and (A.2) are used first to eliminate $\cos(\phi_B - \phi_A)$ and give $\cos\theta_B$, and then either

(A.1) or (A.2) can be used to find the cosine of $(\phi_B - \phi_A)$ and (A.3) the sine, thus determining the difference in longitudes uniquely.

Projections are used to map the surface of a sphere onto a plane piece of paper, which is far more convenient for measurement than a globe. Access to cheap computing means that tasks previously done by pencil and ruler can now be done numerically, so that to some extent the formulae for spherical geometry have replaced the need for projections. Maps sometimes aid our intuition however, something which no computer has. For example, in Chapter 7 we used the Mercator projection to make transform faults lie along straight lines. We could have determined this numerically, but deciding which faults were straight and which were not (because they had a different pole of rotation, for example) is a very difficult computational task, whereas the eye can pick out such things immediately. The important results to know for any projection are the formulae relating (x, y) (or the polar coordinates R, α) on the projection to (θ, ϕ) on the sphere, and the useful properties of the projection. Maps are invariably drawn by computer now, so the geometrical constructs for the projections are unimportant.

Two desirable properties of projections are preservation of angle and area. Unfortunately no projection can offer both these desirable properties simultaneously (if there were such a projection it would be universally accepted as the most useful one). Projections that preserve angle are called *conformal*. The angle between two infinitesimal line segments on the sphere is the same as the angle between their projections. The lengths of each segment will be altered. Thus circles drawn on the sphere will become circles on the projection because adjacent line elements along the circle will remain at the same angle. The radius of the circle will not be preserved because the lengths of the line elements may change. We made use of this property in Practical 2 and in mapping transform faults into straight lines (the limit of a circle of infinite radius), by a Mercator projection about the pole of rotation, in Chapter 7.

The *Mercator* projection is the best-known conformal projection. Mercator maps exaggerate the areas near the poles. Lines of constant latitude and longitude are mapped into straight lines. The formulae defining the projection are

$$\begin{aligned} x &= \phi \\ y &= \log \cot (\theta/2) \end{aligned} \tag{A.4}$$

Mercator projections may be taken about any pole. In Chapter 7 we took the pole to be the pole of relative rotation between two plates.

The *stereographic projection* is also conformal. It was used in Practical 2 to locate earthquakes. It has the advantage over Mercator that points in the vicinity of the pole on the sphere lie close to the pole on the projection. It is a true perspective view of the sphere. The formulae are

$$\begin{aligned} R &= 2\tan(\theta/2) \\ \alpha &= \pi - Z \end{aligned} \tag{A.5}$$

where Z is azimuth. Again any pole can be used for the projection.

The *Lambert equal area* projection was used in Practical 4 to find the fault plane solution. Equal area gives a good indication of the global coverage of seismic stations, since we are concerned with the density of stations per unit area at any particular position. A conformal map would distort the apparent coverage. Note also that in Practical 4 we used apparent positions of the stations to allow for refraction in the earth. The formulae are

$$\begin{aligned} R &= 2\sin(\theta/2) \\ \alpha &= \pi - Z \end{aligned} \tag{A.6}$$

Further information on map projections is to be found in Snyder (1982).

SELECTED BIBLIOGRAPHY

Aki, K & P.G. Richards, 1980. *Quantitative Seismology: Theory and Methods.* W.H. Freeman & Co., San Fransico

Born, M. & E. Wolf, 1975. *Principles of Optics.* Pergammon Press, Oxford.

Batchelor, G.K. 1967. *An Introduction to Fluid Mechanics.* Cambridge University Press, Cambridge.

Bott, M.H.P. 1971. *The Interior of the Earth.* Arnold, London.

Bullen, K.E. 1975. *The Earth's Density.* Chapman & Hall, London.

Bullen, K.E. & B.A. Bolt, 1985. *An Introduction to the Theory of Seismology.* Cambridge University Press, Cambridge.

Bulletin of the International Seismological Centre, Newbury, Berkshire.

Carslaw, H.S. & J.C. Jaeger, 1980. *Conduction of Heat in Solids.* Oxford University Press, Oxford.

Cox, A. 1973. *Plate Tectonics and Geomagnetic Reversals.* W.H. Freeman & Co., San Francisco.

Dziewonski, A.M. & J.H. Woodhouse, 1982. *Studies of the seismic source using normal mode theory,* in: *Proceedings of the Eurico Fermi School on Earthquakes : Observations, theory and Interpretation.* Vol. 85, ed. K. Kanamoni & E. Boschi.

Ewing, W.M., W.S. Jardetsky & F. Press, 1957. *Elastic Waves in Layered Media.* McGraw-Hill, New York.

Gilbert, F., 1979. *An Introduction to Low Frequency Seismology.* Lecture notes: Eurico Fermi School on Physics of the Earth's Interior, Italian Physical Society, Villa Monastero, Verrena, Italy.

Gutenberg, B. & C.F. Richter, 1949. *Seismicity of the Earth.* Princeton University Press, Princeton, N.J.

Holmes, A., 1978. *Principles of Physical Geology.* Nelson, Sunbury-on-Thames, Middlesex.

Hudson, J.A., 1980. *The Excitation and Propagation of Elastic Waves.* Cambridge University Press, Cambridge.

Jeffreys, H., 1957. *Cartesian Tensors.* Cambridge University Press, Cambridge.

Jeffreys, H., 1976. *The Earth.* Cambridge University Press, Cambridge.

Jeffreys, H. & K.E. Bullen, 1970. *Seismological Tables.* Brit. Assoc. Adv. Sci., Gray-Milne trust, London.

Kasahara, K., 1981. *Earthquake Mechanisms.* Cambridge University Press, Cambridge.

Kennett, B.L.N., 1983. *Seismic Wave Propagation in Stratified Media.* Cambridge University Press, Cambridge.

Lapwood, E.R., & T. Usami, 1981. *Free Oscillations of the Earth.* Cambridge University Press, Cambridge.

LePichon, X., J. Francheteau & J. Bournin, 1973. *Plate Tectonics.* Elsevier, Amsterdam.

Landau, L.D. & E.M. Lifshitz, 1975. *Theory of Elasticity,* Vol. 7 of *Course of Theoretical Physics.* Pergammon Press, Oxford.

Landau, L.D. & E.M. Lifshitz, 1975. *Electrodynamics of Continuous Media.* Vol. 8 of *Course of Theoretical Physics.* Pergammon Press, Oxford.

Malvern, L.E., 1969. *Introduction to the Mechanics of a Continuous Medium.* Prentice-Hall Inc., Englewood Cliffs, N.J.

Press, F. & Siever, R., 1978. *Earth.* W.H. Freeman & Co., San Francisco.

Simon, R.B., 1981. *Earthquake interpretations. A Manual for Reading Seismograms.* William Kaufmann Inc., Los Altos, Ca.

REFERENCES

Agnew, D., J. Berger, R. Buland, W.E. Farrell & J.F. Gilbert, 1976. International Deployment of Accelerometers, *EOS Trans. AGU*, **57**, 180–8.

Ansell, J.H. & E.G.C. Smith, 1975. Detailed structure of a mantle seismic zone using the homogeneous station method, *Nature*, **253**, 518–20.

Attwater, T., 1970. Implications of plate for the cenozoic tectonic evolution of western north America. *Bull. Geol. Soc. Am.*, **81**, 3513–36.

Atwater, T. & H. Menard, 1970. Magnetic lineations in the north-east Pacific, *Earth Planet. Sci. Lett.*, **7**, 445–50.

Bessonova, E.N., V.M. Fishman, V.Z. Ryaboyi, & G.A. Sitnikova, 1974. The tau method for the inversion of travel times — I. Deep seismic sounding data. *Geophys. J.R. Astr. Soc.*, **36**, 377–98.

Buland, R., 1976. The mechanics of locating earthquakes. *Bull. Seismol. Soc. Amer.*, **66**, 173–87.

Buland, R., J. Berger & F.Gilbert, 1979. Observations from the IDA network of attenuation and splitting during a recent earthquake. *Nature*, **277**, 358–62.

Buland, R. & C.H. Chapman, 1983. The computation of seismic travel times. *Bull Seism. Soc. Am.*, **73**, 1271–1302.

Bullard, E.C., J.E. Everett & A.G. Smith, 1965. Fit of continents around Atlantic: in Blackett, P.M.S., on continental drift. *Roy. Soc. Lond. Phil. Trans. A.*, **258**, 41–75.

Creager, K.C. & T.H. Jordan, 1984. Slab penetration into the lower mantle. *J. Geophys. Res.*, **89**, 3031–49.

Crossley, D.J. & D. Gubbins, 1975. Static deformation of the Earth's liquid core. *Geophys. Res. Lett.*, **2**, 1–4.

Dahlen, F.A., 1974. On the static deformation of an earth with a fluid core. *Geophys. J.R. Astr. Soc.*, **36**, 461–85.

Doornbos, D.J. & E.S. Husebye, 1972. Array analysis of PKP phases and their precursors. *Phys. Earth Planet Interiors*, **5**, 387–99.

Engdahl, E.R. & S. Billington, 1986. Focal depth determination of central Aleutian earthquakes. *Bull. Seism. Soc. Am.*, **76**, 77–93.

Fitch, T.J., 1975. Compressional velocity in source regions of deep earthquakes: An application of the master earthquake technique. *Earth Planet. Sci. Lett.*, **26**, 156–66.

Gilbert, F. & A.M. Dziewonski, 1975. An application of normal mode theory to the retrieval of structural parameters and source mechanisms from seismic spectra. *Phil. Trans. R. Soc. Lond. A.*, **278**, 187–269.

Gubbins, D., 1980. Source location in laterally varying media, in: *Identi-*

fication of Seismic sources: Earthquake or Underground Explosion? eds. E.S. Husebye & S. Mykkeltreit, 543–73.

Haddon, R.A.W. & J.R. Cleary, 1974. Evidence for scattering of seismic PKP waves near the mantle-core boundary. *Phys. Earth Planet. Interiors.*, **8**, 211–34.

Hiertzler, J.R., G.O.Dickson, E.M.Herron, W.C.Pitman & LePichon, X., 1968. Marine magnetic anomalies, geomagnetic field reversals, and motions of the ocean floor and continents. *J. Geophys. Res.*, **73**, 2119–36.

Isacks, B. & P. Molnar, 1969. Mantle earthquake mechanisms and the sinking of the lithosphere. *Nature*, **223**, 1121–24.

Isacks, B., J. Oliver & L.R. Sykes, 1968. Seismology and the new global tectonics. *J. Geophys. Res.*, **73**, 5855–99.

Johnson, L.E. & F. Gilbert, 1972. Inversion and inference for teleseismic ray data. *Methods in Comp. Phys.*, **12**, 231–66.

Kennett, B.L.N. & J.A. Orcutt, 1976. A comparison of travel time inversions for Marine Refraction Profiles. *J. Geophys. Res.*, **81**, 4061–70.

Masters, T.G. & F. Gilbert, 1981. Structure of the inner core inferred from observations of its spheroidal shear modes. *Geophys. Res. Lett.*, **8**, 569–71.

Masters, G., T.H. Jordan, P.G. Silver & F. Gilbert, 1982. Aspherical earth structure from fundamental spheroidal mode data. *Nature*, **298**, 609–13.

Masters, T.G., J. Park & F. Gilbert, 1983. Observations of coupled spheroidal and toroidal modes. *J. Geophys. Res*, **88**, 10285 – 98.

McKenzie, D.P., 1969. Speculations on the consequences and causes of plate tectonics. *Geophys. J. R. Astr. Soc.*, **18**, 1–32.

McKenzie, D.P., 1988. Edward Crisp Bullard, 1907-1988. *Biographical Memoirs of Fellows of the Royal Society*, **33**, 67–98.

McKenzie, D.P. & W.J. Morgan, 1969. The evolution of triple junctions. *Nature*, **224**, 125–33.

Peterson, J. & N.A. Orsini, 1976. Seismic research observations: Upgrading the worldwide seismic data network. *EOS Trans AGU*, **57**, 548–56.

Raitt, R.W., G.G. Shor, T.J.G. Francis & G.B. Morris, 1969. Anisotropy of the Pacific upper mantle. *J. Geophys. Res.*, **74**, 3095–3109.

Sclater, J.G. & Francheteau, J., 1970. The implications of terrestrial heat flow observations on current tectonic and geochemical models of the crust and upper mantle of the earth. *Geophys. J. R. Astr. Soc.*, **20**, 509–42.

Smylie, D.E. & L. Mansinha, 1971. The elasticity theory of dislocations in real earth models and changes in rotation of the earth. *Geophys. J.R. Astr. Soc.*, **23**, 329–54.

Snyder, J.P., 1982. Map projections used by the US Geological Survey. *Geol. Surv. Bull.*, **1532**, US Govt. Printing Office, Washington.

Vine, F.J., 1966. Spreading of the ocean floor: new evidence. *Science*, **154**, 1405–15.

Ward, S.N., 1980. A technique for the recovery of the seismic moment tensor applied to the Oaxaca, Mexico earthquake of November 1978. *Bull. Seism. Soc. Am.*, **70**, 717–34.

Wunsch, C., 1974. Simple models of the deformation of an earth with a fluid core — I. *Geophys. J.R. Astr. Soc.*, **39**, 413–19.

INDEX

Page numbers in italic indicate definitions